Signal Analysis: Quantum Field Theory Applications in Communication and Intelligence

Signal Analysis: Quantum Field Theory Applications in Communication and Intelligence

Harish Parthasarathy
Professor
Electronics & Communication Engineering
Netaji Subhas University of Technology (NSUT)
New Delhi, Delhi-110078

Manakin PRESS

CRC Press
Taylor & Francis Group
Boca Raton London New York

CRC Press is an imprint of the
Taylor & Francis Group, an **informa** business

First published 2026
by CRC Press
4 Park Square, Milton Park, Abingdon, Oxon, OX14 4RN

and by CRC Press
2385 NW Executive Center Drive, Suite 320, Boca Raton FL 33431

CRC Press is an imprint of Informa UK Limited

British Library Cataloguing-in-Publication Data
A catalogue record for this book is available from the British Library

ISBN: 9781041206866 (hbk)
ISBN: 9781041217121 (pbk)
ISBN: 9781003728627 (ebk)

DOI: 10.1201/9781003728627

Typeset in Minion Pro
by Manakin Press, Delhi

Manakin
PRESS

Preface

This book is about some classical and quantum statistical aspects of theoretical physics as applied to engineering problems in communication theory, statistical signal processing and neural networks. There are several methods to transmit signals over a classical/quantum channel. A typical example is by using a waveguide excited by a classicl/quantum electromagnetic/gauge field at one end and observing the resulting effect on a quantum system like a superstring located at the other end of the guide. Alternately, we can transmit a signal by modulating its effect on a superstring at one end of the guide and observing its effect on the electromagnetic field or a string gauge field at the other end. This method works because superstrings interact with gauge fields just as charged point particles interact with an electromagnetic field. A classical signal can be communicated over a quantum channel by modulating an electromagnetic field with the signal and causing this electromagnetic field to interact with a quantum system thereby altering its state evolution in accordance with the Schrodinger equation. However, during this evolution, quantum noise enters into the dynamics from the bath/environment and hence exact recovery of the state is not possible. We have to make non-demolition measurements on the quantum system and estimate thereby its state on a real time basis using the Belavkin quantum filter. Once an estimate of the state has been recovered, we can decode from it the classical message signal that was transmitted over the channel by means of detection operators for quantum measurement. Moreover, there may be other random classical parameters apart from the electromagnetic field in the quantum system Hamiltonian that evolve according to a Markovian law. The objective then is to estimate not only the evolving quantum state but also the evolving classical parameters via an extended version of the Belavkin filterr. We discuss this problem also here. This problem can be viewed as the quantum generalization of the classical extended Kalman filter wherein both states and system parameters are required to be estimated simultaneously on a real time basis. A comparison of the Belavkin quantum filter with the Kushner-Kallianpur classical stochastic filter can be drawn in the special case when the state evolves according to Hamiltonian dynamics with noise and further, noisy measurements of the state are taken. In such a case, the quantum state can be replaced by a Wigner distribution which has similar properties as a classical joint probability distribution of position and momentum eventhough joint pdf's of position and momentum do not exist in the quantum theory owing to the Heisenberg uncertainty principle. In the absence of measurement noise, the quantum system state evolves according to the Lindblad master equation which in the Wigner distribution domain is just like a classical Fokker-Planck equation but with extra quantum correction terms in powers of Planck's constant. In the presence of non-demolition measurement terms, the evolution of the state estimate using the Belavkin filter is like a classical Fokker-Planck equation with extra measurement terms that make it resemble the canonical Kushner-Kallianpur filter but again it also contains extra quantum correction terms expressed in powers of Planck's constant owing again to the non-commutativity of the observables. It should be noted that the KK classical filter in this case is the classical Fokker-Planck equation with drift terms being given by the Poisson bracket of the Hamiltonian with the probability density, diffusion terms coming from noise in the Hamiltonian dynamics plus measurement noise terms. This book also discusses quantum neural networks based on the Schrodinger equation, the Lindlbad master equation, the Belavkin quantum filter equation and the quantum Boltzmann equation for tuning the parameters in the Hamiltonian and the Lindblad operators so that the probability density corresponding to the wave function or to the mixed state tracks a given family of probability densities or so that the expected value of a family of observables tracks a given set of classical signals. This can be used to synthesize the pdf of a stochastic signal from measurements of another signal that is strongly correlated with the former from measurements of the latter. The quantum Boltzmann equation for a system of N identical particles interacting with external fields and also with a noisy bath can be derived from the Lindblad master equationo for specific separable forms of the Lindlbad operators by partial tracing over other particles and then making the molecular chaos approximation, resulting thereby in a nonlinear evolution equation for one particle density operators. The nonlinearity in the dynamics signifies quantum collison terms. Still more generally, when non-demolition measurements are made, we can partial trace out the Belavkin equation (which is essentially the Lindlbad master equation with measurement noise terms) over the remaining particles to obtain a quantum Boltzmann equation for the one particle filtered state nonlinear evolution equation with measurement terms. We can then cause the pdf corresponding to this one particle filtered state to track a given family of pdf's by controlling the Hamiltonian and Lindblad parameters while simultaneously ensuring that the average value of a set of one particle observables in the filtered state is highly correlated with a corresponding set of classical signals. These problems are discussed in this book. The classical and quantum Boltzmann equations in curved background space-time metric are also discussed here. The essential point in the classical formulation is to use the geodesic equations of a charged particle in an electromagnetic field in general relativity to set up the Boltzmann equation and then to derive the charge, current and energy-momentum tensor of the plasma matter from this Boltzmann distribution function and use these as sources in the Maxwell and Einstein field equations for the electromagnetic and metric tensor fields. For Fermi wave operator fields, the famous Gorkov equations for the temperature Green's functions can be used to derive the dynamics of densities of one particle observables like charge and current densities by averaging w.r.t an initial Gibbs state. This programme can be carried out even in the case when the Hamiltonian is given by Dirac's equation in curved background space-time with electromagnetic interactions. The one particle averaged probability densities of observables can be made to track given families of pdf's by controlling the metric tensor field or other parameters in the Hamiltonian. This book also involves a discussion of certain specialized topics in quantum gravity involving techniques for computing quantum correlations of field observables in coherent state when the observables comprise metric perturbations, fields produced by the Dirac current, the electromagnetic field etc.

when all these fields taking into account their mutual interactions are second quantized. The techniques are based on perturbative solutions to the Einstein-Maxwell-Dirac field equations. The Einstein field equations are driven by the energy-momentum tensors of the Maxwell and Dirac fields, the Maxwell field in curved space-time is driven by the Dirac current and finally the Dirac field in curved space-time interacts with the Maxwell field. The zeroth order solutions to the Einstein, Maxwell and Dirac equations are simply obtained as plane wave superpositions of the corresponding creation and annihilation operators. The higher order perturbative corrections to the metric tensor, the Maxwell potentials and the Dirac wave operator field will therefore be polynomials in the creation and annihilation operators of these fields and hence by using their commutation/anticommutation relations and the action of Bosonic and Fermionic annihilation operators on Bosonic and Fermionic coherent states, one can in principle evaluate the higher order field correlations in a coherent state of the Bosons and Fermions. Symmetry breaking in quantum field theory has been discussed bringing in relation to this concepts such as quantum effective action and the field equations satisfied by it and its relation to the field propagator. The method of deriving the quantum effective action by a Legendre transformation of the quantum momentum generating function w.r.t a classical current field has been introduced. The quantum effective action defines the potential which is gauge invariant. The ground state of the system is at the minimum of this potential. Using these facts, we show that when symmetry is spontaneously broken in the sense that the ground state is non-zero, or equivalently, when the field acquires non-zero vacuum expectation values, then the inverse propagator matrix of the field obtained as the Hessian matrix of the potential in the ground state has some zero eigenvalues with the corresponding eigenvectors being defined by the gauge group generators acting on the ground/vacuum state. These zero eigenvalues of the inverse propagator in the ground state correspond to zero mass Goldstone Bosons and thus we obtain the fundamental priniciple that spontaneous symmetry breaking leads to the generation of massless Bosons. There is another fundamental principle in relation to this which has not been discussed here. It is that if the gauge symmetry is approximately broken by the addition of a small perturbation to the potential, then the ground state gets shifted slightly and the inverse propagator will then no longer have zero eigenvalues, it will have small eigenvalues of magnitude dictated by the strength of the symmetry breaking potential. These massive particles are called pseudo-Goldstone Bosons. Thus, our next fundamental principle is that approximate symmetry breaking leads to the generation of massive pseudo-Goldstone Bosons. Another interesting problem discussed here is Feynman's theory of automata in which the physical world is on a minute scale discrete. We develop a model for discrete quantum mechanics, ie, finite state quantum mechanics in this context based on partitioning physical space into regular hexagons and show how in the continuum limit when the hexagon sidelengths become zero, the continuous state Schrodinger equation is obtained. During our discussion of quantum field theory, we discuss the celebrated Atiyah-Singer Index Theorem in the context of anomalies. Such anomalies arise when although the action is gauge invariant, the path measure for Chiral fields is not gauge invariant and the resulting change in the path integral can be shown to be equivalently obtained by adding to the field Lagrangian a gauge field Lagrangian term which is easily shown to be equal to the four divergence of a four current called the anomalous current. The change in the Lagrangian thus introduced by non-invariance of the path measure has an integral over space-time equal to the index of Dirac operator which is an integer and essentially equal to the difference between the number of left and right Chiral Fermions. This fact makes this integral a topological invariant, ie it cannot change continuously under small variations of the gauge field. This existence of such an anomalous current is based on consideration of global and local gauge symmetry principles. Thus lack of gauge invariance of the path integral manifests itself as the presence of gauge currents that are not conserved. These anomalous currents can actually be derived from triangle vertex Feynman diagrams in which each vertex is connected to a gauge field while the triangle sides are associated with the matter field propagators. We discuss this problem here. The existence of anomalous currents therefore modifies the gauge field propagator which can be controlled by a classical gauge field connected to the third triangle vertex and can be used as a control signal in quantum communication theory. Basically, by encoding a classical message signal in the form of the gauge field connected to the third vertex, we can alter the scattering probabilities of matter particles by controlling the gauge propagator exchanged by the matter particles during their interaction and hence transmit signals. This control of quantum scattering amplitudes via classical message fields is a version of the well known Cq communication problem. We also discuss in this context, various important problems in classical communication systems like amplitude modulation, frequency modulation, band pass noise in these modulation systems and their complex lowpass representations using the Hilbert transform, evaluating the effect of noise on the demodulated output, demodulation of FM signals using the phase locked loop and the effect of noise on the demodulated output, delta modulation based on signal prediction theory, the effect of noise on delta modulation. We have included this discussion because of the important role played by quantum field theory in communication systems. For example, to transmit a classical message over a quantum channel, we can use symmetry breaking principles like consider a quantum matter field ϕ spread out over the region between the transmitter and receiver interacting with a quantum gauge field and also a classical gauge field. In the absence of the classical gauge field, the matter and gauge field have a gauge invariant joint Lagrangian. The introduction of a classical gauge field that interacts with the quantum matter field and the introduction of a classical current source interacting with the quantum gauge field, alters the field pattern of the quantum matter and gauge fields in the Heisenberg picture and hence the state of the quantum matter and gauge fields in the Schrodinger picture. At the receiver end we can either adopt the Heisenberg picture and measure the quantum mean and higher order correlations in the given initial state, say a coherent state or else adopt the Schrodinger picture and take measurements on the evolved state and from the measured outcome using the maximum likelihood method, estimaat the classical gauge field and gauge current at the received end that was used to encode the classical message to be transmitted. Another communication problem in quantum mechanics that can be solved using symmetry breaking principles is to allow the scalar field to acquire vacuum expectation values so that massless Goldstone Bosons are generated and these massless Bosons propagate from the transmitter to the receiver and interact with the other fields causing a changed field pattern whose correlations can be measured and thereby used to infer about the vacuum expectation value of the scalar fields which was used to encode a classical message. More generally, we can consider a set of matter and gauge fields upon which the quantum effective action, ie, the potential depends. We assume that this potential is gauge invariant w.r.t local gauge transformations of the matter and gauge fields. Now when the matter field falls to the ground state acquiring nonzero vacuum expectation values, the gauge field dynamics gets altered and by measuring it at the receiver end, we get to know what was the vacuum expectation value of the matter field which was used to encode the message. A similar circle of ideas is applied to supersymmetry breaking. We consider a supersymmetric action for Chiral and gauge superfields interacting with each other. When we set the auxiliary fields in this action equal to their values at which the action is stationary, then we obtain an action for all the remaining fields, ie, the scalar, the Dirac, the gauge and the gaugino fields. This action functional involves a superpotential and the conditions for unbroken supersymmetry, namely that the change in the vacuum expectation values of the Fermionic fields, namely the Dirac and the gaugino field be zero results in conditions on the vacuum expectation values of the scalar field, the superpotential and its derivative evaluated at the vacuum expectation value of the scalar field. When supersymmetry is unbroken, we can calculate the masses of the Bosons and Fermions appearing in the supersymmetric Lagrangian and when supersymmetry is broken, we can then calculate the change in the masses of the same particles. Thus, by breaking supersymmetry by controlling the superpotential, we can control the masses of the particles and hence the dynamics of the Boson and Fermion fields. This fact can be used to communicate signals from a transmitter to a receiver over which the fields are spread out by controlling the parameters in the superpotential using classical messages. In the context of communication using symmetry breaking we also briefly discuss supersymmetry breaking using messenger superfields propagating from a hidden sector in between the Planckian energy scale and the energy scale of the normally observable particles. The superpotential comprises of a trilinear combination of Chiral superfields one of which has components that acquire non-zero vacuum expectation values thereby inducing supersymmetry breaking. The other two Chiral superfields in the trilinear combination in the superpotential are the messenger superfields which also interact with the gauge superfields in the observable energy sector thereby causing the propagators of the components of the gauge superfield to get altered. This alteration in the gauge field propagators induce further corrections to the matter field propagators in the observable sector with which these gauge fields interact. As a result supersymmetry is broken in the observable sector which is the reason why we do not observe superpartners of the observable particles in the lower energy scale.

The idea that messenger superfields can be used to transmit supersymmetry breaking effects can be exploited in a quantum communication system. Finally, we note that gravity can also used to mediate supersymmetry breaking leading to massive particles. The interaction Lagrangian between the metric tensor field and the energy-momentum tensor of matter can be generalized in the form of an interaction between the metric superfield and the matter current superfield. The matter current superfield is the usual supersymmetry current expressible in terms of the matter Chiral superfields and the superpotential. This supersymmetry current obeys a conservation law by virtue of the superfield equations for the Chiral superfield. It follows that the interaction action between the metric superfield and the matter current superfield will contain bilinear interaction terms in the components of the metric superfield and the corresponding components in the current superfield. When some components of the metric superfield acquire non-zero vacuum expectation values, then supersymmetry is broken and since the current superfield (which is bilineary coupled to the metric superfield) is expressible in terms of the matter Chiral superfield and the superpotential (which is a function of the matter superfield) the matter field action will acquire supersymmetry breaking terms expressible in terms of the superpotential acting on the scalar components of the Chiral field and perhaps other Chiral superfields like the gauge superfield used to construct the gauge field action in the form of its "F term" just as the F term of the superpotential acting on the Chiral superfield is also a component of the total action. It is precisely this breaking of supersymmetry that gives the gaugino (which is a component of the gauge superfield) its mass. The gaugino mass appears at a lower energy than the Planckian scale of gravity. At a higher energy, we have gravitinos replacing gauginos and since the gravitinos and gauginos, both interact with the supercurrent, their masses are related. The vacuum energy of the matter fields is balanced by the negative energy of gravitational field with the latter arising because of broken supersymmetry when some components of the metric superfield acquire nonzero vacuum expectation values. Hence, combining these facts, it follows that the expression for the gaugino mass involves both the superpotential as a function of the scalar components of the Chiral matter supefield and the gaugino field with a coefficient being proportional to the gravitino mass. It should be noted that in the bilinear interaction terms between the supercurrent and the metric superfield, one of the components is the bilinear interaction between the metric field and the matter energy-momentum tensor that is present in conventional general relativity and another component is the bilinear interaction between the gravitino field and the supersymmetry current. The action corresponding to the latter term yields an amplitude for goldstino exchange in terms of the matrix elements of the supersymmetry current between two goldstino states and the gravitino propagator. At lower energies, the gravitino is absent and the same reaction involving goldstino exchange has an amplitude in terms of the matrix elements of the supersymmetry current without the presence of gravitino propagator. This latter amplitude can be computed from basic principles of Lorentz invariance and its expression by virtue of supersymmetry current conservation involves a coefficient F whose square is proportional to the vacuum energy of matter because the sum of squared of the matrix elements of the supersymmetry charge between vacuum and one Goldstino states is proportional to the vacuum energy density by virtue of the supersymmetry algebra anticommutation relations which states that the anticommutator between two Fermionic generators is proportional to the Bosonic generator of four momentum. Equating these expressions for the amplitude of Goldstino exchange with and wihout gravitinos, yields an expression for the gravitino mass in terms of the vacuum energy. However since as noted earlier, the positive vacuum energy must be canceled out by the negative energy of the gravitational field, it follows that the coefficient F is related to the vacuum energy of gravitation and hence the gravitino mass is also related to the vacuum energy of gravitation. We also saw that supersymmetry breaking by gravity in the ground state having nonzero expectation values leads to gaugino mass terms and hence the gaugino mass must be proportional to the square root of the vacuum energy of gravitation which in turn as we saw is proportional to the gravitino mass. This sort of supersymmetry breaking is called anomaly mediated supersymmetry breaking because the supercurrent which is bilinearly coupled to the metric superfield has a conservation law with terms dependent upon a Chiral superfield that is nonzero iff the trace of the energy-momentum tensor of the matter field is non-zero iff an "R-current" defined in terms of the supercurrent is not conserved. This phenomenon of mediating supersymmetry breaking via control of the vacuum expecation values of some of the metric superfield components can be exploited in communication theory by controlling gravity at the transmitter end and observing its effect on gaugino masses at the receiver end. This book contains many other such interesting discussions and will be useful for the mathematical physicist and electronics engineer who wishes to use quantum field theory in the design of communication systems. This book also includes some basic problems on algebra and algebraic geometry because of the results on abstract complexes with abstract differentials which enable one to cast the problem of calculating the dimensions of the cohomological groups occurring in quantum field theory as the number zero modes of the Dirac-Yang-Mills operator and the Atiyah-Singer index theorem giving a formula for the Chiral anomaly, namely the difference of the dimensions of the zero eigenspaces of the Dirac operator corresponding to positive and negative chirality in terms of integrals of forms in form of abstract algebraic theorems. The second volume of this set will discuss more problems in this field and will emphasize on how quantum neural networks can be designed using string theory. The basic mechanism in this design involves defining a string amplitude for propagation between an initial and a final string state in the presence of externally interacting strings carrying definite momentum fields or more generally when these externally interacting strings are in definite states defined by excited energy eigenfunctions of the String Hamiltonian obtained in terms of the eigenfunctions of an infinite sequence of quantum mechanical harmonic oscillators. The book also contains some material related to nonlinear, inhomogeneous and anisotropic waveguide and cavity resonator theory based on perturbative computations of the shift in the oscillation frequencies or modes of a confined electromagnetic field when the medium within the confining region has permittivity and permeability tensor that depends upon both space, frequency of the field and more generally, even upon the field itself. Numerical computations of the modes are then based either on perturbation theory involving expanding the inhomogeneous and anisotropic permeability and permittivity tensors as a Fouirer series in the z coordinate and simultaneously also expanding the fields as a Fourier series in the z coordinate and then applying the first stage of perturbation theory in which we treat the higher order spatial Fourier modes as small thereby reducing the problem to an infinite sequence of pde's in the xy coordinates alone to which we then apply the second stage of perturbation theory in which we regard the dependence of the pemittivity and permeability on the xy coordinates as being small so that we can use standard techniques in eigenvalue perturbation theory to calculate the modal shifts as is done in time independent perturbation theory in quantum mechanics for calculating the shift in the energy levels of an atomic system when the Hamiltonian gets perturbed by a small time independent potential. The second numerical method is more frequently used and involves formulating the basic waveguide/cavity resonator modal field problem as a variational problem and then applying the finite element method wherein the spatial region is partitioned into pixels and the field values at the vertices of the pixels are interpolated to give the field values within the pixels using standard basis functions. The variational integral is thus approximated by a function of the vertex node field values and optimizing this energy functionals w.r.t the nodal field values then results in a finite dimensional nonlinear eigenvalue problem which can be solved using standard nonlinear algebraic perturbation theory in a finite dimensional vector space. There is yet another method for solving the original nonlinear infinite dimensional eigenmode problem using the variational/direct method which called the method of moments in which the field within the walls is expanded as a linear combination of basis functions that satisfy the appropriate boundary conditions and then this expansion is substituted into the pde and inner products on both sides with the same basis functions are formed resulting once again in a nonlinear finite dimensional eigenvalue problem.

Author

Table of Contents

Chapter 1

Parameter Estimation, Probability Density, Quantum Mechanics, Blackhole Physics, and Robot Tracking

Calculation of the capacity of a multivariate Gaussian channel.

A quantum neural network based on Schrodinger's equation for synthesizing a family of probability densities that tracks a given family of densities and simultaneously whose mean is forced to be strongly correlated with another signal.

Selected topics in statistical signal and image processing

Information theoretic data compression as an application of Shannon's noiseless coding theorem, estimating dynamically the velocity field of a fluid forced by random terms from discrete spatial velocity measurements using the extended Kalman filter, estimating the metric tensor field, the matter density and velocity field and the electromagnetic field in space-time from discrete noisy spatial measurements of some functions of these–an application of Einstein's general relativity with noisy forces, group theoretic image processing, processing classical image fields by transforming them into quantum state vectors and designing unitary processors.

The expectation-maximization algorithm as a sequence of approximations of a parameter estimate converging to the true maximum likelihood estimate but easier to implement in most cases.

Tracking a desired trajectory by robot subject to stochastic noise using error feedback, tracking a desired wave function/state by a quantum robot by controlling the parameters adaptively in its Hamiltonian and Lindblad operators. When the dynamics of the quantum robot is described by means of a Hudson-Parthasarathy noisy Schrodinger equation, then we take non-demolition measurements to obtain filtered estimates of its state and we control the parameters of the original Hudson-Parthasarathy equation adaptively by matching the filtered system state to a desired state trajectory.

Proof that exponential equivalence of two families of random variables implies that the large deviation rate function of the first family can be derived as a limit of the latter.

Predicting families of probability density functions from measurements of correlated signals using the Schrodinger equation as a natural tool that generates families of probablity densities parametrized by the time variable.

Principal component analysis as an optimization problem in linear algebra: Construct a matrix of rank $\leq r$ that best approximates a given matrix of full rank in the Frobenius norm sense by using the standard factorization of any rank r matrix as a product of two matrices, the first having full column rank r and the second having full row rank r.

DOI: 10.1201/9781003728627-1

A problem in latin square channels on general relativity with emphasis on quantum blackhole physics.

The metric tensor has random or even quantum random fluctuations which means that the metric tensor can be a quantum field described in terms of bosonic spin two creation and annihilation operators or even quantum noisy fields described in terms of creation and annihilation fields in space-time. We solve the Klein-Gordon, or the Dirac equation or the Maxwell equation or non-Abelian matter and gauge field equations in such a background metric and the resulting wave field statistics in a coherent state of the gravitational field can be computed. If we take a wave functional field calculated using an action functional for the field in the background random metric, we can calculate the resulting mixed state of the field after time t after averaging the pure state obained by the Feynman path integral method in the coherent state. A pure state after evolution in the background gravitational field becomes mixed owing to the randomness (or quantum randomness) of the metric and hence entropy is generated by the gravitational field.

1.1 Calculation of the capacity of a multivariate Gaussian channel

Let N be a Gaussian random vector of size $d \times 1$ with zero mean and covariance matrix K_N. Let Y=X+N where X is a random vector independent of N. The capacity of this Gaussian channel is $sup_{p(X)}(H(X+N) - H(N)) = (sup_{p(X)}H(X+N)) - H(N)$. This supremum is to be calculated over all probability distributions p(X) of X with a power constraint
$mathbbEX^2 \leq P$. It is clear that under this power constraint, the supremum is attained when X is a normal random vector with covariance K_X chosen so as to maximize $(1/2)log(|K_X + K_N|/|K_N|)$ subject to the constraint $Tr(K_X) \leq P - Tr(K_N)$ where it is assumed that $P \geq Tr(K_N)$.

1.2 A quantum neural network based on Schrodinger's equation for synthesizing a family of probability densities that tracks a given family of densities and simultaneously whose mean is forced to be strongly correlated with another signal

Consider the motion of an electron bound to the nucleus with a potential $V(r)$. This electron is made to interact with an external electromagnetic field $E(t), B(t)$ which are functions of only time and not the spatial coordinates. This electromagnetic field can be derived from an electric potential $\Phi(t,r) = -E(t).r$ and a magnetic vector potential $A(t,r) = B(t) \times r/2$. The total Hamiltonian of the electron is then

$$H(t) = (p+eA(t,r))^2/2m+V(r)-e\Phi(t,r)$$
$$= H_0-(ie/2m)(divA(t,r)+2(A(t,r),\nabla))+e^2A(t,r)^2/2m$$
$$= H_0 + eV_1(t) + e^2V_2(t)$$

wy.here

$$H_0 = p^2/2m + V(r)$$

is the Hamiltonian of the electron bound to its nucleus without any electromagnetic perturbation and $V_1(t) = (-i/2m)(B(t) \times r, \nabla) - E(t).r$ and $V_2(t) = (B(t) \times r)^2/8m$. Note that $V_1(t)$ is a linear function of the electromagnetic field components while $V_2(t)$ is a quadratic function of the magnetic field. We wish to make the output wave function $\psi(t,r)$ yield a mean value of some function $f(r)$ of the position at time t given by

$$\mu(\psi,t) = \int f(r)|\psi(t,r)|^2d^3r$$

that is close to a given signal z(t) while at the same time, cause the pdf $|\psi(t,r)|^2$ track a given pdf $p_0(t,r)$. This problem is therefore something like training our quantum neural network (qnn) to track a high dimensional probability density $p_0(t,r)$ while taking as input a signal $z(t)$ (lower dimensional quantity) that is presumed to be correlated with the given high dimensional density. Thus, our adaptive algorithm for updating $E(t), B(t)$ in time must be designed based on an adaptive gradient algorithm to minimize some instantaneous error energy of the form

$$c_1(z(t) - \mu(\psi,t))^2 + c_2(p_0(t,r) - |\psi(t,r)|^2)^2$$

1.3 Selected topics in statistical signal and image processing

Special issue on applications of statistical image processing to engineering, physics and astronomy.

Scope of the journal.

The papers in in this journal involve theoretical developments in statistical image processing like parameter and field estimation from noisy measurements, modeling image fields using partial differential equations, real time stochastic filtering of imaged fields based on noisy measured data, enhancement and denoising of images using quantum image processing techniques, group representation theoretic techniques for filtering image fields on a surface, for feature extraction and for statistical pattern recognition and for estimating the group transformation element acting on an image field. The journal will also publish applications of the above techniques to physics and astronomy especially to problems of determining cosmic expansion parameters for evolving 3-D galaxies based on noisy 2-D images, for estimating the parameters involving scattering, absorption and emission of elementary particles during scattering processes based on Feynman diagrammatic computations of the probabilities of the process involved, estimating the wave function and density operator of a quantum system using classical and quantum stochastic filtering techniques.

1.4 EM algorithm for implementing the maximum likelihood estimator

$p(x|\theta)$ is the probability density function of the observation x given the parameter θ. z is the latent r.v.

Let q(z) be any density.

$$log(p(x|\theta)) = \int log(p(z,x|theta)/q(z))q(z)dz + \int log(q(z)/p(z|x,\theta))q(z)dz$$

since

$$p(z,x|\theta) = p(z|x,\theta)p(x|\theta)$$

We start with a guess estimate θ_0 for θ and set $q(z) = p(z|x,\theta_0)$. We write the two terms above as

$$T_1(x,\theta,\theta_0) = \int log(p(z,x|theta)/q(z))q(z)dz = log(p(z,x|theta)/p(z|x,\theta_0))p(z|x,$$

and

$$T_2(x,\theta,\theta_0) = \int log(q(z)/p(z|x,\theta))q(z)dz = \int log(p(z|x,\theta_0)/p(z|x,\theta))p(z|x,\theta_0)dz$$

We maximize $T_1(x,\theta,\theta_0)$ w.r.t θ and denote the resulting value of θ by θ_1. It is obvious that

$$T_2(x,\theta_0,\theta_0) = 0 \leq T_2(x,\theta_1,\theta_0)$$

by the non-negativity of the relative entropy function. It follows that since

$$T_1(x,\theta_1,\theta_0) \geq T_1(x,\theta_0,\theta_0)$$

that

$$log(p(x|\theta_1)) = T_1(x,\theta_1,\theta_0)+T_2(x,\theta_1,\theta_0) \geq T_1(x,\theta_0,\theta_0)+T_2(x,\theta_0,\theta_0) = log(p(x|\theta_0)$$

and hence θ_1 is an improvement on θ_0 towards obtaining the maximum likelihood estimate of θ.

1.5 Tracking a desired trajectory by robot subject to stochastic noise using error feedback, tracking a desired wave function/state by a quantum robot by controlling the parameters adaptively in its Hamiltonian and Lindblad operators. When the dynamics of the quantum robot is described by means of a Hudson-Parthasarathy noisy Schrodinger equation, then we take non-demolition measurements to obtain filtered estimates of its state and we control the parameters of the original Hudson-Parthasarathy equation adaptively by matching the filtered system state to a desired state trajectory

Consider two robots, a master robot and a slave robot. The Master robot has the Hamiltonian $H_1(q_1, \theta_1(t))$ while the slave robot has the Hamiltonian $H_2(q_2, \theta_2(t))$. Here, q_1 is a complete set of commuting variables for the master robot while q_2 is a complete set of commuting variables for the slave robot. $\theta_1(t)$ are control parameters for the master robot while $\theta_2(t)$ are the control parameters for the slave robot. These control parameters are to be updated according to a feedback adaptive control law. Let $\psi_1(t, q_1)$ denote the wave function of the master robot and $\psi_2(t, q_2)$ the wave function for the slave robot both in the absence of noise. Thus these two wave functions satisfy the Schrodinger equation:

$$i\partial_t\psi_1(t, q_1) = H(q_1, \theta_1(t))\psi_1(t, q_1), i\partial_t\psi_2(t, q_2) = H_2(q_2, \theta_2(t))\psi_2(t, q_2)$$

where the adaptive control law for $\theta_k(t), k = 1, 2$ may be taken as

$$\theta_1(t+dt) = \theta_1(t) - \mu_1 dt.\partial/\partial\theta_1(t) \sum_{k=1}^{p} (< \psi_1(t+dt)|X_k|\psi_1(t+dt) > - < \psi_2(t+dt)|X_k|\psi_2(t+dt) >)^2$$

$$\theta_2(t+dt) = \theta_2(t) - \mu_2 dt.\partial/\partial\theta_2(t) \sum_{k=1}^{p} (< \psi_1(t+dt)|X_k|\psi_1(t+dt) > - < \psi_2(t+dt)|X_k|\psi_2(t+dt) >)^2$$

with $X_k, k = 1, 2, ..., p$ being a set of observables. The idea here is that q_1 and q_2 are identical variables but acting in different components of the tensor product of the master robot and slave robot Hilbert spaces and we adapt the control parameters at each time instant so as to minimize the sum of the error energies between the average value of observables in for the master and slave robots.

1.6 Proof that exponential equivalence of two families of random variables implies that the large deviation rate function of the first family can be derived as a limit of the latter

A problem in large deviation theory: Let $X(\epsilon), \epsilon \to 0$ be a family of random variables. Suppose $Y(\epsilon, m), \epsilon \to 0, m \to \infty$ is another family of random variables satisfying for each m an LDP with rate function I_m and suppose that the latter family is exponentially equivalent to the former, ie,

$$limsup_{m\to\infty} limsup_{\epsilon\to 0}\epsilon.log(P(|Y(\epsilon, m) - X(\epsilon)| > \delta)) = -\infty$$

Then, $X(\epsilon), \epsilon \to 0$ satisfies the LDP with rate function

$$I(x) = lim_{\delta\to 0} lim_m inf_{y\in B(x,\delta)} I_m(y)$$

To see this, we observe that

$$\{X(\epsilon) \in B(x, \delta)\} \subset \{Y(\epsilon, m) \in B(x, 2\delta))\} \cup \{\| X(\epsilon) - Y(\epsilon, m) \| > \delta\}$$

because

$$|Y(\epsilon, m) - x| \le |X(\epsilon) - x| + |Y(\epsilon, m) - X(\epsilon)|$$

and therefore,

$$P(X(\epsilon) \in B(x, \delta)) \le P(Y(\epsilon, m) \in B(x, 2\delta)) + P(|X(\epsilon) - Y(\epsilon, m)| > \delta)$$

Thus, by exponential equivalence,

$$limsup_{\epsilon\to 0}\epsilon.log(P(X(\epsilon) \in B(x, \delta)))$$

$$\le liminf_m limsup_\epsilon \epsilon.log(P(Y(\epsilon, m) \in B(x, 2\delta))$$

and

$$liminf_\epsilon \epsilon.log(P(X(\epsilon) \in B(x, \delta)))$$
$$\le liminf_m liminf_\epsilon \epsilon.log(P(Y(\epsilon, m) \in B(x, 2\delta))$$

and likewise,

$$\{Y(\epsilon, m) \in B(x, \delta)\} \subset \{X(\epsilon) \in B(x, 2\delta)\} \cup \{|Y(\epsilon, m) - X(\epsilon)| > \delta\}$$

so that once again by exponential equivalence,

$$limsup_m limsup_\epsilon \epsilon.log(P(\{Y(\epsilon, m) \in B(x, \delta)\}$$

$$\le limsup_\epsilon P(\{X(\epsilon) \in B(x, 2\delta)\})$$

and

$$limsup_m liminf_\epsilon \epsilon.log(P(\{Y(\epsilon, m) \in B(x, \delta)\}$$
$$\le liminf_\epsilon P(\{X(\epsilon) \in B(x, 2\delta)\})$$

From these equations, the stated result is immediate.

1.7 Predicting families of probability density functions from measurements of correlated signals using the Schrodinger equation as a natural tool that generates families of probablity densities parametrized by the time variable

Predicting higher dimensional signals from lower dimensional signals. A typical example of this is predicting EEG signals from speech signals. Speech is a low dimensional signal since it is highly correlated and hence its prediction error variance is very small. We can then use Kolmogorov's formula for the entropy rate of a Gaussian process to show that speech has a very low information rate. On the other hand, the EEG signals of a person are high dimensional signals especially when the person has a brain disease since these signals collected at different points on the head surface are very erratic and display a high degree of uncorrelatedness. This causes their prediction error variance to be very high and hence can be regarded as high dimensional signals. Moreover, EEG signals are collected at different spatial points on the head surface and hence even their physical dimnension is very large. Prediction of a high dimensional signal from a low dimensional one therefore involves making additional use of some sort of a parametric model for the low and high dimensional signals and then predicting the parameters in the latter model from the parameters of the former. Another important fact of such prediction theory is the situation in which we are more interested in the statistical properties of the signal to be predicted rather than in the actual sample waveforms. In this case we must then attempt to predict the joint probability density of the high dimensional signal at a set of time points whose successive time difference lags have been specified. This is because a stationary process is characterized completely by its finite dimensional probability distributions at times whose lags, ie, successive time differences alone are specified. In order to predict the joint probability distributions of the high dimensional signal from low dimnensional signal data, we must exploit some sort of correlatedness between the two data. Such a prediction is most easily accomplished using a quantum neural network (qnn) in which Schrodinger's equation generates a one parameter family of joint probability density functions naturally as the modulus square of the wave function with the potential energy in Schrodinger's equation being adapted according to the error between the low dimensional signal amplitude and its predicted value based on quantum averages of the high dimenional pdf calculated using the wave function. The qnn is trained using this error and also the error between the true pdf of the high dimensional signal and that obtained using the modulus square of the wave function generated by Schrodinger's equation in the qnn. For example, we can take our qnn to be a high dimensional version of the Hydrogen atom Hamiltonian with the inverse square potential law and perturb this Hamiltonian by a control time varying electromagnetic field interacting with the electron and adapted in such a way as to cause the wave function modulus square to track the given high dimensional pdf with the scale factor of the electromagnetic field being proportional to the square of the prediction error energy between the low dimensional signal and its predicted value based on quantum averages of the high dimensional pdf generated from the qnn. The Hydrogen atom model is particularly useful because all atoms are described by this poential apart from small perturbations coming from the interaction energy between the different electrons. Perturbation theory gives us the corrections to the Hydrogenic atom wave function caused by mutual electron interactions and thus qnn's can be designed based on the behaviour of actual atoms in the laboratory by introducing control electromagnetic fields. A lot of work in the existing literature has gone into predicting high dimensional signals using fuzzy neural networks but not much work has gone into predicting pdf's, the primary reason being that a pdf has several constraints like non-negativity and integrability to unity which are hard to impose using conventional neural networks. QNN's based on the Schrodinger equation are naturally able to accomplish these two constraints. Further, QNN's can naturally be implemented in the laboratory using atoms and molecules or even a crystal lattice of atoms and molecules.

1.8 Principal component analysis as an optimization problem in linear algebra

Construct a matrix of rank $\leq r$ that best approximates a given matrix of full rank in the Frobenius norm sense by using the standard factorization of any rank r matrix as a product of two matrices, the first having full column rank r and the second having full row rank r

Linear algebra in principal component analysis: Let X be an $n \times m$ matrix. The aim is to choose for a given $r < min(n, m)$, an $n \times m$ matrix Y of rank $\leq r$ such that $\| X - Y \|$ is minimized where $\| . \|$ denotes Frobenius norm. The answer to this problem is as follows. Let

$$X = UDV^*$$

be the singular value decomposition of X. We can always choose the unitary matrices U, V (after an appropriate permutation of the columns of U and the corresponding rows of V) so that the $n \times m$ diagonal matrix D has the form

$$D = \begin{pmatrix} D_1 & 0 \\ 0 & 0 \end{pmatrix}$$

where

$$D_1 = diag(d_1, d_2, ..., d_p), d_1 \geq d_2 \geq ... \geq d_p > 0$$

where $p = rank(X)$. Then, the optimal Y is given by

$$Y = UD_0V^*$$

where

$$D_0 = \begin{pmatrix} D_2 & 0 \\ 0 & 0 \end{pmatrix}$$

with

$$D_2 = diag[d_1, ..., d_r]$$

In other words, D_0 is obtained from D by setting to zero all the diagonal values $d_{r+1}, ..., d_p$. The proof of this result is obtained by noting that any $n \times m$ matrix of rank $\leq r$ can be expressed as $Y = EF$ where E is $n \times r$ and F is $r \times m$. Then expand $\| X - EF \|^2$ as

$$U(E, F) = \| X \|^2 + Tr(F^*E^*EF) - 2Re(Tr(F^*E^*X)))$$

set the variational derivative of this w.r.t F^* to zero to get

$$E^*EF - E^*X = 0$$

to get

$$F = (E^*E)^{-1}E^*X$$

Substitute this expression into U to get

$$U = Tr(X^*(X - EF)) = Tr(X^*X) - Tr(X^*PX)$$

where

$$P = E(E^*E)^{-1}E^*$$

is the orthogonal projection onto $Range(E)$. As E varies over all $n \times r$ matrices, P will vary over all $n \times n$ orthogonal projections of rank $\leq r$. It is then clear that $Tr(X^*PX) = Tr(PXX^*)$ is a maximum when

$$P = UZU^*$$

where

$$Z = \begin{pmatrix} I_r & 0 \\ 0 & 0 \end{pmatrix}$$

and substitituting this into the expression

$$Y = EF = PX$$

gives us the required solution to this optimization problem.

1.9 Latin square channels

Let A be a finite alphabet and let $\theta : A \times A \to A$ be a Latin square, ie, for each $x \in A$, $y \to \theta(x,y)$ is a permutation of A and $y \to \theta(y,x)$ is also a permutation of A. Let ν be a probabiliity distribution on A and define $\nu_x(y) = \nu(\theta(x,y))$,$x, y \in A$. Then $x \to_x$ is a transition probability distribution on A. Define

$$N(n,\nu,\delta) = min(|E| : E \subset A^n, \nu^n(E) > 1-\delta\})$$

and $M(n,\nu,\delta)$ to be the maximum size of a code for strings in A^n for which there exists a decoder with error probability smaller than δ. Note that A code of size M for A^n with error probability smaller than δ means a set of M distinct elements $u_k, k = 1,2,...,M$ in A^n along with mutually disjoint subsets $V_k, k = 1,2,...,M$ of A^n such that $\nu_{u_k}(V_k) > 1-\delta, k = 1,2,...,M$ where by ν_u, we mean the product probability distribution $\nu_{u(1)} \times ... \times \nu_{u(n)}$ on A^n where $u = (u(1),...,u(n)) \in A^n$.

Problem: For a Latin square channel, what is the relationship between $N(n,\nu,\delta)$ and $M(n,\nu,\delta)$? Note that from the standard noiseless coding theorem of Shannon, we have the result that

$$lim_\delta limsup_n log(N(n,\nu,\delta))/n = lim_\delta liminf_n log(N(n,\nu,\delta))/n = H(\nu)$$

and from the standard noisy coding theorem of Shannon,

$$lim_\delta limsup_n log(M(n,\nu,\delta))/n = lim_\delta liminf_n log(M(n,\nu,\delta))/n = C(\nu)$$

where $C(\nu)$ is the capacity of the channel $(x,y) \to \nu_x(y)$ For this Latin square channel since $H(\nu_x) = H(\nu)$ since ν_x is simply a permutation of ν, it follows that

$$C(\nu) = sup_\mu(H(\mu) - H(\nu)) = max_\mu H(\mu) - H(\nu) = log(a) - H(\nu)$$

Suppose $n = 1$. Consider $M(\nu,\delta) = M(1,\nu,\delta)$. Thus, we can choose distinct $x_1,...,x_M$ in A and disjoint subsets $V_1,...,V_M$ in A such that $\nu_{x_k}(V_k) = \nu(\theta(x_k,V_k)) > 1-\delta, k = 1,2,...,M$.

Likewise considering N$(\nu,\delta) = N(1,\nu,\delta), wecanfindan$E$\subset A$ such that $|E| = N$ and $\nu(E) > 1-\delta$. Now $\nu(\theta(x_k,V_k)) > 1-\delta$ implies

$$N(\nu,\delta) \leq |\theta(x_k,V_k))| = |V_k|$$

for each k. Summing this over all k then results in

$$NM \leq \sum_k |V_k| \leq |A|$$

since the $V_k's$ are disjoint. Thus,

$$M(\nu,\epsilon) \leq |A|/N(\nu,\epsilon)$$

Intuitively this means that if $N(\nu,\epsilon)$ is very large, then we require a large set to obtain a probability $> 1-\epsilon$ which means we cannot find to many disjoint sets each having a probability $> 1-\epsilon$ and therefore, we cannot find a large set of codewords which can be decoded with high probability.

1.10 general rel-ativity and cosmology with emphasis on quantum blackhole physics

The Lorentz group of special relativity as the group of linear transformations that preserves the Minkowski metric of flat space-time or equivalently as the group of linear transformations of the space-time manifold that preserves the speed of light.

Proper time as a Lorentz invariant.

Four vectors in special relativity as quantities that transform as the space-time coordinates under a Lorentz transformation.

The Lorentz transformation between inertial systems.

On the existence of an absolute inertial frame.

Special relativity mechanics, four velocity and four momentum vectors. Energy and momentum as the components of the four momentum vector.

Tensors in special relativity and their transformation laws. The energy momentum tensor of a collection of particles and of a fluid. Fluid dynamics in special relativity as the vanishing of the four divergence of the energy-momentum tensor.

Viscous fluids with thermal conduction in special relativity. The contribution of viscosity and thermal conduction to the energy-momentum tensor of a fluid.

Invariance of the Maxwell equations under Lorentz transformations, the Lorentz transformation laws of the electromagnetic field.

Special relativity quantum mechanics: The Klein-Gordon equation of the scalar field in the presence of electromagnetic interactions, Factorization of the energy-momentum relationship of special relativity into linear factors using the Dirac matrices, Dirac's relativistic equation for the electron in the presence of an electromagnetic field.

The spinor representation of the Lorentz group, invariance of the Dirac equation under Lorentz transformations.

Entropy evolution of the density operator in Dirac's equation coupled to a noisy bath.

Evolution of a system of identical Dirac particles with mutual interaction, the quantum Boltzmann approximation.

Special relativistic motion of an elastic body. The Lagrangian of an elastic body, quantization of the elastic equations of motion.

$$S = \int rho_0 g_{\mu\nu}(x) u^{\mu}_{,t} u^{\nu}_{,t} d^4x - (1/2) \int C(\mu\nu\rho\sigma) u^{\mu\nu} u^{\rho\sigma} d^4x$$

where

$$u^{\mu\nu} = \partial^{\nu} u^{\mu} + \partial^{\mu} u^{\nu}$$

is the strain tensor.

The entropy of a relativistic Langevin particle in stochastic classical mechanics.

$$d(v(t)/\sqrt{1 - v(t)^2/c^2}) = -bv(t)dt + \sigma.dB(t)$$

Solve using perturbation theory with $1/c$ as the small perturbation parameter.

Chapter 2

Tracking Pdfs Through Control, Quantum Systems, Neural Networks, and Belavkin Filters

2.1 Design of classical and quantum neural networks for tracking probability densities using elastic materials by adaptation of control parameters like the anisotropic elastic constants

Consider an elastic body with elastic coefficient tensor $\{C(abcd)\}$. Let $u_a(t,r)$ be the displacement field of the elastic body so that its strain tensor field is given by

$$u_{ab} = u_{a,b} + u_{b,a}$$

Letting ρ denote the mass density of the elastic body, its total kinetic energy is given by

$$T = (1/2)\sum_a \int \rho u_{a,t}^2 d^3r$$

and its potential energy by

$$U = (1/2)\int C(abcd)u_{ab}u_{cd}d^3r$$

The contribution to the action functional from the interaction of the charged elastic body with the external electromagnetic field is given by

$$L_{em} = -\rho_q \int \Phi(t, r + u(t,r))d^4x$$

$$+\rho_q \int u_{a,t}(t,r)A_a(t,r)d^4x$$

Since the displacement of the elastic body is small, we can after neglecting a quantity that is independent of the displacement field write

$$L_{em} = \rho_q E_a(t,r)u_a(t,r)d^3r + \rho_q \int u_{a,t}A_a d^3r$$

where

$$E_a = -\Phi_{,a}$$

is the electrostatic field. The complete equation of the elastic body in the presence of electromagnetic and random forces is then

$$\rho_m u_{a,tt}(t,r) = \rho_q(E_a(t,r) + \epsilon(abc)u_{b,t}(t,r)B_c(t,r))$$

$$+C(abcd)u_{cd,b}(t,r) + w_a(t,r)$$

DOI: 10.1201/9781003728627-2

The idea now is to control the electromagnetic fields and the elastic constants (by allowing the latter to be functions of time) so that the empirical probability density functional of the displacement field of the elastic body tracks a given pdf with time. This tracking can be done by making use of the large deviation principle for evaluating the probability that the empirical density of the displacement field deviates from a desired one by an amount more than a given threshold.

2.2 Designing a quantum neural network based on Feynman's path integrals for fields from a Lagrangian density with time varying adaptive control parameter tuned so that the probability density functional of the field calculated using the path integral at time t tracks a given probability density functional

Using Einstein field equations for a classical and a quantum neural network.

Assume that we are given a Lagrangian density $L(\phi, \phi_{,\mu}, f)$ where $\phi = \{\phi_k\}$ is a set of fields driven by a set $f = \{f_k\}\}$ of inputs. We also assume that the system is subject to some random forces g so that the field equations can be expressed as

$$\partial_\mu \frac{\partial L}{\partial_{,\mu}} - \frac{\partial L}{\partial \phi} = g$$

These are nonlinear stochastic pde's. The aim is to choose the parameters θ in the control inputs f so that the empirical pdf of the fields matches as closely as possible a given pdf. Assuming that the noise g is small, we can calculate using perturbation theory, rate functional of the empirical pdf of the fields as the size of the space-time set based on which the empirical pdf of the fields is being computed becomes very large. For example. Then we compute this rate function evaluated at the desired density and minimize this w.r.t the control parameters. In this way, the desired pdf has been compressed into the set of control parameters. Now suppose that we pass over to the quantum scenario wherein the field equations are quantized. This can be done for example using the Feynman path integral. The pdf of the field a time t can be computed using the path integral:

$$P(\phi_t, t) = |\int_{\phi_0, \phi_t} exp(i \int S_t(\phi, theta)) D\phi|^2$$

where the path integral is w.r.t the fields over all space and over the time interval $[0, t]$ with the initial field over all space being ϕ_0 and the final field at time t being $\phi_t = \phi$, the field configuration at which the pdf is being evaluated. Here, we are assumign that

$$S_t(\phi, \theta]) = \int_{\mathbb{R}^3 \times [0,t]} L(\phi(s, x), \phi_{,\mu}(s, x), \theta(s)) d^3x ds$$

is the action functional of the fields based on data upto time t and θ are the control parameters of the Lagrangian. These parameters must be adapted in such a way so that the Schrodinger pdf $P(\phi, t)$ tracks a given probability density functional $P_0(\phi, t)$ with time. Such an adaptive control law may be based on the gradient descent algorithm such as

$$d\theta(t)/dt = -\mu.\frac{\partial}{\partial\theta(t)}(P_t(\phi, t) - P_{0t}(\phi, t))^2$$

Implementing this algortithm thus amounts to calculating

$$\partial P_t(\phi, t)/\partial\theta(t)$$

More generally, we can adopt a neural firing approach in which the dependence of the error energy on the past values of the parameters is also taken into account. Such an algorithm will amount to adopting the iteration

$$d\theta(t)/dt = -\sum_{t_k \leq t} \mu(k)\frac{\partial}{\partial\theta(t_k)}(P_t(t, \phi)) - P_{0t}(\phi, t))^2$$

Still more generally, we can consider an observable X in field space, ie, in the ϕ representation, ie, the field poistion representation, X is represented by a "continuous matrix"

$$< \phi|X|\phi' >= X(\phi, \phi')$$

and then the average value of the field at time t is computed as

$$< X > (t) = \int \bar{\psi}(\phi, t)X(\phi, \phi')\psi(\phi', t)D\phi.D\phi'$$

where

$$\psi(\phi_t, t) = \int_{\phi_0, \phi_t} exp(i\int S_t(\phi, theta))D\phi$$

We choose several such observables $X_k, k = 1, 2, ..., N$ and then design our adaptive gradient search algorithm so as to minimize

$$\sum_{k=1}^{N}(< X_k > (t) - X_d(t))^2$$

at each time using the approximate gradient search algorithm:

$$d\theta(t)/dt = -\mu\frac{\partial}{\partial\theta(t)}\sum_{k=1}^{N}(< X_k > (t) - X_d(t))^2$$

2.3 Quantum nonlinear neural networks based on the quantum Boltzmann equation

The quantum Boltzmann equation can be derived by partial tracing the Schrodinger evolution equation for the mixed state of a quantum mechanical system comprising N identical particles assuming that the joint density operator is invariant under permutations in the position representation. This evolution equation has the form

$$i\partial_t \rho(t) = [\sum_{k=1}^{N} H_k(t), \rho(t)] + [\sum_{1\le k<j\le N} V_{kj}, \rho(t)]$$

where

$$H_k(t) = (P_k + eA_k(t, r_k))^2/2m - e\Phi(t, r_k) - eV_0(r_k) = H_{0k} + eV_{1k}(t) + e^2V_{2k}(t)$$

with r_k, P_k being respectively the position and momentum of the kth particle, $V_0(r_k)$ is the interaction potential energy of the kth particle, ie, electron with its nucleus. H_{0k} is the unperturbed Hamiltonian of the kth electron:

$$H_{0k} = P_k^2/2m + V_0(r_k)$$

$$V_{1k}(t) = -\Phi(t, r_k) + (1/2m)(P_k.A(t, r_k) + A(t, r_k).P_k)$$

$$= \Phi(t, r_k) - (ie/2m)(2(A(t, r_k), \nabla_k) + divA(t, r_k))$$

$$V_{2k}(t) = A(t, r_k)^2/2m$$

Let

$$\rho_1(t) = Tr_{23...N}\rho(t), \rho_{12}(t) = Tr_{34...N}\rho(t) = \rho_1(t) \otimes \rho_1(t) + g_{12}(t)$$

where g_{12} is assumed to be small. Here we are assuming that for each $r = 1, 2, ..., N$ all the marginals of $\rho(t)$ obtained by taking partial traces over the remaining $N - r$ Hilbert spaces are identical copies of each other but acting on different component tensor product Hilbert space. The different component Hilbert spaces are all identical copies of $L^2(\mathbb{R}^3)$. Thus we get by partial tracing the exact equations

$$i\partial_t \rho_1(t) = [H_1(t), \rho_1(t)] + (N - 1)Tr_2[V_{12}, \rho_{12}(t)]$$

$$i\partial \rho_{12}(t) = [H_1(t) + H_2(t) + V_{12}, \rho_{12}(t)] + (N - 2)[V_{13} + V_{23}, \rho_{123}(t)]$$

When we make approximations, by regarding the interaction potentials $V_{ij}, i < j$ as small just as g_{12}, then we get from these equations,

$$= i(\rho_1' \otimes \rho_1 + \rho_1 \otimes \rho_1' + g_{12}') = [H_1, \rho_1] \otimes \rho_1 + \rho_1 \otimes [H_1, \rho_1] + ig_{12}'$$

$$= [H_1 + H_2 + V_{12}, \rho_1 \otimes \rho_1]$$

or equivalently,

$$ig_{12}' = [H_1 + H_2 + V_{12}, \rho_1(t) \otimes \rho_1(t)]$$

so that

$$g_{12}(t) = -i\int_0^t [H_1(s) + H_2(s) + V_{12}, \rho_1(s) \otimes \rho_1(s)]ds$$

Exercise:Indicate the approximations made in deriving this equation. We thus get

$$i\partial_t \rho_1(t) = [H_1(t), \rho_1(t)] + (N - 1)Tr_2[V_{12}, g_{12}(t)]$$

$$= [H_1(t), \rho_1(t)] + (N-1)\int_0^t Tr_2(ad(V_{12}).ad(H_1(s)+H_2(s)+V_{12})(\rho_1(s)\otimes\rho_1(s))ds$$

This has the approximate solution

$$\rho_1(t) = T(t,0)(\rho_1(0)) + (N-1)\int_{0<s<t_1<t} T(t,t_1)(Tr_2(ad(V_{12}).ad(H_1(s)+$$

$$H_2(s)+V_{12})(T(s,0)\otimes T(s,0)(\rho_1(0)\otimes\rho_1(0))dsdt_1)$$

where

$$T(t,s) = T\{exp(-i\int_s^t ad(H_1(u)))du)$$

Assume now that there are parameters θ in H_1 and V_{12} that can be made time dependent. The aim is to adapt these parameters so that the average value of a set of observables $X_a, a \in I$ given by $Tr(\rho_1(t)X_a), a \in I$ are highly correlated with given corresponding signals $s_a(t), a \in I$ and simultaneously, the probability density of another observable Y given by $Tr(\rho(t)dE_Y(\lambda)/d\lambda)$ where E_Y is the spectral measure of Y, tracks a given probability density $p_0(t,\lambda)$. The idea herein is to predict $Tr(\rho_1(t)X_a)$ based on $s_a(t-m), m = 0,1,...,D$ for each a and to make the adaptation constant an increasing function of the resultant prediction error energies while using the gradient algorithm to adapt θ to minimize the instantaneous error energy in the pdf $\int(p_0(t+dt,\lambda) - Tr(\rho_1(t+dt)E'_Y(\lambda)))^2 d\lambda$. The gradient of this error energy w.r.t $\theta(t)$ is calculated by substituting for $\rho_1(t+dt)$ its expression in terms of $\rho_1(t)$ and $\theta(t)$ given by the quantum Boltzmann equation.

2.4 Probabilistic neural networks. The weights of the neural network are made random with a joint probability density defined by the Schrodinger wave equation whose potential parameters are adapted so that the output of the neural network tracks a given output process for a given input

Probabilistic neural networks: The weights do not describe the neural network but rather the probability distribution of the weights. If for example, a neural network is trained with an input stochastic process and another output stochastic process, then the weights will be random and dependent upon the statistics of the input an output stochastic process. In this case, the limiting converged weights will have a probability distribution or if the input and output stochastic processes are jointly non-stationary, the weight process will be characterized by its evolving probability density. Such an evolving probability density can in turn be characterized by the modulus square of a quantum mechanical wave function evolving according to Schrodinger's wave equation.

2.5 Belavkin extended Kalman filter in which the Hudson-Parthasarathy noisy Schrodinger equation contains time varying classical random parameters in its Hamiltonian and Lindblad operators that follow a classical Markovian evolution with known infinitesimal generator

Simultaneous estimation of quantum observables and classical parameters of a quantum system defined by the Hudson-Parthasarathy noisy Schrodinger equation based on non-demolition observables (Quantum version of the extended Kalman filter).

The Hudson-Parthasarathy noisy Schrodinger evolution equation is

$$dU(t) = -(iH(\theta(t)) + P(\theta(t))dt + L(\theta(t))dA(t) - L(\theta(t))^*dA(t)^*)U(t)$$

where

$$P(\theta) = L(\theta)L(\theta)^*/2$$

$\theta(t)$ is a finite classical parameter vector satisfying the classical sde

$$d\theta(t) = d\epsilon(t)$$

where $\epsilon(t)$ is a vector valued Brownian motion process with covariance $\sigma^2 min(t,s)$. The non-demolition measurement process is given by

$$Y_o(t) = U(t)^*Y_i(t)U(t), Y_i(t) = cA(t) + \bar{c}A(t)^*$$

let $\eta_o(t) = \sigma(Y_o(s) : s \leq t)$ the measurement algebra upto time t. Assume that the bath is in the vacuum coherent state $|e(0)>$. Let f be a complex valued function on the parameter space. Let X be an observable in the system Hilbert space. Thus the direct sum (X, f) is our extended observable in system and parameter space. Let $j_t(X, f)$ denote the evolution of this observable after time t. Thus,

$$j_t(X, f) = (U(t)^*XU(t), f(\theta(t)) = (k_t(X), f(\theta(t)) = (k_t(X), l_t(f))$$

We have by quantum Ito's formula,

$$dk_t(X) = k_t(\phi_0(X, \theta(t))dt + k_t(\phi_1(X, \theta(t))dA(t) + k_t(\phi_2(X, \theta(t)))dA(t)^*$$

where

$$\phi_0(X, \theta) = i[H(\theta), X] - P(\theta)X/2 - XP(\theta)/2 + L(\theta)XL(\theta)^*$$

$$= i[H(\theta), X] - (1/2)(L(\theta)L(\theta)^*X + XL(\theta)L(\theta)^* - 2L(\theta)XL(\theta)^*))$$

$$\phi_1(X, \theta) = XL(\theta) - L(\theta)X, \phi_2(X, \theta) = L(\theta)^*X - XL(\theta)^*$$

We also have by the classical Ito formula,

$$dl_t(f) = df(\theta(t)) = f'(\theta(t))d\epsilon(t) + (\sigma^2/2)\Delta f(\theta(t))dt$$

where

$$\Delta = \sum_k \frac{\partial^2}{\partial \theta_k^2}$$

is the Laplacian in parameter space. We let

$$D_i = \partial/\partial\theta_i$$

and then we can write

$$dl_t(f) = \sum_i l_t(D_i f)d\epsilon_i(t) + l_t(\sigma^2 \Delta f/2)dt$$

The two equations can be combined and expressed as

$$dj_t(X,f) = j_t(\psi_0(X,f))dt + j_t(\psi_1(X,f))dA(t) + j_t(\psi_2(X,f))dA(t)^* \\ + j_t(\psi_{3k}(X,f))d\epsilon_k(t)$$

where summation over the repeated index k is implied. Here,

$$\psi_0(X,f) = (\phi_0(X,\theta), \sigma^2\Delta f/2)$$

regarded as direct sums of two observables,

$$\psi_1(X,f) = (\phi_1(X,\theta), 0), \psi_2(X,f) = (\phi_2(X,\theta), 0),$$

$$\psi_{3k}(X,f) = (0, D_k f)$$

In all these expressions, $(X,f) = diag(X,f)$ is the operator direct sum of the two observables X, f. We can equivalently write

$$(X(t), f(\theta(t))) = j_t(X,f) = (k_t(X), l_t(f))$$

Now we are in a position to develop quantum filtering theory in the language of Belavkin for estimating both $X(t) = k_t(X)$ and $f(\theta(t)) = l_t(f)$. Let

$$\pi_t(X,f) = \mathbb{E}(j_t(X,f)|\eta_o(t)) \\ = (\mathbb{E}(k_t(X)|\eta_o(t)), \mathbb{E}(l_t(f)|\eta_o(t)) = (\mathbb{E}(X(t)|\eta_o(t)), \mathbb{E}(l_t(f)|\eta_o(t)))$$

We have from the reference probability approach of Gough the orthogonality relations

$$\mathbb{E}(dj_t(X,f) - d\pi_t(X,f)|\eta_o(t)) = 0,$$

$$\mathbb{E}(dj_t(X,f) - d\pi_t(X,f))dY_o(t)|\eta_o(t)) + \mathbb{E}(j_t(X,f) - \pi_t(X,f))dY_o(t)|\eta_o(t)) = 0$$

Now we can write

$$dY_o(t) = dY_i(t) - k_t(\bar{c}L + cL^*)dt$$

Write

$$M = -(\bar{c}L + cL^*) = M(\theta)$$

So

$$\mathbb{E}(dj_t(X,f)dY_o(t)|\eta_o(t)) =$$
$$\mathbb{E}(dj_t(X,f)diag(dY_i(t)+k_t(M)dt,0)|\eta_o(t))$$
$$dj_t(X,f)dY_i(t) = \bar{c}j_t(\psi_1(X,f))dt = \bar{c}.dt.\pi_t(\psi_1(X,f))$$
$$= \bar{c}.dt\mathbb{E}(diag(\phi_1(X(t),\theta(t))|\eta_o(t))$$
$$= \bar{c}.dt.Tr(diag(\phi_1(X,\theta),0)\rho_B(t))$$

where now θ is regarded as an observable and $\rho_B(t)$ is the Belavkin filtered state at time t. Note that if we choose a representation q in which the $Y_o(t)'s$ are all diagonal (q represents the classical stochastic degrees of freedom), then we can represent the Belavkin filtered state as a block diagonal matrix $\rho_B(t,q), q \in Q$, where for each q, $\rho_B(t,q)$ is a system space operator. Using this representation, we can write

$$\mathbb{E}(dj_t(X,f)dY_o(t)|\eta_o(t)) = \bar{c}.dt.diag(\int Tr(\rho_B(t,q)\phi_1(X,\theta(q)))dq,0)$$

Note that θ will be a diagonal matrix in the q-representation because it commutes with both the system and bath operators. We note that

$$\pi_t(X,f) = \mathbb{E}(diag(X(t),f(\theta(t))|\eta_o(t))$$

$$= diag(\int Tr(\rho_B(t,q)X)dq, \int f(\theta(q))Tr(\rho_B(t,q))dq)$$

Note that the normalization condition

$$\int Tr(\rho_B(t,q))dq = 1$$

holds good. We write down the dynamics of $\pi_t(X,f)$ as

$$d\pi_t(X,f) = F_t(X,f)dt + G_t(X,f)dY_o(t)$$

where $F_t(X,f)$ and $G_t(X,f)$ are functions of the Abelian family of operators $Y_o(s), s \le t$ and hence mutually commute. Then, we find that

$$\mathbb{E}(d\pi_t(X,f).dY_o(t)|\eta_o(t)) = G_t(X,f)|c|^2dt,$$

$$\mathbb{E}(j_t(X,f)dY_o(t)|\eta_o(t))$$
$$= \mathbb{E}(j_t(X,f)diag(k_t(M),0)|\eta_o(t))dt = \mathbb{E}(diag(k_t(XM),0)|\eta_o(t))dt$$
$$= diag(\int Tr(\rho_B(t,q)XM(\theta(q)))dq,0)dt$$

Putting all these pieces together, we get

$$0 = \mathbb{E}((dj_t(X,f) - d\pi_t(X,f))dY_o(t)|\eta_o(t)) + \mathbb{E}((j_t(X,f) - \pi_t(X,f))dY_o(t)|\eta_o(t))$$

$$= \bar{c}.dt.diag(\int Tr(\rho_B(t,q)\phi_1(X,\theta(q)))dq, 0) - G_t(X,f)|c|^2 dt$$

$$+diag(\int Tr(\rho_B(t,q)XM(\theta(q)))dq, 0)dt - \pi_t(X,f)\pi_t(M)dt$$

where

$$\pi_t(X,f)\pi_t(M) = diag[(\int Tr(\rho_B(t,q)X)dq)) \int Tr(\rho_B(t,q)M(\theta(q)))dq)), 0]$$

Further,

$$0 = \mathbb{E}(dj_t(X,f) - d\pi_t(X,f)|\eta_o(t))$$

$$= \pi_t(\psi_0(X,f))dt - F_t(X,f)dt - G_t(X,f)\pi_t(M)dt$$

or equivalently,

$$F_t(X,f) = \pi_t(\psi_0(X,f)) - G_t(X,f)\pi_t(M) =$$

$$diag(\int Tr(\rho_B(t,q)\phi_0(X,\theta(q))dq, (\sigma^2/2)\int Tr(\rho_B(t,q))\Delta f(\theta(q))dq)$$

$$-G_t(X,f)diag[\int Tr(\rho_B(t,q)M(\theta(q)))dq, 0]$$

An alternative approach based on tensor products of observables rather than direct sums: Let B_1 denote the Banach algebra of all bounded Hermitian operators in the system Hilbert space $\mathfrak{h}$ and let B_2 denote the commutative Banach algebra of all bounded complex valued functions on the parameter space $\mathbb{R}^p$. We define

$$B = B_1 \otimes B_2$$

so that the system observable $X]inB_1$ is represented as $X \otimes 1$ and the parameter observavble $f \in B_2$ is represented as $1 \otimes f$. The evolution of observables in B_1 is k_t and the evolution of observables in B_2 is l_t. Thus $j_t = k_t \otimes l_t$ defines the evolution of observables in B. We have

$$j_t(X \otimes f) = k_t(X) \otimes l_t(f) = f(\theta(t))X(t) = l_t(f)k_t(X)$$

$$dj_t(X,f) = dk_t(X) \otimes l_t(f) + k_t(X) \otimes dl_t(f)$$

$$= f(\theta(t))(\phi_0(X(t),\theta(t))dt + \phi_1(X(t),\theta(t))dA(t) +$$

$$\phi_2(X(t),\theta(t))dA(t)^*) + D_k f(\theta(t))d\epsilon_k(t) + (\sigma^2/2)\Delta f(\theta(t))X(t)dt$$

We define

$$\pi_t(X \otimes f) = \mathbb{E}(f(\theta(t))X(t)|\eta_o(t)) = \mathbb{E}(j_t(X \otimes f)|\eta_o(t)) = \int Tr(\rho_B(t,q)X)f(\theta(q))dq$$

so that

$$\mathbb{E}(j_t(X \otimes f)dY_o(t)|\eta_o(t)) = \mathbb{E}(f(\theta(t))(k_t(X)k_t(M)|\eta_o(t))dt$$

$$= \mathbb{E}(f(\theta(t))k_t(XM)|\eta_o(t))dt$$

$$= dt \int f(\theta(q))Tr(\rho_B(t,q)XM(\theta(q)))dq$$

Note that $\rho_B(t,q)$ describes simultaneously the filtered density matrix on the system Hilbert space as well as the conditional probability density of the classical parameter $\theta(t)$ given the measurement algebra $\eta_o(t)$ upto time t. $Tr(\rho_B(t,q))$ equals the conditional probability density of $\theta(t,q)$ while $\int \rho_B(t,q)dq$ equals the filtered state on the system Hilbert space $\mathfrak{h}$. Then,

$$0 = \mathbb{E}(dj_t(X \otimes f) - d\pi_t(X \otimes f)|\eta_o(t)) =$$

$$= dt \int f(\theta(q))Tr(\rho_B(t,q))\phi_0(X,\theta(q)))dq + (\sigma^2/2)\int Tr(\rho_B(t,q)X)\Delta f(\theta(q))dq$$

$$-F_t(X,f)dt - G_t(X,f)\pi_t(M)dt - - - (1)$$

where

$$\pi_t(M) = \int Tr(\rho_B(t,q)M(\theta(q)))dq$$

Note that (1) is the same as

$$F_t(X,f) = -G_t(X,f)\pi_t(M(\theta)) + \pi_t(\phi_0(X,\theta)\otimes f) + (\sigma^2/2)\pi_t(X\otimes\Delta f) - - - (1')$$

We also consider the equation

$$0 = \mathbb{E}(dj_t(X\otimes f) - d\pi_t(X,f))dY_o(t)|\eta_o(t)) + \mathbb{E}(j_t(X\otimes f) - \pi_t(X,f))dY_o(t)|\eta_o(t))$$

This is the same as

$$0 = dt\bar{c}\int f(\theta(q))Tr(\rho_B(t,q)\phi_1(X,\theta)M(\theta(q)))dq$$

$$-G_t(X,f)|c|^2dt + \pi_t(XM(\theta)\otimes f)dt - \pi_t(X\otimes f)\pi_t(M)dt$$

or equivalently,

$$G_t(X,f)|c|^2 = \bar{c}\int f(\theta(q))Tr(\rho_B(t,q)\phi_1(X,\theta)M(\theta(q)))dq$$

$$+\pi_t(XM(\theta)\otimes f) - \pi_t(X\otimes f)\pi_t(M) - - - (2)$$

This is the same as

$$|c|^2 G_t(X,f) = \bar{c}\pi_t(\phi_1(X,\theta)M(\theta)\otimes f) + \pi_t(XM(\theta)\otimes f) - \pi_t(X\otimes f)\pi_t(M)$$

(1) and (2) can be solved for $F_t(X,f)$ and $G_t(X,f)$ thereby obtaining the filter in the form

$$d\pi_t(X,f) = F_t(X,f)dt + G_t(X,f)dY_o(t)$$

Note that in the tensor product formalism, $j_t(X,f)$ is the same as $j_t(X\otimes f)$ and $\pi_t(X,f)$ is the same as $\pi_t(X\otimes f)$. In the special case when $c = 1$ our simultaneous filtering and parameter estimation algorithm is given by

$$d\pi_t(X\otimes f) = G_t(X,f)(dY_o(t) - \pi_t(M(\theta))dt) + (\pi_t(\phi_0(X,\theta)\otimes f) + (\sigma^2/2)\pi_t(X\otimes\Delta f))dt$$

$$= (\pi_t(\phi_0(X,\theta)\otimes f) + (\sigma^2/2)\pi_t(X\otimes\Delta f))dt$$

$$+(\pi_t(\phi_1(X,\theta)M(\theta)\otimes f) + \pi_t(XM(\theta)\otimes f) - \pi_t(X\otimes f)\pi_t(M))(dY_o(t) - \pi_t(M(\theta))dt)$$

From this qsde which is in fact a classical Schrodinger equation, we can derive a classical Schrodinger equation for $\rho_B(t,q)$ which gives the conditional state and probability density of the quantum system and its parameters respectively the conditioning being on the non-demolition measurement process upto time t.

The case when the coherent state is not vacuum and when measurements in include classical measurements which are functions of the classical stochastic parameter $\theta(t)$. The coherent state of the bath is taken as $|phi(u)>=|e(u)>exp(-|u|^2/2)$. The measurement model is assumed to be

$$dY_{o1}(t) = dY_i(t) + j_t(M(\theta))dt, dY_{o2}(t) = h(\theta(t))dt + dv(t)$$

Let $Y_o(t) = (Y_{o1}(t), Y_{o2}(t))$ and let $\eta_o(t) = \sigma(Y_o(s) : s \leq t)$.

$$\mathbb{E}(j_t(X,f)dY_{o1}(t)|\eta_o(t)) = \pi_t(X,f)(cu(t) + \bar{c}\bar{u}(t))dt + \pi_t((X \otimes f)M)dt)$$

where

$$\pi_t((X \otimes f)M) = \int Tr(\rho_B(t,\theta)XM(\theta))f(\theta)d\theta$$

$$\mathbb{E}(j_t(X,f)dY_{o2}(t)|\eta_o(t)) = \mathbb{E}(f(\theta(t))X(t)h(\theta(t))|\eta_o(t))$$

$$= \int Tr(\rho_B(t,\theta)X)f(\theta)h(\theta)d\theta$$

$$\mathbb{E}[dj_t(X \otimes f)dY_{01}(t)|\eta_o(t)] = \mathbb{E}[f(\theta(t))(\phi_1(X(t),\theta(t))dt + \phi_2(X(t),\theta(t))dA(t) +$$

$$\phi_3(X(t),\theta(t))dA(t)^*)(cdA(t) + \bar{c}dA(t)^* + j_t(M(\theta(t))dt)|\eta_o(t))$$

$$+\mathbb{E}[X(t)(f'(\theta(t))d\epsilon(t) + (\sigma^2/2)\Delta f(\theta(t))dt)(cdA(t) + \bar{c}dA(t)^* + j_t(M(\theta(t))dt)|\eta_o(t)]$$

$$= \pi_t(f(\theta)\phi_2(X,\theta))\bar{c}dt = dt\bar{c}\int f(\theta)Tr(\rho_B(t,\theta)\phi_2(X,\theta))d\theta$$

Further,

$$\mathbb{E}[dj_t(X \otimes f)dY_{02}(t)|\eta_o(t)] = \mathbb{E}[f(\theta(t))(\phi_1(X(t),\theta(t))dt + \phi_2(X(t),\theta(t))dA(t) +$$

$$\phi_3(X(t),\theta(t))dA(t)^*)(h(\theta(t))dt + dv$$

$$+\mathbb{E}[X(t)(f'(\theta(t))d\epsilon(t) + (\sigma^2/2)\Delta f(\theta(t))dt)(h(\theta(t))dt + dv(t))|\eta_o(t)]$$

$$= \int Tr(\rho_B(t,\theta)X)f'(\theta)R_{\epsilon v}(t)dt$$

where

$$R_{v\epsilon}(t)dt = d\epsilon(t).dv(t)$$

$$\mathbb{E}(dj_t(X \otimes f)|\eta_o(t)) = \pi_t(\phi_0(X,\theta))dt + \pi_t(\phi_1(X,\theta))u(t)dt$$

$$+\pi_t(\phi_2(X,\theta))\bar{u}(t)dt + (\sigma^2/2)\pi_t(X \otimes \Delta f)dt$$

We leave it to the reader to complete the calculations from this stage onwards, ie, substitute into the optimal normal equations

$$\mathbb{E}(dj_t(X,f) - d\pi_t(X,f)|\eta_o(t)) = 0,$$

$$\mathbb{E}((j_t(X,f) - \pi_t(X,f))dY_{ok}(t)|\eta_o(t)) +$$

$$\mathbb{E}(dj_t(X,f)) - d\pi_t(X,f))dY_{ok}(t)|\eta_o(t)) = 0, k = 1,2$$

with the assumption

$$d\pi_t(X,f) = F_t(X,f)dt + \sum_{k=1,2} G_{kt}(X,f)dY_{ok}(t)$$

where

$$F_t(X,f), G_{kt}(X,f) \in \eta_o(t), k = 1,2$$

and obtain the equations for the operators $F_t(X,f), G_{kt}(X,f), k = 1,2$ in terms of the structure maps $\phi_k(X,f), k = 0,1,2$ and the measured observable $M(\theta)$ after time t. Finally, substitute

$$\pi_t(X,f) = \int Tr(\rho_B(t,\theta)X)f(\theta)d\theta$$

to derive a stochastic Schrodinger equation for $\rho_B(t,\theta)$.

2.6 Construction of fractals using cubes filling regions of space

The paper proposes an algorithm for generating planar and three dimensional fractals using at each stage the introduction of a set of squares forming annuli. At a given stage, if $N(\epsilon)$ new squares are introduced in the form of annuli and if the linear dimension of a square is ϵ, then from standard dimension theory, it follows that the dimension of the region covered by the squares is the limit as $\epsilon \to 0$ of the quantity $log(N(\epsilon))/log(1/\epsilon)$. A short explanation of this dimension formula may be provided in the paper. Further, the authors suggest a scheme for generating fractals using iterative affine linear maps of the form $x(n+1) = A(n)x(n) + b(n+1)$ where $A(n)$ is a sequence of square matrices and $b(n)$ is a sequence of column vectors. Some explanation of how the algorithm proposed in the paper using squares filling annular regions can be cast in the form of such a dynamical system may be provided. Further, it is a well known fact in chaotic dynamics, that apart from the fractional dimension of the region filled by the trajectory of a dynamical system, the other parameters that characterize chaotic/random behaviour of a system are (a) Spectrum of the process calculated using the periodogram, (b) Lyapunov exponents of the dynamical system which measure sensitivity to initial conditions these exponents being evaluated using ergodicity of the invariant measure and the Jacobian of the iteration map, (c) the invariant measure of the dynamical system which gives information about the limiting/equilibrium probability distribution of the dynamical system trajectory, ie, after several iterations, the proportion of time spent by the trajectory in each Borel subset of the region. It would be nice if the authors could indicate how their iterative algorithm can be used to evaluate these parameters. In order to do this, the authors would have to first cast their algorithm in the analytical language of dynamical systems, for example, they could indicate that if $S(n)$ is the spatial region occupied by the set of squares after the nth iteration, then how to express $S(n+1)$ in terms of $S(n)$, in short, a mathematical algorithm would faciliate the understanding of the process described by the authors. One way to cast the dynamics of the spatial region would be to use approximations to the indicator functions of the region and then to look at the dynamics of these approximate indicators. Another way to describe the dynamics of the fractal region would be to define a measure at each iteration, where the measure is calculated based on an approximation of indicators. The paper is interesting but apart from a dimensional analysis, in order for the results to be useful in engineering applications, the other important parameters of the fractal sequence generated as mentioned above should be included at least in a brief way.

2.7 Quantum neural networks for optimal mixed state evolution with entropy and relative entropy cost functions. The mixed state evolution depends on control parameters which are adapted in such a way as to minimize a weighted combination of the entropy and the relative entropy distance w.r.t a given evolving state

The state is assumed to evolve in accordance with the Schrodinger-Von-Neumann-Liouville-Lindblad equation, indeed, it is assumed to describe the dynamics of an open quantum system:

$$\rho'(t) = -i[H(\theta(t)), \rho(t)] - (1/2)(L(\theta(t))L(\theta(t))^* \rho(t) + \rho(t)L(\theta(t))L(\theta(t))^*$$
$$-2L(\theta(t))^* \rho(t) L(\theta(t)))$$

where

$$H(\theta) = H_0 + \sum_k \theta_k H_k, L(\theta) = L_0 + \sum_k \theta_k L_k$$

The aim is to minimize a cost function equal to a weighted linear combination of the negative of the Von-Neumann entropy of the state and its relative entropy distance w.r.t another (desired) evolving state $\sigma(t)$. The adaptive itertative algorithm is therefore given by

$$\theta_k(t+h) = \theta_k(t) - \mu_k(t)h.\frac{\partial}{\partial \theta_k(t)})C(t+h, \rho(t+h))$$

where

$$C(t, \rho(t)) = a.H(\rho(t)) + b.D(\rho(t)|\sigma(t))$$

with

$$H(\rho) = -Tr(\rho.log\rho), D(\rho|\sigma) = Tr(\rho(log\rho - log(sigma))), a, b > 0$$

For calculating the partial derivatives w.r.t $\theta_k(t)$, we must discretize the above evolution equation to get

$$\rho(t+h) = \rho(t) - ih[H(\theta(t)), \rho(t)] - (h/2)E(\theta(t))(\rho(t))$$

where

$$E(\theta)(\rho) = (L(\theta)L(\theta)^*\rho + \rho L(\theta)L(\theta)^* - 2L(\theta)^*\rho L(\theta))$$

Then,

$$\partial\rho(t+h)/\partial\theta_k(t) = -ih[\partial H(\theta(t))/\partial\theta_k(t), \rho(t)] - (h/2)(\partial E(\theta(t))/\partial\theta_k(t)))(\rho(t))$$

Using this formula, we can evaluate the partial derivatives of $C(t+h, \rho(t+h))$ w.r.t $\theta_k(t)$. Sometimes, in order to improve the convergence speed of the iterative algorithm, it is useful to make use of the partial derivatives of the cost $C(t+h, \rho(t+h))$ w.r.t $\theta_k(t-t_m), m = 1, 2, ..., r$. In other words, we would use a delay firing algorithm of the kind

$$\theta_k(t+h) = \theta_k(t) - h\sum_{m=1}^{r} \mu_{km}(t)\partial C(t+h, \rho(t+h))/\partial\theta_k(t-t_m)$$

In order to evaluate these partial derivatives, we have to express $C(t, \rho(t))$ in terms of all the past values of the parameters. To this end, we express the Schrodinger-Lindblad evolution equation as

$$\rho'(t) = T(\theta(t))(\rho(t))$$

where $T(\theta)$ is the θ-dependent generator of a quantum dynamical semigroup. It is given by

$$T(\theta) = -iad(H(\theta)) + F(\theta), F(\theta) = -(1/2)E(\theta)$$

We can express the solution as follows. Let

$$\rho(t) = U(t)(\rho(0)), U(t) = U_0(t)W(t)$$

where

$$U_0'(t) = -iad(H(\theta(t))).U_0(t)$$

and hence,

$$U_0'(t)W(t) + U_0(t)W'(t) = -i.ad(H(\theta(t)))U_0(t)W(t) + F(\theta(t))U_0(t)W(t)$$

giving

$$W'(t) = U_0(t)^*F(\theta(t)U_0(t)W(t)$$

where $U_0(t)$ is the unitary operator defined on the space of Hilbert Schmidt operators with inner product $< X, Y >= Tr(X^*Y)$ given by

$$U_0(t) = T\{exp(-i\int_0^t ad(H(\theta(s)))ds)\}$$

$$= I + \sum_{n\geq 1}\int_{0<t_n<...<t_1<t} exp(-iad(H(\theta(t_1))))...exp(-iad(H(\theta(t_n))))dt_1...dt_n$$

Problem: Prove that $U_0(t)$ is a unitary operator.

Hint: Consider $X(t) = U_0(t)X, Y(t) = U_0(t)Y$. Then,

$$(d/dt) < U_0(t)X, U_0(t)Y >= i < [H(t), X(t)], Y(t) > -i < X(t), [H(t), Y(t)] >$$

Now use the fact that $H(t)$ is Hermitian and that

$$< [H, X], Y >= Tr([H, X]^*Y)$$

$$= Tr((X^*H - HX^*)Y) = Tr(X^*(HY - UH)) =< X, [H, Y] >$$

for H Hermitian. Thus $< U_0(t)X, U_0(t)Y >=< X, Y > \forall t$.

Problem: Compute $U_0(t)^{-1} = U_0(t)^*$.

hint: If H is Hermitian, then prove that $exp(-i.ad(H))$ is unitary on the above mentioned Hilbert space. To prove this you may observe that this is a special case of the above mentioned unitarity result. Thus, conclude that

$$U_0(t)^{-1} = U_0(t)^* = I + \sum_{n\geq 1}\int_{0<t_n<...<t_1<t} exp(i.ad(H(\theta(t_n))))...exp(iad(H(\theta(t_1))))dt_1...dt_n$$

Now observe that we can write

$$W(t) = I + \sum_{n\geq 1}\int_{0<t_n<...<t_1<t} U_0(t_1)^*F(\theta(t_1))U_0(t_1, t_2)$$

$$F(\theta(t_2)))U_0(t_2, t_3)...U_0(t_{n-1}, t_n)F(\theta(t_n))U_0(t_n)dt_1...dt_n$$

where for $t > s$, we have defined

$$U_0(t, s) = U_0(t)U_0(s)^*$$

$U_0(t, s)$ describes the free evolution of the mixed state from time s to time t under the Hamiltonian $H(t)$, where by free evolution, we mean in the absence of the Lindblad noise operator terms that couple the system to the noisy bath.

Some remarks: In a quantum neural network, we would typically expect the state of the evolving system $\rho(t)$ to be close to another state $\sigma(t)$ but we would also require that the average of a set of observables $X_1, ..., X_n$ in the evolving state to be correlated with a vector valued signal $\xi(t) \in \mathbb{R}^n$. This means that we must make the adaptive parameter update equation faster by increasing the adaptation constants $\mu_k(t)$ when the prediction error square involved in predicting the process $(Tr(\rho(t)X_k))_{k=1}^n$ based on $\xi(t - m\delta), m = 0, 1, ..., p$ is large. The prediction error process $e(t)$ is given by

$$e(t) = Tr(\rho(t)X) - \sum_{m=0}^{p} B(m)\xi(t - m\delta)$$

or in component form,

$$e_k(t) = Tr(\rho(t)X_k) - \sum_{m,j} B_{kj}\xi_j(t - m\delta), k = 1, 2, ..., n$$

where the coefficient matrices $B(m), m = 0, 1, ..., p$ are chosen so that

$$\sum_{r=0}^{N} |e(t - r\delta)|^2$$

is a minimum. One way to select the adaptive coefficients $\mu_k(t)$ would be to increase its value at the next iteration if $e(t)^2$ increases rapidly with increase in $\theta_k(t)$. That will cause the value of $\theta_k(t)$ at the next iteration to either increase or decrease at a rapid rate depending on whether the distance of the generated state from the reference state decreases or increases with the value of $\theta_k(t)$.

On the statistical performance of a quantum neural network based on mixed state evolution: We are adapting the qnn parameters in the Hamiltonian and the Lindblad operators so that the evolving state $\rho(t)$ continuously tracks another evolving state $\sigma(t)$. Typically, there will be measurement errors while reading the state $\sigma(t)$ for the training process. We can model these measurement errors either as a classically random family of TPCP maps S_t so that the measured state is $S_t(\sigma_t)$ and then calculate the mean square fluctuation of the qnn output state $\rho(t)$ w.r.t the desired state $\sigma(t)$. Or else, we model $\sigma(t)$ as the output of a Belavkin filter based on non-demolition measurments and then calculate the mean square deviation. In order to use the latter approach, we must assume that $\sigma(t)$ is a solution of a Lindblad equation, and then dilate it so that $\sigma(t)$ becomes the "system component" of the joint state of the system and bath for which the Belavkin equation gives its estimate as the solution to a stochastic Schrodinger equation.

2.8 Quantum neural networks based on the Hudson-Parthasarathy noisy Schrodinger equation with Belavkin filtered state matched to given processes

A problem in quantum neural networks: Given the Hudson-Parthasarathy-Schrodinger qsde

$$dU(t) = -(iH(\theta(t)) + P(\theta(t)))dt + L(\theta(t))dA(t) - L(\theta(t))^*dA(t)^*)U(t)$$

with the parameter vector $\theta(t)$ satisfying the classical sde $d\theta(t) = f(\theta(t))dt + g(\theta(t))dB(t) where f : \mathbb{R}^d \to \mathbb{R}^d, g : \mathbb{R}^d \to \mathbb{R}^{d \times p} and \mathrm{B(t)} an \mathrm{R}^p$ valued Brownian motion, we take non-demolition measurements

$$Y_o(t) = U(t)^*Y_i(t)U(t), Y_i(t) = A(t) + A(t)^*$$

we calculate using Belavkin's filter, the conditional expectation

$$\pi_t(X \otimes F) = \mathbb{E}(k_t(X)F(\theta(t))|\eta_o(t)) = \mathbb{E}(j_t(X \otimes F)|\eta_o(t))$$

where

$$j_t(X \otimes F) = k_t(X)F(\theta(t)), k_t(X) = U(t)^* XU(t)$$

Note that

$$j_{t+s}(X \otimes F) = j_{t+s,s} o j_s(X \otimes F) = j_{t+s,s}(X(s)F(\theta(s))), X(s) = k_s(X)$$

We have

$$dj_t(X \otimes F) = (j_t(\phi_0(X,\theta)F)dt$$
$$+j_t(X \otimes KF))dt + j_t(\phi_1(X,\theta)F)dA(t) + j_t(\phi_2(X,\theta)F)dA(t)^*$$
$$+j_t(X.D_iF)dB_i(t)$$

where

$$D_iF(\theta) = \partial_j F(\theta).g_{ji}(\theta), DF(\theta) = g(\theta)^T \nabla F(\theta)$$

summation over the repeated index j being implied and

$$KF(\theta) = f(\theta)^T \nabla F(\theta) + (1/2)Tr(g(\theta)g(\theta)^T \nabla\nabla^T F(\theta_k L))$$

ie, K is the infinitesimal generator of the diffusion process $\theta(t)$, ie,

$$KF(\theta(t))dt = \mathbb{E}(dF(\theta(t))|\theta(t))$$

Note that $\theta \to \phi_k(X,\theta)F(\theta)$ is an element of the Banach algebra $\mathcal{B}(\mathfrak{h}) \otimes B_1$ wjhere B_1 is the commutative Banach space of all bounded continuous functions on $\mathbb{R}^d$. We can define the joint generator of the quantum-classical Markov process $(X(t), \theta(t))$ as

$$K_1(X \otimes F)(\theta) = \phi_0(X,\theta)F(\theta) + X.KF(\theta)$$

and then write

$$dj_t(X \otimes F) = j_t(K_1(X \otimes F))dt + j_t(\phi_1(X,\theta)F)dA(t) + j_t(\phi_2(X,\theta)F)dA(t)^*$$

$$+j_t(X.D_iF)dB_i(t)$$

Then,

$$\mathrm{E}(\mathrm{d}j_t(X \otimes F)|\eta_o(t))$$
$$= \pi_t(K_1(X \otimes F))dt + \pi_t(\phi_1(X,\theta)F)dA(t) +$$
$$\pi_t(\phi_2(X,\theta)F)dA(t)^* + \pi_t(X.D_iF)dB_i(t)$$

Likewise performing the computation of the other conditional expectations, we obtain a classical Schrodinger equation

$$d\pi_t(X \otimes F) = F_t(X,F)dt + G_t(X,F)dY_o(t)$$

This filtering equation can be expressed in the form

$$d\rho(t,\theta)) = K_1^* \rho(t,\theta)dt + H_1(\rho(t,\theta))(dY_o(t) - H_2(\rho(t,\theta))dt)$$

Now we incorporate control parameters η in the drift coefficient $f(\theta)$ of the process $\theta(t)$. Thus these control parameters will appear in the generator K_1 of the quantum-classical Markov process $(X(t), \theta(t))$ which is represented by j_t. We can thus express the above quantum-classical Belavkin-Kushner-Kallianpur filter as

$$d\rho(t,\theta) = K_1(\eta)^* \rho(t,\theta)dt + H_1(\rho(t,\theta)))(dY_o(t) - H_2(\rho(t,\theta))dt)$$

The parameters η can be controlled so that $\rho(t,\theta)$ at each time instant is close to a desired state, or more generally so that the expected values of joint observables $X_k \otimes F_k, k = 1, 2, ..., M$ at each time t are close to given functions of time $\xi_k(t), k = 1, 2, ..., M$. This is achieved by the gradient adaptive algorithm:

$$\eta(t+dt) = \eta(t) - \mu.dt.(\partial/\partial\eta(t)) \sum_{k=1}^{M} (\xi_k(t+dt) - \int Tr(\rho(t+dt,\theta)X_k)F_k(\theta)d\theta)^2$$

where

$$\rho(t+dt,\theta) = \rho(t,\theta) + K_1(\eta(t))^* \rho(t,\theta)dt + H_1(\rho(t,\theta))(dY_o(t) - H_2(\rho(t,\theta))dt)$$

2.9 MATLAB problems in digital image processing

Simulate an image field $X(n,m), 1 \leq n, m \leq N$ so that these are all iid random variables with each $X(n,m)$ being of the form $\xi.N(u(n,m)/\xi)+W(n,m)$ where $N(\lambda)$ is a random variable having the Poisson distribution with mean λ and $W(n,m)$ is an $N(0,\sigma^2)$ random variable independent of $N(\xi)$. Show that

$$\mathbb{E}(X(n,m)) = u(n,m), Var(X(n,m)) = \xi.u(n,m) + \sigma^2$$

so that in the limit as $\xi \to 0$, $X(n,m)$ always has the mean value of $u(n,m)$ and a decreasing variance of σ^2. Evaluate the negative log-likelihood function of the image field and add to this an edge gradient energy term $A.\sum_{n,m}[(u(n+1,m) - u(n,m))^2 + (u(n,m+1) - u(n,m))^2]$ and compute the denoised estimate of the image field as that $\{u(n,m)\}\}$ which minimizes this sum. Evaluate the approximate mean and covariance of this image field estimate using perturbation theory.

Simulate a time varying image field using the algorithm

$$X_{t+1}[n,m] = \sum_{(k,l)\in D} a[k,l]X_t[n-k, m-l] + W_t[n,m], 1 \leq n, m \leq N$$

with $W_t, t = 1, 2, ..$ being an iid Gaussian process with values in $\mathbb{R}^{N \times N}$ having mean zero and a prescribed covariance matrix. Calculate the optimal Wiener filter and Kalman filter for this model with the measurement process

$$Z_t[n, m] = (HX_t)[n, m] + V_t[n, m]$$

where V_t is another iid zero mean Gaussian process independent of the W_t process. Here, H is a linear transformation of the image field. Note that such a linear transformation has the general form

$$HX = \sum_{k=1}^{p} E_k X F_k$$

where E_k, F_k are $N \times N$ matrices.

Given an image field $U(x, y)$, its edges can be smoothened out and in fact, its contrast can be reduced by a diffusive evolution. This means that after time t, the diffused image field is given by

$$U_t(x, y) = P_t(x, y) * U(t, x, y) = \int P_t(x', y')U(x - x', y - y')dx'dy'$$

where $P_t(x, y)$ is a Gaussian point spread function whose covariance matrix increases proportionally with time. Show that such a $P_t(x, y)$ will satisfy the diffusion equation

$$\partial P_t(x, y)/\partial t = D_{xx}\partial^2 P_t(x, y)/\partial x^2 + D_{yy}\partial^2 P_t(x, y)/\partial y^2 + 2D_{xy}\partial^2 P_t(x, y)/\partial x \partial y$$

where

$$\mathbf{D} = \begin{pmatrix} D_{xx} & D_{xy} \\ D_{xy} & D_{yy} \end{pmatrix}$$

is a positive definite matrix. Hence, deduce that $U_t(x, y)$ also satisfies the same diffusion equation with initial condiition $U_0(x, y) = U(x, y)$. Implement this diffusion evolution for an image field with sharp edges and display the original image and the smoothened image.

Generate an $N \times N$ image field $X(n, m)$. At each point (n, m) calculate its gradient magnitude along the directions making angles $2\pi k/M, k = 0, 1, ..., M - 1$ with the horizontal using the formula

$$|\nabla_k X(n, m)|^2$$
$$= ((X(n+1, m) - X(n, m))cos(2\pi k/M) + (X(n, m+1) - X(n, m))sin(2\pi k/M))^2$$

Set a threshold ϵ and then calculate the points (n, m) and the directions k associated to these points at which

$$|\nabla_k X(n, m)| > \epsilon$$

If this happens for a triplet (n, m, k), then there is an edge at (n, m) in the image that is perpendicular to the direction $2\pi k/M$. In this case, replace $X(n, m)$ by its average taken around its neighbours. Display the original image with its edges and its smoothened version.

Generate an image $X(n,m), -N \leq n,m \leq N$. Rotate this image field by an angle $\theta \in [0,2\pi)$ so that the transformed image is

$$Y(n,m,\theta) = X([n.cos(\theta) + m.sin(\theta)], [-n.sin(\theta) + m.cos(\theta)])$$

where (n,m) now vary over $[-[N\sqrt{2}], [N\sqrt{2}]]$, $[x]$ denoting the integer part of x. Plot the original and the rotated image fields over the region $[-[N\sqrt{2}], [N\sqrt{2}]] \times [-[N\sqrt{2}], [N\sqrt{2}]]$ where if any index goes out of the range of the image field, we set the image field value at that index equal to zero.

Generate an image $X(n,m), 1 \leq n,m \leq N$. Calculate its SVD: $X = UDV^T$. Retain the first p largest diagonal entries only in D and synthesize this compressed image Y. Calculate the SNR for this compression defined by $\| X \|^2 / \| X - Y \|^2$. Plot this SNR with p increasing from 1 to N.

This problem deals with clustering. Generate N data points $X(n) \in \mathbb{R}^p, n = 1,2,...,N$. Let $q < N$. Determine via an iterative algorithm vectors $\mu(k) \in \mathbb{R}^p, k = 1,2,...,q$ and a partition $E_1,...,E_q$ of $\{1,2,...,N\}$ so that

$$\sum_{k=1}^{p} \sum_{n \in E_k} \| X(n) - \mu(k) \|^2$$

is a minimum. For solving this problem, first observe that if $E_k, k = 1,2,...,p$ is a partition of $\{1,2,...,N\}$, then the $\mu(k)'s$ that minimize the above error energy are given by

$$\mu(k) = \frac{sum_{n \in E_k} X(n)}{\mu(E_k)}$$

where $\mu(E)$ denotes the cardinality of the set E. Next observe that if $\mu(k), k = 1,2,...,q$ are points in $\mathbb{R}^p$, then the partition $E_k, k = 1,2...,q$ that minimizes the above error energy is given by

$$E_k = \{n : 1 \leq n \leq N, \| X(n) - \mu(k) \| \leq \| X(n) - \mu(j) \| \forall j = 1,2...,q\}$$

By repeatedly alternating these two processes, for an initial set of data points and an initial partition $E_k, k = 1,2,...,q$ verify convergence.

Generate two classical image fields $X(n,m), Y(n,m) = X(n,m) + W(n,m)$ where $X(n,m)$ is the signal field and $W(n,m)$ is the noise field. The pixel index range is $1 \leq n,m \leq N$. Transform these image fields into quantum state vectors $|\psi_X>$ and $|\psi_Y>$ of size $2^{N^2} \times 1$. This is most easily done by replacing $X(n,m)$ by the qubit

$$|e_X(n,m)> = X(n,m)exp(i\phi_1(n,m))|1> + \sqrt{1 - X(n,m)^2}exp(i\phi_0(n,m))|0>$$

and $Y(n,m)$ by the qubit

$$|e_Y(n,m)>=Y(n,m)exp(i\chi_1(n,m))|1>+\sqrt{1-X(n,m)^2}exp(i\chi_0(n,m))|0>$$

where we assume that $X(n,m)$ and $Y(n,m)$ have been normalized so that their magnitudes fall in the range $[0,1]$. Now order the pixel indices $(n,m), 1 \leq n,m \leq N$ in lexicographic order and define

$$|\psi_X>=\bigotimes_{n,m}|e_X(n,m)>, |\psi_Y>=\bigotimes|e_Y(n,m)>$$

Write a program to express $|\psi_X>$ as

$$|\psi_X>=\sum_{k=0}^{2^{N^2}-1} c_X(k)|k>$$

where

$$|k>=|x(M-1)x(M-2)...x(0)>, M=N^2, k=\sum_{k=0}^{M-1} x(k)2^k, x(k)\in\{0,1\}$$

In other words, you must find a formula for $c_X(k)=c_X(x(M-1)...x(0))$. Also write a program to recover $X(n,m)$ from $|\psi_X>$ for each (n,m).

hint:

$$c_X(x(M-1)...x(0))=c_X(k)=U(M-1,x(M-1))...U(0,x(0))$$

where for each (n,m), we write

$$N(n,m)=(n-1)N+m$$
$$=M-j(n,m), j(n,m)=M-N(n-1)-m=N^2-N(n-1)-m$$
$$U(j(n,m),x(M-j(n,m)))=X(n,m)exp(i\phi_1(n,m)), x(M-j(n,m))=1,$$
$$U(j(n,m),x(M-j(n,m)))=\sqrt{1-X(n,m)^2}exp(i\phi_0(n,m)), x(M-j(n,m))=0$$

Using this, write a program to recover $X(n,m)$ from $|\psi_X>$.

Generate a three dimensional random image field $X(n,m,k)$ for $-N \leq n,m,k \leq N$. Choose a set of Euler angles α, β, ψ and generate the corresponding 3-D rotation matrix

$$R=R(\alpha,\beta,\psi)=R_z(\alpha)R_x(\beta)R_z(\psi)$$

Apply this rotation to the above 3-D image field and add noise to it obtaining thereby the rotated and noise corrupted 3-D image field

$$Y(n,m,k)=X(R^{-1}(n,m,k))+W(n,m,k)$$

Using the notion of group theoretic Fourier transform on the rotation group $SO(3)$, estimate the rotation Euler angles from measurements of the 3-D image fields X, Y.

hint: Pick a positive real number r and consider the image fields on a sphere of radius r defined by $X_1(\theta,\phi) = X([r.cos(\phi)sin(\theta)], [r.sin(\phi)sin(\theta)], [r.cos(\theta)])$ and Likewise for Y, W. Express thus the image statistical model on a sphere of radius r as

$$Y_1(\theta,\phi) = X_1(R^{-1}(\theta,\phi)) + W_1(\theta, phi)$$

Note that to compute the field $X_1(R^{-1}(\theta,\phi))$ you must calculate first the cartesian coordinates of the point $R^{-1}(\theta,\phi)$ and then evaluate X at this point. Now take the spherical-harmonic transform on both sides as

$$X_2(l,m) = \int X_1(\theta,\phi)\bar{Y}_{lm}(\theta, phi)sin(\theta)d\theta d\phi$$

$$Y_2(l,m) = \int Y_2(\theta,\phi)\bar{Y}_{lm}(\theta, phi)sin(\theta)d\theta d\phi$$

and likewise for W. Show that the model can be expressed as

$$Y_2(l) = \pi_l(R)X_2(l) + W_2(l), l = 0, 1, 2, ...$$

where

$$X_2(l) = (X_2(l,m)))_{m=-l}^{l}, Y_2(l) = (Y_2(l,m))_{m=-l}^{l}$$

and π_l is the irreducible representation of $SO(3)$ corresponding to the subspace $span\{Y_{lm} : |m| \leq l\}$. Specifically, its matrix elements are given by

$$[\pi_l(R)]_{m'm} = \int Y_{lm}(R^{-1}(\theta,\phi))\bar{Y}_{lm'}(\theta,\phi)sin(\theta)d\theta.d\phi$$

By considering the explicit form of $\pi_l(R)$ for the first few l, estimate R as

$$\hat{R} = argmin_R \sum_l \| Y_2(l) - \pi_l(R)X_2(l) \|^2$$

[9] Generate an image field according to the recursive algorithm

$$X(n,m) = a(1,,0)X(n-1,m) + a(0,1)X(n,m-1) + a(1,1)X(n-1,m-1) + W(n,m)$$

starting with $X(0,0), X(1,0), X(0,1)$ where $W(n,m)$ is an independent collection of N^2 r.v's and the distribution of $W(n,m)$ is $\xi.P(u(n,m)/\xi) + w$ where $P(\lambda)$ has the Poisson distribution with mean λ and w is $N(0,\sigma_w^2)$ independent of $P(u(n,m)/\xi)$. Plot both the image fields $W(n,m)$ and $X(n,m)$. Which is smoother ? Now calculate the likelihood function of $\{X(n,m)\}$ and hence design estimates of $a(0,0), a(1,0), a(0,1)$ as well as $\{u(n,m)\}$.

2.10 Using a quantized field theory for implementing a quantum neural network

Let $L(\phi_k(x), \phi_{k,\mu}(x), \theta(x))$ be a Lagrangian density for the fields $Q_k(x) = \phi_k(x), k = 1, 2, ..., N$ and let $\theta(x)$ be control parameters dependent upon the space-time argument $x = (t, r)$. The $Q_k's$ are canonical position fields and

$$P_k(x) = \partial L/\partial\phi_{k,0}$$

are the canonical momenta. The Hamiltonian density is obtained by a Legendre transformation:

$$H(Q_k(x), Q_{k,r}(x), P_k(x)|\theta(x)) = P_k(x)Q_{k,0}(x) - L$$

Schrodinger's equation for the wave functional $\psi(t, Q)$ reads

$$i\partial_t\psi(t, Q) = [\int H(Q_k(t, r), Q_{k,m}(t, r), -i\delta/\delta Q_k(t, r)|\theta(t, r))d^3r]\psi(t, Q)$$

We wish to control the parameters θ adaptively in time so that the wave functional $\psi(t, Q)$ tracks a desired wave functional $\chi(t, Q)$ or more accurately, so that the probability density functional $|\psi(t, Q)|^2$ tracks the given probability density functional $f(t, Q)$. To this end, let μ be a measure on the space of position functionals of space $\{Q(r) : r \in \mathbb{R}^3\}$. Then the algorithm for adapting θ would read

$$\partial_t\theta(t, r) = -dt^{-1}\mu(t, r)\frac{\delta}{\delta\theta(t, r)}\int(f(t + dt, Q) - |\psi(t + dt, Q)|^2)^2 d\mu(Q)$$

where

$$\psi(t+dt, Q) = \psi(t, Q) - idt.[\int H(Q_k(t, r), Q_{k,m}(t, r), -i\delta/\delta Q_k(t, r)|\theta(t, r))d^3r]\psi(t, Q)$$

Note that

$$\delta\psi(t+dt, Q)/\delta\theta(t, r) = -idt\frac{\partial H(Q_k(t, r), Q_{k,m}(t, r), -i\delta/\delta Q_k(t, r)|\theta(t, r))}{\partial\theta(t, r)}\psi(t, Q)$$

Exercise: Cast the above parameter update equation in the language of stochastic differential field equations when there are random fluctuations in the given pdf $f(t, Q)$. Explain how you would linearize these stochastic differential field equations to obtain linear stochastic differential field equations for the parameter field fluctuations $\delta\theta(t, r)$. Further improve upon this adaptation by considering the average value of an observable $X(Q, Q')$ in the position representation

$$< X > (t) = \int X(Q, Q')\psi(t, Q)\psi(t, Q')^* d\mu(Q)d\mu(Q')$$

computing the linear prediction error for $< X > (t)$ based on a finite memory past samples $S(t - \tau), 0 \leq \tau \leq T$ of a given signal and causing the adaptation coefficient $\mu(t, r)$ to be an increasing function of the corresponding prediction error square. What is the meaning of such an adaptation ?

Chapter 3

Statistical Image Processing, Quantum Mechanics Parameters, Metric Estimation, String Theory, Stochastic Calculus, Group Representations

3.1 Examination problems in digital image processing

Let $f_1(x)$ and $f_2(x)$ be statistically independent random functions defined on $\mathbb{Z}$. Consider the separable random image field $f(x,y) = f_1(x)f_2(y)$. Calculate the autocorrelation function of $f(x,y)$ in terms of those of f_1 and f_2. Show that it can be expressed using the Kronecker tensor product as

$$R(x,y|x',y') = R_1(x|x').R_2(y|y')$$

or equivalently, in terms of matrices,

$$R = R_1 \otimes R_2$$

Using this result, express the KL expansion of R in terms of the KL expansions of R_1 and R_2. Show specifically that if $\{\lambda_n\}$ and $\{\mu_n\}$ are respectively the eigenvalues of R_1 and R_2 with corresponding orthonormal eigenfunctions $\{\phi_n\}\}$ and $\{\psi_n\}$ respectively, then $\{\lambda_n\mu_m : n,m \in \mathbb{Z})\}$ are the eigenvalues of R with corresponding orthonormal eigenfunctions $\{\phi_n \otimes \psi_m : n,m \in \mathbb{Z}\}$. Explain how you would perform the PCA of f using the PCA's of f_1 and f_2 respectively.

Given a random image field $((X(n,m)))_{1\leq n,m\leq N}$, declare that there is an edge at the pixel (n,m) that makes an angle θ with the x axis if the gradient of the image field at (n,m) along the perpendicular direction to $(cos(\theta), sin(\theta))$ has a magnitude greater that a given threshold ϵ, ie if

$$|-(X(n+1,m)-X(n,m))sin(\theta)+(X(n,m+1)-X(n,m))cos(\theta)| > \epsilon$$

Given that $((X(n,m)))$ is a random Gaussian field with mean $\mu(n,m) = \mathbb{E}(X(n,m))$ and autocorrelation $R(n,m|n',m') = \mathbb{E}(X(n,m)X(n',m'))$, derive an expression for the probability that the image field has an edge at (n,m) along the direction θ. Also derive an expression for the joint probability that there are edges at the pixels $(n_k,m_k), k = 1,2,...,p$ along the directions $\theta_k, k = 1,2,...,p$ respectively. Specialize your result to the case when $X(n,m), n,m = 1,2,...,p$ are iid $N(\mu,\sigma^2)$ random variables. Also derive an expression for the total edge energy in the entire image field along a given direction θ. This total edge energy is defined by

$$\sum_{n,m} |-(X(n+1,m)-X(n,m))sin(\theta)+(X(n,m+1)-X(n,m))cos(\theta)|^2$$

DOI: 10.1201/9781003728627-3

Given a random image field $X(n,m), n,m = 1,2,...,N$ and a real number I, segment the image field pixles into two disjoint sets E_1 and $E_2 = E_1^c$, where E_1 is the set of those pixels (n,m) for which $X(n,m) > I$, so that E_2 is the set of those pixels $X(n,m) \leq I$. Let $W_1(n,m)$ and $W_2(n,m)$ be two non-random window functions. Process the pixels $(n,m) \in E_1$ by multiplication with the window $W_1(n,m)$ and the pixels $(n,m) \in I_2$ by multiplication with the window $W_2(n,m)$. Assuming the $X(n,m)$ are iid with probability density $f(X)$, calculate the joint moment generating function of the processed image field $Y(n,m), n,m = 1,2,...,N$. Note that

$$Y(n,m) = W_1(n,m)X(n,m)\chi_{X(n,m)>I} + W_2(n,m)X(n,m)\chi_{X(n,m)\leq I}$$

Calculate the moment generating function of the total sum of the processed pixel values in E_1 and also of those processed pixel values in E_2 and also the joint moment generating function. Note that these sums are respectively given by

$$X_1 = \sum_{(n,m):X(n,m>I} W_1(n,m)X(n,m), X_2 = \sum_{(n,m):X(n,m)\leq I} W_2(n,m)X(n,m)$$

Also compute the moment generating function of the number $N_1(I)$ of pixels in E_1 and of the number $N_2(I)$ of pixels in E_2 and the joint moment generating function of $(N_1(I), N_2(I))$. Note that

$$N_1(I) = \sum_{n,m} \chi_{X(n,m)>I}, N_2(I) = \sum_{n,m} \chi_{X(n,m)\leq I}$$

Make use of the iid property of the pixel values. Derive an expression for the joint distribution of $(X_1, N_1(I))$ and hence of the average $X_1/N_1(I)$ of the processed pixel values in E_1 and likewise of $X_2/N_2(I)$, the average of the processed pixels in E_2.

Given a vector valued image field $X(n,m) \in \mathbb{R}^d, n,m = 1,2,...,N$ (for example if $d = 3$, we have an RGB coloured image) and disjoint Borel sets $B_1,...,B_M \subset \mathbb{R}^d$ such that $\bigcup_{k=1}^{M} B_k = \mathbb{R}^d$, we segement the image pixels into M disjoint segments $E_1,...,E_M$ where $(n,m) \in E_k$ iff $X(n,m) \in B_k, k = 1,2,...,M$. Assuming that $X(n,m), n,m = 1,2,...,M$ are iid random variables and N_k is the number (random) of pixels in E_k, derive a formula for the joint moment generating function of $N_k, k = 1,2,...,M$. Further apply the window $W_k(n,m)$ to all pixel values in E_k for each $k = 1,2,...,M$. The average of the processed pixel values in the kth segment E_k is given by

$$X_k = \frac{\sum_{n,m} W_k(n,m)X(n,m)\chi_{\{X(n,m)\in B_k\}}}{N_k}, N_k = \sum_{n,m} \chi_{\{X(n,m)\in B_k\}}$$

Calculate the joint moment generating function of the r.v's $(N_1,...,N_M, N_1X_1,...,N_MX_M)$ in term of the pdf $f(X)$ of a given $X(n,m)$. Note that we are assuming the pixel values to be iid.

Let $X(n,m), n,m = 1,2,...,N$ be an iid random image field with pdf $f(X)$. Choose a pixel (n,m) with probability p and reject it with probability $q = 1-p$. Calculate the Histogram of the chosen pixels by partitioning $\mathbb{R}$ into disjoint intervals $I_1,...,I_M$, say $I_k = (a_k, a_{k+1}], k = 1,2,...,M, a_1 = -A, a_{M+1} = A$ and $f(X)$ is assumed to be concentrated over $[-A, A]$. Assign the histogram value of $f_k = N_k/(\sum_j N_j)$ to the interval I_k where N_k denotes the number of chosen pixels (n,m) with $X(n,m) \in I_k$. Calculate the average of the rth order moment μ_r of this Histogram using the formula $\mu_r = \sum_k b_k^r f_k$ where $b_k \in I_k$ is fixed for each k, for example, we can take $b_k = (a_k + a_{k+1})/2$. Suppose, we choose to smoothen out this historgram by the maximum entropy method, ie, maximize $-\int g(X).log(g(X))dX$ subject to the constraints $\int X^k g(X)dX = \mu_k, k = 1,2,...,r$. Then, calculate the optimum $g(X)$ and its statistical average.

Given two $N \times M$ image field matrices X, Y, how would you choose a rank r matrix Z so that $\| X - Z \|^2 + \lambda. \| Z - Y \|^2$ is a minimum. Note that the significance of this problem is that we are finding the best rank r (compressed image) approximation to the given image field X subject to a constraint that this approximation retains some of the major features in another image field Y. In order to do this problem, write $Z = EF$ where E is $N \times r$ and F is $r \times M$. Make use of the SVD of X, Y.

3.2 Problems in functional analysis and operator theory with applications to image processing

Given a Banach algebra X, let Δ denote the space of all homomorphisms from X into $\mathbb{C}$. Show that an ideal I of A is a maximum proper ideal in A iff $h(I) = 0$ for some $h \in \Delta$. Deduce that an element $x \in A$ is not invertible iff $h(x) = 0$ for some $h \in \Delta$. An element $\lambda \in \mathbb{C}$ is said to belong to the spectrum $\sigma(x)$ of $\in A$ iff $\lambda e - x$ is not invertible. Show that $\sigma(x) = \{h(x) : h \in \Delta$. For any $x \in A$, define $\hat{x} : \Delta \to \mathbb{C}$ by $\hat{x}(h) = h(x)$. Show that $\hat{x} \in C(\Delta)$. Show that the set $G(X)$ of invertible elements in X is an open set. Specifically, show that if x is invertible, then so is $x + y$ whenever $\| y \| < \| x^{-1} \|^{-1}$. Show that if $\rho(x) = sup|\sigma(x)|$, then $\rho(x) \leq \| x \|$ by first demonstrating that if $|\lambda| > \| x \|$, then $\lambda - x$ is invertible with inverse given by

$$(\lambda - x)^{-1} = \lambda^{-1} \sum_{n \geq 0} \lambda^{-n} x^n$$

which is an absolutely convergent series and hence represents an element of X. Show that $\sigma(x)$ is a closed subset of $\mathbb{C}$. Deduce that $\sigma(x)$ is a compact subset of $\mathbb{C}$. Show that if $x \in A$, then $x^n \in A, n = 1,2,...$ by using the factorization

$$\lambda^n - x^n = (\lambda - x)(\lambda^{n-1} + \lambda^{n-2} x + ... + x^{n-1})$$

and hence deducing that if $\lambda - x$ is not invertible, then so is $\lambda^n - x^n$. Thus show that if $\lambda \in \sigma(x)$, then $|\lambda| \leq \| x^n \|^n \; \forall n = 1,2,....$ Hence deduce that

$$\rho(x) \leq inf_{n \geq 1} | \; \| x^n \|^{1/n}$$

Show using the Cauchy residue theorem that if Γ is any circle of radius $> \rho(x)$ in the complex plane with the origin as centre, then

$$x^n = \int_\Gamma \lambda^n (\lambda - x)^{-1} d\lambda, n = 1,2,...$$

Hence deduce that

$$liimsup \parallel x^n \parallel^{1/n} \leq \rho(x)$$

Conclude from these two facts that

$$\rho(x) = limsup \parallel x^{n 1/n} = inf_n \parallel x \parallel^{1/n}$$

3.3 Application of the functional analysis of infinite dimensional matrices to image processing problems

Let $X = ((X(n,m)))$ be an complex matrix of infinite dimmension, ie, $x, m \in N = \{1, 2, ...\}$. Define the norm on such image fields by

$$\parallel X \parallel = sup_{n,m}|X(n,m)|$$

Denote the set of all such image fields with this norm by B. Show that B is a Banach algebra. Describe principal component analysis, eigendecomposition of normal, unitary and Hemitian matrices, singular value decompostion of such infinite matrices in an appropriate way. Specifically, consider the Hilbert space $L^2(\mathbb{N})$ and an image field X to be a bounded operator in this Hilbert space. Describe the spectral theorem for Hermitian matrices in this framework. Show that for the spectral norm,

$$\parallel X \parallel \leq sum_n \sum_m |X(n,m)|^2$$

If instead, the image field X is regarded as an operator from $L^p(\mathbb{N})$ for some $p \geq 1$ into $L^r(\mathbb{N})$ for some $r \geq 1$, then

$$|Xf(n)| \leq (\sum_m |X(n,m)|^q)^{1/q}. \parallel f \parallel_p, 1/p + 1/q = 1$$

and hence

$$\parallel Xf \parallel_r \leq [\sum_n (\sum_m |X(n,m)|^q)^{r/q}]^{1/r}. \parallel f \parallel_p$$

and hence that the norm induced on image fields from $L^p(\mathbb{N})$ into $L^r(\mathbb{N})$ satisfies the inequality

$$\parallel X \parallel_{p,r} = sup_{f \in L^p} \parallel Xf \parallel_r / \parallel f \parallel_p \leq [\sum_n (\sum_m |X(n,m)|^q)^{r/q}]^{1/r}$$

3.4 Problems in image processing for infinite dimensional image fields

Let $\phi(t,x,y)$ be the quantum Klein-Gordon field in the plane $\mathbb{R}^2$. Its equation of motion is obtained from the Lagrangian density

$$L(\phi, phi_t, \partial_x\phi, \partial_y\phi) =$$

$$(1/2)(\partial_t\phi)^2 - (1/2)(\partial_x\phi)^2 - (1/2)(\partial_y\phi)^2 - m^2\phi^2/2$$

The equations of motion

$$\partial_t^2\phi - \partial_x^2\phi - \partial_y^2\phi + m^2\phi = 0$$

can be solved to give

$$\phi(t,x,y) = C. \int [E(p)^{-1/2}a(p_x,p_y)exp(-i(E(p)t - p_x x - p_y y)) +$$

$$E(p)^{-1/2}a(p_x,p_y)^* exp(i(E(p)t - p_x x - p_y y))]dp_x dp_y$$

where

$$E(p) = E(p_x,p_y) = p_x^2 + p_y^2 + m^2$$

The canonical commutation relations

$$[\phi(t,x,y), \partial_t\phi(t,x',y')] = i\delta(x-x')\delta(y-y')$$

and

$$[\phi(t,x,y), \phi(t,x',y')] = 0$$

imply

$$[a(p_x,p_y), a(p'_x,p'_y)] = 0,$$

$$[a(p_x,p_y)^*, a(p'_x,p'_y)^*] = 0,$$

$$[a(p_x,p_y), a(p'_x,p'_y)^*] = \delta(p_x - p'_x)\delta(p_y - p'_y)$$

provided that we choose

$$C = 1/2\pi$$

When we perform a 2-D filtering of this quantum Klein-Gordon field, we get another quantum field given by

$$\psi(t,x,y) = h(x,y) * \phi(t,x,y) = \int h(x',y')\phi(t,x-x',y-y')dx'dy'$$

$$= C \int [E(p)^{-1/2}H(p_x,p_y)a(p_x,p_y)$$

$$exp(-i(E(p)t - p.r) + E(p)^{-1/2}a(p_x,p_y)^* H(p_x,p_y)^* exp(i(E(p)t - p.r))]d^2p$$

where $r = (x,y)$ and $p = (p_x,p_y)$. In this expression,

$$H(p_x,p_y) = H(p) = \int h(r)exp(-ip.r)d^2r$$

We wish to design our filter h so that this KG field is at closest distance from another quantum field given by

$$\chi(t,x,y) = \int f(t,p)a(p)exp(-ip.r) + f(t,p)^* a(p)^* exp(ip.r) d^3 r$$

when the two fields are in a state ρ, for example, we can take ρ as a cohrent pure state $-\phi(u) >$ that satisfies

$$a(p)|\phi(u) >= u(p)|\phi(u) >, a(p)^*|\phi(u) >= -\delta|\phi(u) > /\delta u(p)$$

Note that the coherent state is defined formally by

$$|\phi(u) >= \sum_{\{} n(p)\|\Pi_p(u(p)^{n(p)} a(p)^{*n(p)}/n(p)!)|0 >$$

$$= \sum_{\{n(p)\}\Pi_p(u(p)^{n(p)}/\sqrt{n(p)!})|\{n(p)\}>}$$

This filter design problem can be cast in the form of an optimization problem involving minimizing

$$\int Tr(\rho(h(r)^*\phi(t,r) - \psi(t,r))^2) dt d^3 r$$

w.r.t h. or equivalently, if ρ is the above pure coherent state, then we must minimize

$$\int < \phi(u)|(h(r) * \phi(t,r) - \psi(t,r))^2|\phi(u) > dt d^2 r$$

w.r.t h.

3.5 A real time version of filtering for quantum image fields passed through a classical random window

Let the quantum Klein-Gordon field interact with a fan window $W(t,x,y)$ whose dynamics is defined by an sde

$$dW(t,x,y) = \sum_{n,m} a(n,m)\partial_x^n \partial_y^m W(t,x,y)dt + \sigma dB(t,x,y)$$

where B is a Brownian motion field, ie, a zero mean Gaussian field with

$$\mathbb{E}(dB(t,x,y)dB(t',x',y')) = Q(x,y|x',y')dt.\delta_{t,t'}$$

or equivalently,

$$\mathbb{E}(B(t,x,y)B(t',x',y')) = min(t,t')Q(x,y|x',y')$$

Let $\phi_n(x,y), n = 1,2,...$ and $\lambda_n, n = 1,2,...$ be the complete set of orthonormal eigenfunctions and eigenvalues of Q. Then, we define

$$\int B(t,x,y)\phi_n(x,y)dxdy = B_n(t)$$

Then

$$\mathbb{E}(B_n(t)B_m(t')) = \lambda_n \delta_{n,m} min(t,t')$$

implying thereby that $B_n(.)/\sqrt{\lambda_n}, n = 1,2,...$ are independent standard Brownian motion processes. We can write

$$B(t,x,y) = \sum_n B_n(t)\phi_n(x,y)$$

When the quantum Klein-Gordon field $\phi(t,x,y)$ passes through this random classical window, the output quantum field is

$$\psi(t,x,y) = \phi(t,x,y)W(t,x,y)$$

and when this field interacts with an atom, with H_0 as the unperturbed Hamiltonian of the atom, then the resulting perturbed atomic Hamiltonian becomes $H_0 + \int \psi(t,x,y)X(x,y)dxdy = H_0 + V(t)$ where $X(x,y)$ are atomic observables. The probability of the atom making a transition from a state $|n>$ to a state $|m>$ in the time duration $[0,t]$ under this perturbation with the field being in the coherent state $|\phi(u)>$ is then given upto first order perturbation theory by

$$P_t(m|n) = |\int_0^t < m \otimes \phi(u)|V(t)|n \otimes \phi(u) > exp(i(E_n - E_m)t)dt|^2$$

Note that

$$< m \otimes \phi(u)|V(t)|n \otimes \phi(u) >=$$
$$< m|X(x,y)|n >< \phi(u)|\psi(t,x,y)|\phi(u) >$$

The problem is to estimate using the EKF, the window function $W(t,x,y)$ on a real time basis from noisy measurements of $P_t(m|n)$. Note that the sde for the window can be expressed in the following alternate way in terms of the Brownian motion processes $B_n(t), n \geq 1$:

$$dW(t,x,y) = \sum_n dW_n(t)\phi_n(x,y) = \sum_{n,m,k} a(n,m)W_k(t)\partial_x^n \partial_y^m \phi_k(x,y)dt$$
$$+\sum_n dB_n(t)\phi_n(x,y)$$

so that

$$dW_k(t) = \sum_{k,l} b(k,l)W_l(t)dt + dB_k(t), k = 1,2,...$$

where

$$b(k,l) = \sum_{n,m} a(n,m) < \phi_k, \partial_x^n \partial_y^m \phi_l >$$

In other words the continuous space-continuous time sde for the original window function has been transformed into an infinite sequence of coupled linear sde's for the generalized Fourier series coefficients of the window driven by an infinite sequence of independent standard Brownian motion processes.

3.6 Design of quantum neural networks using quantum scattering theory:The Lippmann-Schwinger equations

Let $H_0, H = H_0 + V$ be a pair of Hamiltonians. The wave operators are defined as

$$W_{\pm} = slim_{t\rightarrow\pm\infty} exp(itH).exp(-itH_0)$$

over the appropriate domain. This formula is based on the following principle: A free particle with Hamiltonian H_0 comes from $-\infty$. Its state evolves according to the free particle Hamiltonian H_0, so that its state at time t evolves as $exp(-itH_0)|\phi_i>$, $|\phi_i>$ being the initial free particle state. This particle comes nea the scattering centre having a potential of V, interacts with this centre and then gets scattered to the state $|\psi_i>$ which evolves according to the interacting particle Hamiltonian H. Thus after this interaction, the state of the particle evolves as $exp(-itH_i)|\psi>$. In the remote past, these two state evolutions coincide because there was no interaction in the remote past, ie

$$lim_{t\rightarrow-\infty} \parallel exp(-itH)|\psi_i> - exp(-itH_0)|\phi_i> \parallel = 0$$

This amounts to saying that

$$|\psi_i> = lim_{t\rightarrow-\infty} exp(-itH)exp(itH_0)|\phi_i> = W_-|\phi_i>$$

$|\phi_i>$ is the incoming free particle state at time $t = 0$ and $|\psi_i>$ is the incoming scattered state at time $t = 0$. After the scattering process, the particle goes to the outgoing scattered state $|\psi_f>$ which evolves according to the interacting particle Hamiltonian H, so that its state at time t is given by $exp(-itH)|\psi_f>$. In the distant future, ie, as $t \rightarrow \infty$, this evolves according to a final free particle state $|\phi_f>$, so that

$$lim_{t\rightarrow\infty} \parallel exp(-itH)|\psi_f> - exp(-itH_0)|\phi_f> \parallel = 0$$

which is the same as saying that

$$|\psi_f> = lim_{t\rightarrow\infty} exp(-itH)exp(itH_0)|\phi_f> = W_+|\phi_f>$$

The scattering matrix element $<\phi_f|S|\phi_i>$ between the initial free particle state $|\phi_i>$ and the final free particle state $|\phi_f>$ is defined by

$$<\phi_f|S|\phi_i> = <\psi_f|\psi_i> = <\phi_f|W_+^* W_-|\phi_i>$$

or equivalently,

$$S = W_+^* W_-$$

S is called the scattering matrix or the scattering operator. It is a unitary operator with domain and range equal to the $E_{ac}(H_0)$, the space of all vectors $f \in D(H_0)$ for which $\lambda \rightarrow \parallel E_0(\lambda)f \parallel$ is absolutely continuous w.r.t the Lebesgue measure on $\mathbb{R}$. Here, E_0 is the spectral measure of H_0.

Note: W_- maps $E_{ac}(H_0)$ onto $E_{ac}(H)$. W_+ also maps $E_{ac}(H_0)$ onto $E_{ac}(H)$. Therefore, W_+^* maps $E_{ac}(H)$ onto $E_{ac}(H_0)$, from which, it follows that $S = W_+^* W_-$ maps $E_{ac}(H_0)$ onto $E_{ac}(H_0)$.

We see that

$$W_+ = I + \int_0^\infty \frac{d}{dt}(exp(itH)exp(-itH_0))dt = I + i\int_0^\infty exp(itH)V.exp(-itH_0)dt$$

$$= I + i\int_{\mathbb{R}_+ \times \mathbb{R}} E(d\lambda)V.exp(it(\lambda - H_0 + i0))dtd\lambda$$

$$= I - \int_{\mathbb{R}} E(d\lambda)V.(\lambda - H_0 + i0)^{-1} = I + \int_{\mathbb{R}} E(d\lambda)V.(H_0 - \lambda - i0)^{-1}$$

Thus,

$$|\psi_f >= W_+|\phi_f >= |\phi_f > + \int_{\mathbb{R}} E(d\lambda)V.(H_0 - \lambda + i0)^{-1}|\phi_f >$$

Yet another formula for W_+ is given by

$$W_+ = I + i\int_{\mathbb{R} \times \mathbb{R}_+} exp(it(H - \lambda + i0))V.E_0(d\lambda)dt$$

$$= I - \int_{\mathbb{R}} (H - \lambda + i0)^{-1}V.E_0(d\lambda)$$

which gives the following expression for the outgoing scattered state in terms of the outgoing free particle state:

$$|\psi_f >= W_+|\phi_f >= |\phi_f > - \int (H - \lambda + i0)^{-1}V.E_0(d\lambda)|\phi_f >$$

so that if for some small $\epsilon > 0$, we have that $|\phi_f >\in R(E_0(\lambda - \epsilon, \lambda + \epsilon))$, then we have approximately,

$$|\psi_f >= |\phi_f > -(H - \lambda + i0)^{-1}V|\phi_f >$$

This is the first Lippman-Schwinger equation. Likewise,

$$W_- = I - \int_{-0\infty}^0 \frac{d}{dt}(exp(itH).exp(-itH_0))dt$$

$$= I - i\int_{-\infty}^0 exp(itH)V.exp(-itH_0)dt$$

$$= I - i\int_{-\infty}^0 dt \int_{\mathbb{R}} exp(it(H - \lambda - i0))V.E_0(d\lambda)$$

$$= I - \int (H - \lambda - i0)^{-1}V.E_0(d\lambda)$$

from which we get the result that if $|\phi_i> \in R(E_0(\lambda-\epsilon,\lambda+\epsilon))$ for some small ϵ, then we have approximately,

$$|\psi_i>=|\phi_i>-(H-\lambda-i0)^{-1}V|\phi_i>$$

This is the second Lippman-Schwinger equation. Now we can ask the question that if $V=V(t)$ changes with time or is rather controlled with the help of a set of parameters to change with time, then how will we describe the scattering matrix as a function of time ?

If the potential $V(t)$ is switched on upto the duration T, then the wave operators will be given by

$$W_{\pm}=lim_{t\to\pm\infty}(T\{exp(-i\int_0^t H(s)ds)\})^{-1}.exp(-itH_0)$$

where

$$H(t)=H_0+V(t)$$

Writing

$$V_1(t)=exp(itH_0)V(t).exp(-itH_0)$$

we get

$$T\{exp(-i\int_0^t H(s)ds)\}=exp(-itH_0)T\{exp(-i\int_0^T V_1(t)dt)\},t\geq T$$

Thus defining

$$W(T)=T\{exp(-i\int_0^T V_1(t)dt)\}$$

we get

$$W_+=T\{exp(-i\int_0^T V_1(t)dt)\}^*=W(T)^*$$

Assume more generally that $V(t)$ is non-zero for $-T\leq t\leq T$ and is zero outside this interval. Then we write $U(t,s)$ for the unitary evolution operator from time s to time $t>s$. Then,

$$U(t,s)=T(exp(-i\int_s^t H(u)du)$$

Write

$$U(t,s)=U_0(t-s)W(t,s),U_0(u)=exp(-iH_0u)$$

Then,

$$i\partial_t U(t,s)=H(t)U(t,s)=(H_0+V(t))U(t,s),U(s,s)=I$$

gives

$$U_0(t-s)(i\partial_t W(t,s)+H_0W(t,s))=(H_0+V(t))U_0(t-s)W(t,s)$$

which implies

$$i\partial_t W(t,s) = U_0(t-s)^* V(t) U_0(t-s) W(t,s)$$

We define $W_1(t,s)$ by

$$U(t,s) = U_0(t) W_1(t,s), t \geq s$$

Then

$$U(s,s) = I = U_0(s) W_1(s,s), W_1(s,s) = U_0(s)^* = U_0(-s)$$

$$i\partial_t W_1(t,s) = V_1(t) W_1(t,s), V_1(t) = U_0(-t) V(t) U_0(t)$$

gives

$$W_1(t,s) = U_0(-s) - i\int_s^t V_1(u_1) W_1(u,s) d_1 u$$

$$= U_0(-s) - i\int_s^t V_1(u_1) U_0(-s) du_1 + (-i)^2 \int_{s<u_2<u_1<t} V_1(u_1) V_1(u_2) W_1(u_2,s) du_1 du_2$$

$$= ... = T(exp(-i\int_s^t V_1(u)du)) U_0(-s) = W_2(t,s) U_0(-s)$$

where

$$W_2(t,s) = T(exp(-i\int_s^t V_1(u)du))$$

So finally, we can write

$$U(t,s) = U_0(t) T(exp(-i\int_s^t V_1(u)du)) U_0(-s)$$

$$= U_0(t-s) + \sum_{n\geq 1} (-i)^n \int_{s<t_n<...<t_1<t} U_0(t-t_1) V_1(t_1) U_0(t_1-t_2)$$

$$V_1(t_2) ... U_0(t_{n-1}-t_n) V_1(t_n) U_0(t_n-s) dt_1 ... dt_n$$

Note that

$$U_0(t-s) W(t,s) = U_0(t) W_1(t,s)$$

gives

$$W(t,s) = U_0(s) W_1(t,s) = U_0(s) W_2(t,s) U_0(-s)$$

so we can write

$$W(t,s) = T(exp(-i\int_s^t U_0(s) V_1(u) U_0(-s) du))$$

$$= I + \sum_{n\geq 1} (-i)^n \int_{s<u_n<...<u_1<t} U_0(s-u_1)$$

$$V(u_1) U_0(u_1-u_2) V(u_2) ... V(u_n) U_0(u_n-s) du_1 ... du_n$$

$$= I + \sum_{n\geq 1} (-i)^n \int_{0<u_n<...<u_1<t-s} U_0(-u_1) V(s+u_1)$$

$$U_0(u_1-u_2) V(s+u_2) . U_0(u_2-u_3) ... V(s+u_n) U_0(u_n) du_1 ... du_n$$

$$= T(exp(-i\int_0^{t-s} V_2(s,u)du))$$

We have

$$U(t,0)|\psi_f> - U_0(t)|\phi_f> \to 0, t \to \infty$$

so that

$$|\psi_f> = lim_{t\to\infty} U(t,0)^* U_0(t)|\phi_f> = lim_{t\to\infty} W_1(t,0)^*|\phi_f> = W_1(T,0)^*|\phi_f>$$

or

$$|\psi_f> = W_+|\phi_f>, W_+ = W_1(T,0)^* = W_2(T,0)^*$$

Likewise,

$$U(0,-t)|\psi(-t)> = |\psi_i>, t>0$$

gives

$$|\psi(-t)> = U(0,-t)^*|\psi_i>$$

and

$$|\psi(-t)> - U_0(-t)|\phi_i> \to 0, t \to \infty$$

gives

$$|\psi_i> = lim_{t\to\infty} U(0,-t)U_0(-t)|\phi_i> = W_-|\phi_i>$$

which implies that

$$W_- = lim_{t\to\infty} U(0,-t)U_0(-t) = lim_{t\to\infty} U(0,-t)U_0(-t) =$$

Now recalling the formula $U(t,s) = U_0(t)W_1(t,s) = U_0(t)W_2(t,s)U_0(-s)$, we get

$$W_- = lim_{t\to\infty} W_1(0,-t)U_0(-t) = W_2(0,-T)$$

The scattering operator is then

$$S = W_+^* W_- = W_2(T,0)W_2(0,-T) = S(T)$$

This expression gives an explicit formula for the scattering operator when the interacting potential $V(t)$ is switched for the duration $-T \leq t \leq T$. Note that by definition of the time ordering operator T, we can write

$$S(T) = T(exp(-i\int_{-T}^{T} V_1(t)dt)) = W_2(T,-T)$$

3.7 Neural network application of this result

Assume that $V(t) = V(\theta(t))$ is function of some control parameters $\theta(t)$ that can be made to vary with time. We start with an input free particle state $|\phi_i>$ at time $-T$ and switch on the scattering potential over the time interval $[-T,T]$, so at time T, the output free particle state becomes $S(T)|\phi_i>$. We wish to control the parameter variation $\theta(t)$ that determines the potential V_1 over the time interval $[-T,T]$ so that the final scattered state $S(T)|\phi_i>$ yields a probability density of scattered particles $f(T,q) = |<q|S(T)|\phi_i>|^2$ that is close to a given probability density $f_0(T,q)$. This can be done using the adaptive control law with delay firing

$$\theta(T+dT) = \theta(T) - dT.\sum_k \mu(k)\frac{\partial}{\partial\theta(T-t_k)}(\int (f_0(T,q) - f(T,q))^2 dq)$$

$$= \theta(T) + 2.dT.\sum_k \mu(k)\int (f_0(T,q) - f(T,q))\partial f(T,q)/\partial\theta(T-t_k)$$

Note that

$$\partial f(T,q)/\partial\theta(T-t_k) = 2Re(<\phi_i|S(T)^*|q><q|(\partial S(T)/\partial\theta(T-t_k))|\phi_i>)$$

with

$$\partial S(T)/\partial\theta(T-t_k) = \partial W(T,-T)/\partial\theta(T-t_k)$$

Now for $t \in [-T,T]$, we have

$$\partial W(T,-T)/\partial\theta(t) = \frac{\partial}{\partial\theta(t)}T(exp(-i\int_{-T}^{T} V_1(\theta(s))ds))$$

3.8 Problem on multipole expansion of the electromagnetic radiation field with neural network application

Consider the propagation of multipole electromagnetic fields at frequency $\omega = Kc$. The relevant Maxwell equations are

$$(\nabla^2 + K^2)(E,H) = 0, divE = 0, divH = 0$$

We try a solution

$$E_1 = f(r)LY_{lm}(\hat{r}), L = -ir\times\nabla, r = |\mathbf{r}|, \hat{r} = (\theta,\phi)$$

Then

$$\hat{r}.E_1 = 0$$

since

$$\hat{r}.L = 0$$

so that E_1 is polarized orthogonal to the radial direction. Further, E_1 will satisfy the Helmholtz equation provided that

$$r^{-1}(rf)'' - l(l+1)f/r^2 + K^2 f = 0$$

ie

$$f'' + 2f'/r - l(l+1)f/r^2 + K^2 f = 0$$

or

$$r^2 f'' + 2rf' + (K^2r^2 - l(l+1))f = 0$$

The solution is a linear combination of $r^{-1/2}J_{l+1/2}(Kr)$ and $r^{-1/2}Y_{l+1/2}(Kr)$ where $J_{l+1/2}(x)$ and $Y_{l+1/2}(x)$ are respectively the Bessel and modified Bessel functions of order $l+1/2$. The appropriate linear combination that corresponds to waves propagating radially outward for large r, ie having the phase factor proportional to $exp(iKr)$ is denoted by $g_l(Kr)$. We can thus write

$$E_1 = E_{lm}(\mathbf{r}) = g_l(Kr)LY_{lm}(\hat{r}), H_1 = curlE_1/-jKc\mu$$

Further we must check the condition $divE_1 = 0$. But

$$div(f(r)LY_{lm}(\hat{r})) = f'(r)\hat{r}.LY_{lm} + f(r)div(LY_{lm}) = f(r)div(LY_{lm})$$

$$div(r \times \nabla Y(\hat{r})) = (\nabla \times r).\nabla Y - r.\nabla \times \nabla Y = 0$$

thereby verfying the required property. Likewise we define

$$H_{lm} = gl(Kr)LY_{lm}(\hat{r})$$

The idea is to expand the electric and magnetic field as linear combinations of the vectors $[E_{lm}, curlE_{lm}/(-j\omega\mu]$ and $[curlH_{lm}/j\omega\epsilon, H_{lm}]$ with the linear combination coefficients determined by the boundary conditions of the fields on a spherical surface of radius R. The boundary conditions can be regarded as the sources of information or equivalently the inputs. Alternately since the linear combination coefficients are uniquely determined by the boundary conditions, they can also be regarded as the input sources while the propagating multipole electromagnetic fields constitute the output information. By controlling the boundary conditions or equivalently the linear combination coefficients, we can constrain the output fields to be such that they cause classical charged particles to move along specified trajectories or the wave functions of electrons to be such that their modulus square tracks specified joint probability density functions.

3.9 Problems in quantum stochastic calculus

The Gisin-Percival equation: Start with the Hudson-Parthasarathy-noisy Schrodinger equation

$$dU(t) = (-(iH + LL^*/2)dt + LdA - L^*dA^*)U(t)$$

Apply both sides of this equation to the pure state

$$|f(0) > \otimes |\phi(u) >$$

where $|\phi(u) >$ is a coherent state of the bath assuming (ie making the ansatz)

$$U(t)(|f(0) > \otimes |\phi(u) >) = |f(t) > \otimes |\phi(u) >$$

to derive the equation

$$d|f(t)>=(-(iH+LL^*/2)dt+u(t)Ldt-L^*(dB(t)-u(t)dt))|f(t)>$$

where

$$B(t)=A(t)+A(t)^*$$

is a classical Brownian motion, ie, commutative Brownian motion in the coherent state $|\phi(u)>$ with drift given by

$$<\phi(u)|B(t)|\phi(u)>=u(t)+\bar{u}(t)$$

Develop a quantum neural network that will take as input a given signal $\xi(t)$ and a probability density $p_0(t,x)$ and will output a wave function $<x|f(t)>$ with probability density $p(t,x)=|<x|f(t)>|^2$ that tracks $p_0(t,x)$, by applying at each time instant the stochastic gradient algorithm to the error energy $\int(p_0(t,x)-p(t,x))^2dx$ after incorporating time varying control parameters $\theta(t)$ into the Hamiltonian H and the Lindblad operator L. The adaptation coefficient μ in this stochastic gradient algorithm must be made to increase in accordance with the prediction error energy $e(t)^2$ where $e(t)$ is the prediction error obtained when estimating $\int xp(t,x)dx=\int x|<x|f(t)>|^2dx$ linearly based on $\xi(t),\xi(t-\delta),\xi(t-2\delta),...\xi(t-k\delta)$. It follows then that the output pdf $p(t,x)$ will track the given pdf $p_0(t,x)$ while simultaneously ensuring that its mean $\int xp(t,x)dx$ remains strongly correlated with the reference signal $\xi(t)$ and its past. More generally, we consider a qsde of the form

$$dU(t)=(L_b^a d\Lambda_a^b(t))U(t)$$

where $a,b\geq 0$, the quantum Ito formula reads

$$d\Lambda_b^a.d\Lambda_d^c=\epsilon_d^a.d\Lambda_b^d$$

We can realize these quantum noise processes in terms of creation and annihilation processes alone:

$$\Lambda_a^0(t)=A_a(t),a\geq 1,\Lambda_0^a(t)=A_a(t)^*,a\geq 1,d\Lambda_b^a(t)=dA_b(t)^*.dA_a(t)/dt,\Lambda_0^0(t)=t$$

Then assuming that the joint state of the system and bath is a pure state of the form

$$|\psi(t)>=|\psi_1(t)\otimes\phi(u)>=U(t)(|\psi_1(0)\otimes\phi(u)>$$

$$L_b^a.d\Lambda_a^b(t)|\psi(t)>=$$

$$=L_0^0dt|\psi_1(t)>|\phi(u)>+L_0^a|\psi_1(t)>dA_a(t)|\phi(u)>$$
$$+L_a^0|\psi_1(t)>dA_a(t)^*|\phi(u)>+L_b^a|\psi_1(t)>dt^{-1}.dA_b(t)^*.dA_a(t)|\phi(u)>$$

where now on the rhs, the summation indices a,b range over only $a,b\geq 1$. Defining the classical Brownian motion processes

$$B_a(t)=A_a(t)+A_a(t)^*,a\geq 1$$

we then get for the rhs

$L^0_0|\psi_1(t)>|\phi(u)> + L^a_0|\psi_1(t)> u_a(t)dt|\phi(u)>$

$+L^0_a|\psi_1(t)> (dB_a(t) - u_a(t)dt)|\phi(u)>$

$+L^a_b|\psi_1(t)> .(dB_b(t) - u_b(t)dt)u_a(t)|\phi(u)>$

where we have used the fact that

$$dA_a(t)|\phi(u)>= u_a(t)dt|\phi(u)>$$

Hence $|\psi_1(t)>$ obeys the classical sde

$$d|\psi_1(t)>= (L^0_0 + u_a(t)L^a_0 - u_a(t)L^0_a - u_a(t)u_b(t)L^a_b)dt + (L^0_b - u_a(t)L^a_b)dB_b(t))|\psi_1(t)>$$

and this can be used in a quantum neural network by regarding the coherent state parameter function $u(t)$ as a control parameter, ie we keep adapting this parameter with time so that control is exerted on the bath state in such a way that the system pdf corresponding to the state $|\psi_1(t)>$ tracks a given pdf.

Let $\phi_k(x), k = 1, 2, ...$ be a sequence of complex classical fields defined on a manifold M and $\bar{\phi}_k(x)$ their complex conjugates. Let $a(k), k = 1, 2, ...$ be a sequence of annihilation operators and $a(k)^*, k = 1, 2, ...$ the corresponding sequence of creation operators satisfying the standard CCR

$$[a(k), a(m)] = 0 = [a(k)^*, a(m)^*], [a(k), a(m)^*] = \delta(k, m)$$

Consider now the quantum Hermitian field

$$\phi(x) = \sum_{k\geq 1}(a(k)\phi_k(x) + a(k)^*\bar{\phi}_k(x))$$

Let $L(\psi(X), \partial_\mu\psi(X))$ be a Lagrangian for another field $\psi(X)$ defined on a manifold N in which $\phi_k(x)$ takes values. Let $|\phi(u)>$ be a coherent state of the creation and annihilation operators:

$$a(k)|\phi(u). = u(k)|\phi(u)>$$

and then evaluate the average action functional

$$S[\psi] =< \phi(u)| \int d^4X \int dx.L(\psi(X + \phi(x)), \psi_{,\mu}(X + \phi(x))|\phi(u)>$$

The problem is to evaluate this as a power series in $\psi(X)$ and it s partial derivatives $\psi_{,\mu_1,...,\mu_k}(X), \mu_1, \mu_2, ..., \mu_k = 1, 2, ..., d, k = 1, 2, ...$ This quantum averaged actional functional for ψ is a generalization of string field theory in which the actional functionals of quantum field theory are derived as lower energy approximations of quantum averages of actional functionals of string theory.

Real time quantum filtering and parameter estimation in the Hudson-Parthasarathy-noisy Schrodinger equation.

The state model is

$$dU(t) = -(iH(\theta(t)) + P(\theta(t)))dt + L(\theta(t)dA(t) + L(\theta(t))^* dA(t)^*)U(t),$$

where $\theta(t)$ is a classical Markov process with generator K. Let X be a system observable. Consider

$$j_t(Xf) = X(t)f(\theta(t) = j_t(X)f(\theta(t))$$

) We can write

$$dj_t(Xf) = f(\theta(t))dj_t(X) + j_t(X)df(\theta(t))$$

The input measurement process is

$$dY_i(t) = h_0(\theta(t))dt + h_1(\theta(t))dA(t) + \bar{h}_1(\theta(t))dA(t)^* + dv(t)$$

The output measurement process is

$$Y_o(t) = U(t)^* Y_i(t)U(t)$$

Note that $Y_o(t)$ is a non-demolition process because of the fact that

$$Y_o(t) = U(T)^* Y_i(t)U(T), T \geq t$$

We have

$$dY_o(t) = dY_i(t) - j_t(h_1(\theta)L(\theta)^* + \bar{h}_1(\theta)L(\theta))dt$$

Here, we are assuming that $h_k, k = 0, 1$ are q-vector valued functions of the parameter so that actually, we mean that

$$dY_{ik}(t) = h_{0k}(\theta(t))dt + h_{1k}(\theta(t))dA(t) + \bar{h}_{1k}(\theta(t))dA(t)^* + dv(t), k = 1, 2, ..., q,$$

$$dY_{ok}(t) = dY_{ik}(t) - j_t(h_{1k}(\theta)L(\theta)^* + \bar{h}_{1k}(\theta)L(\theta))dt, k = 1, 2, ..., q$$

Also $v(t)$ is an q-vector valued Brownian motion.

Remark: If $X(\theta)$ is a system observable valued function of θ, then $j_t(X(\theta) = U(t)^* X(\theta(t)))U(t).\theta \to X(\theta)$ can be viewed as an element of $B(\mathfrak{h}) \otimes B_1$ where B_1 is the Banach space of all complex valued functions on the parameter space $\mathbb{R}^d$ that fall in the domain of the generator K of the Markov process $\theta(t)$.

We can also write

$$dj_t(Xf) = j_t(\phi_0(X, \theta)f(\theta) + XK_0 f(\theta))dt$$

$$+ j_t(\phi_1(X, \theta)f(\theta))dA(t) + j_t(\phi_2(X, \theta)f(\theta))dA(t)^* + J_t(XD_i f(\theta))dB_i(t)$$

where

$$\phi_0(X, \theta) = i[H(\theta), X], \phi_1(X, \theta) = -L(\theta)X + XL(\theta), \phi_2(X, \theta) = -XL(\theta)^* + L(\theta)^* X$$

and we are assuming that the Markov process $\theta(t)$ is actually a diffusion process of the form

$$d\theta_i(t) = \mu_i(\theta(t))dt + \sigma_{ij}(\theta(t))dB_j(t)$$

with

$$D_i = \sigma_{ji}(\theta)\partial/\partial\theta_j$$

and

$$K_0 = \mu_i(\theta)\partial/\partial\theta_i + (1/2)(\sigma(\theta)\sigma(\theta)^T)_{ij}\partial^2/\partial\theta_i\partial\theta_j$$

is the generator of the diffusion process θ. Note that if G is any operator acting on the space B_1 of complex valued functions of the parameter θ, we regard $\phi_k(X,\theta)Gf(\theta)$ as the $B(h)$ valued function $\theta \to \phi_k(X,\theta)Gf(\theta)$ on the parameter space or equivalently as an element of $B(\mathfrak{h}) \otimes B_1$. The Cq filter is defined as the conditional expectation operator acting on $B(\mathfrak{h}) \otimes B_1$ and taking values in $\eta_o(t) = \sigma(Y_o(s) : s \leq t)$ defined by

$$\pi_t(Xf) = \mathbb{E}(j_t(Xf)|\eta_o(t)) = \mathbb{E}(X(t)f(\theta(t))|\eta_o(t)) = \int Tr(\rho_B(t,\theta)X(t))f(\theta)d\theta$$

where for each parameter $\theta \in \mathbb{R}^p$,$\rho_B(t,\theta)$ is a function of $\eta_o(t)$ ie of $Y_o(s), s \leq t$ with values in $B(\mathfrak{h})$, so that $\rho_B(t,\theta) \geq 0$ and $\int Tr(\rho_B(t,\theta))d\theta = 1$. The Cq filter satisfies a stochastic differential equation of the form

$$d\pi_t(Xf) = F_t(X,f)dt + G_t(X,f)dY_o(t) = F_t(X,f)dt + G_{tk}(X,f)dY_{ok}(t)$$

summation over the repeated index k being implied. We assume that generally the classical measurement noise $v(t)$ can be correlated with the classical parameter process noise $B(t)$. Thus, we assume that $(v(t), B(t)), t \geq 0$ is a Brownian motion process (ie and independent increment zero mean Gaussian process) with

$$dB(t)dB(t)^T = Qdt, dv(t)dv(t)^T = Rdt, dB(t)dv(t)^T = Sdt, dv(t)dB(t)^T = S^T dt$$

Using the orthogonality principle in filtering theory, we derive the filter equations

$$\mathbb{E}(dj_t(Xf) - d\pi_t(X,f))|\eta_o(t)) = 0 - - - (1)$$

$$\mathbb{E}((dj_t(Xf) - d\pi_t(X,f))dY_{ok}(t)|\eta_o(t)) + \mathbb{E}((j_t(Xf) - \pi_t(X,f))dY_{ok}(t)|\eta_o(t)) = 0$$

or equivalently, in matrix notation,

$$\mathbb{E}((dj_t(Xf) - d\pi_t(X,f))dY_o(t)^T|\eta_o(t)) +$$
$$\mathbb{E}((j_t(Xf) - \pi_t(X,f))dY_o(t)^T|\eta_o(t)) = 0 - - - (2)$$

Now

$$dj_t(Xf)dY_o(t)^T =$$
$$(j_t(\phi_0(X,\theta)f(\theta) + XK_0f(\theta))dt + j_t(\phi_1(X,\theta)f(\theta))dA(t) + j_t(\phi_2(X,\theta)f(\theta))dA(t)^*$$
$$+ j_t(XD_if(\theta))dB_i(t))(h_1(\theta(t))dA(t) + \bar{h}_1(\theta(t))dA(t)^* + dv(t))^T$$
$$= j_t(\phi_1(X,\theta)f(\theta)h_1(\theta)^*)dt + j_t(XDf(\theta)^T S)dt$$

where

$$h_1(\theta)^* = \bar{h}_1(\theta)^T$$

is a row vector and and $Df(\theta) = ((D_i f(\theta)))_i$ is regarded as a column vector so that $Df(\theta)^T$ is a rwo vector. Hence deduce that

$$\mathbb{E}(dj_t(Xf)dY_o(t)^T|\eta_o(t)) = \pi_t(\phi_1(X,\theta)f(\theta)h_1(\theta)^*)dt + \pi_t(XDf(\theta)^T S)dt$$

Also,

$$d\pi_t(X,f)dY_o(t)^T = G_t(X,f)dY_o(t)dY_o(t)^T$$
$$= G_t(X,f)dY_i(t)dY_i(t)^T = G_t(X,f)(R+h_1(\theta)h_1(\theta)^*)dt$$

and hence

$$\mathbb{E}(d\pi_t(X,f)dY_o(t)^T|\eta_o(t)) = G_t(X,f)(R+\pi_t(hh^*(\theta)))dt$$

Note that $G_t(X,f)$ is regarded as a row vector. Also,

$$(j_t(Xf) - \pi_t(Xf))dY_o(t)^T =$$

$$(j_t(Xf) - \pi_t(Xf))(dY_i(t)^T - j_t(h_1(\theta)L(\theta)^* + \bar{h}_1(\theta)L(\theta))dt)$$
$$= (j_t(Xf) - \pi_t(Xf))(h_0(\theta(t))^T dt + h_1(\theta(t))^T dA(t) + h_1(\theta(t))^* dA(t)^*$$
$$+dv(t)^T - j_t(h_1(\theta)^T L(\theta)^* + h_1(\theta)^* L(\theta))dt)$$

thereby yielding

$$\mathbb{E}[(j_t(Xf) - \pi_t(Xf))dY_o(t)^T|\eta_o(t)] =$$
$$[\pi_t(Xfh_0(\theta)^T) - \pi_t(Xfh_0(\theta)^T)]dt + [\pi_t(Xfh_1(\theta)^T) - \pi_t(Xf)\pi_t(h_1(\theta)^T)]u(t)dt$$
$$+[\pi_t(Xfh_1(\theta)^*) - \pi_t(Xf)\pi_t(h_1(\theta)^*)]\bar{u}(t)dt$$
$$-[\pi_t(X(L(\theta)^* h_1^T f + L(\theta)h_1^* f) - \pi_t(Xf)\pi_t(h_1(\theta)^T L(\theta)^* + h_1(\theta)^* L(\theta)))]dt$$

Thus (2) gives us

$$\pi_t(\phi_1(X,\theta)f(\theta)h_1(\theta)^*) + \pi_t(XDf(\theta)^T S) - G_t(X,f)(R+\pi_t(hh^*(\theta)))$$
$$+[\pi_t(Xfh_0(\theta)^T) - \pi_t(Xfh_0(\theta)^T)]dt + [\pi_t(Xfh_1(\theta)^T) - \pi_t(Xf)\pi_t(h_1(\theta)^T)]u(t)$$
$$+[\pi_t(Xfh_1(\theta)^*) - \pi_t(Xf)\pi_t(h_1(\theta)^*)]\bar{u}(t)$$
$$-[\pi_t(X(L(\theta)^* h_1^T f + L(\theta)h_1^* f) - \pi_t(Xf)\pi_t(h_1(\theta)^T L(\theta)^* + h_1(\theta)^* L(\theta)))] = 0 - - - (3)$$

Further, (1) gives us

$$\pi_t(\phi_0(X,\theta)f(\theta) + XK_0 f(\theta)) + \pi_t(\phi_1(X,\theta)f(\theta))u(t) + \pi_t(\phi_2(X,\theta)f(\theta))\bar{u}(t)$$

$$= F_t(X,f) + G_t(X,f)[\pi_t(h_1(\theta))u(t) + \pi_t(\bar{h}_1(\theta))\bar{u}(t)) - \pi_t(\bar{h}_1(\theta)L(\theta) + h_1(\theta)L(\theta)^*)]$$

where we have used the identity

$$\mathbb{E}(dY_o(t)|\eta_o(t)) = \mathbb{E}(dY_i(t)|\eta_o(t) - \pi_t(\bar{h}_1(\theta)L(\theta) + h_1(\theta)L(\theta)^*)dt$$

$$= [\pi_t(h_1(\theta))u(t) + \pi_t(\bar{h}_1(\theta))\bar{u}(t)) - \pi_t(\bar{h}_1(\theta)L(\theta) + h_1(\theta)L(\theta)^*)]dt$$

Combining all these equations, we get the filter equation as

$$d\pi_t(Xf) =$$

$$(\pi_t(\phi_0(X,\theta)f(\theta)+XK_0f(\theta))+\pi_t(\phi_1(X,\theta)f(\theta))u(t)+\pi_t(\phi_2(X,\theta)f(\theta))\bar{u}(t))dt$$
$$+G_t(X,f)(dY_o(t)-[\pi_t(h_1(\theta))u(t)+\pi_t(\bar{h}_1(\theta))\bar{u}(t))-\pi_t(\bar{h}_1(\theta)L(\theta)+h_1(\theta)L(\theta)^*)]dt)$$
$$=(\pi_t(\phi_0(X,\theta)f(\theta)+XK_0f(\theta))+\pi_t(\phi_1(X,\theta)f(\theta))u(t)+\pi_t(\phi_2(X,\theta)f(\theta))\bar{u}(t))dt$$
$$+[\pi_t(\phi_1(X,\theta)f(\theta)h_1(\theta)^*)+\pi_t(XDf(\theta)^TS)$$
$$+[\pi_t(Xfh_0(\theta)^T)-\pi_t(Xfh_0(\theta)^T)]dt+[\pi_t(Xfh_1(\theta)^T)-\pi_t(Xf)\pi_t(h_1(\theta)^T)]u(t)$$
$$+[\pi_t(Xfh_1(\theta)^*)-\pi_t(Xf)\pi_t(h_1(\theta)^*)]\bar{u}(t)$$
$$-[\pi_t(X(L(\theta)^*h_1^Tf+L(\theta)h_1^*f)-\pi_t(Xf)\pi_t(h_1(\theta)^TL(\theta)^*$$
$$+h_1(\theta)^*L(\theta)))](R+\pi_t(hh^*(\theta)))^{-1}(dY_o(t)$$
$$-[\pi_t(h_1(\theta))u(t)+\pi_t(\bar{h}_1(\theta))\bar{u}(t))-\pi_t(\bar{h}_1(\theta)L(\theta)+h_1(\theta)L(\theta)^*)]dt)$$

3.10 Additional problems in group representation theory

Let G be a compact group and let $\hat{G}$ be the collection of all inequivalent irreducible representations of G. Prove that $\hat{G}$ is countable, that each element $\pi\in\hat{G}$ is a finite dimensional irreducible representation of G and can be chosen to be a unitary representation and that if L denotes the left regular representation of G, then

$$L=\bigoplus_{\pi\in\hat{G}}d(\pi)\pi$$

Show that if

$$\chi_\pi(g)=Tr(\pi(g))$$

is the character of any irreducible representation π of G, then

$$\chi_L(g)=\sum_{\pi\in\hat{G}}d(\pi)\chi_\pi(g)$$

Show that if U is any representation of G, then U can be decomposed as (ie, U is equivalent to)

$$U=\bigoplus_{\pi\in E(U)}m(\pi)\pi$$

where $E(U)\subset\hat{G}$ and for each $\pi\in\hat{G}$, $m(\pi)$ is a finite positive integer. To show the latter, first define an inner product on $\mathcal{H}$ that is invariant w.r.t U by the group averaging process (the method works only because G is a compact group), then show that U is a unitary representation of G w.r.t this inner product and then prove that if W is any subspace of the representation space $\mathcal{H}$ in which U acts that is U-invariant, then $\mathcal{H}^\perp$ is also U-invariant. Hence, deduce that U is completely reducible, ie, decomposible into a direct sum of irreducible representations.

Show that if $\pi, \pi' \in \hat{G}$ (ie, π, π' are inequivalent irreducible unitary representations of G) then

$$< \chi_\pi, \chi_{\pi'} >= 0, < \chi_\pi, \chi_\pi >= 1$$

where the inner product is the standard $L^2(G)$ inner product taken w.r.t the Haar measure:

$$< f_1, f_2 >= \int_G \bar{f}_1(g) f_2(g) dg$$

Hence deduce that if U, V are two representations of G with decompositions

$$U = \bigoplus_{\pi \in \hat{G}} m(\pi)\pi, V = \bigoplus_{\pi \in \hat{G}} n(\pi)\pi$$

then

$$< \chi_U, \chi_V >= \sum_{\pi \in \hat{G}} m(\pi) n(\pi)$$

Here $m(\pi), (\pi), \pi \in \hat{G}$ are non-negative integers. Show that U and V are equivalent representations, iff $m(\pi) = n(\pi) \forall \pi \in \hat{G}$ iff $\chi_U = \chi_V$. For this, you must use the fact that

$$\chi_U = \sum_{\pi \in \hat{G}} m(\pi) \chi_\pi$$

is an orthogonal decomposition of χ_U and likewise

$$\chi_V = \sum_{\pi \in \hat{G}} n(\pi) \chi_\pi$$

and hence that $m(\pi) = n(\pi) \forall \pi \in \hat{G}$ iff $\chi_U = \chi_V$

3.11 Using the Einstein field equations in the design of a classical neural network

The Einstein field equations are derived from a Lagrangian $L(q_{ab}, q'_{ab}, N^a, N)$ with q_{ab} a three dimensional (spatial) metric and N^a a three dimensional vector field and N a scalar field. This is called the ADM action. The equations of motion are

$$\frac{d}{dt} \partial L / \partial q'_{ab} - \partial L / \partial q_{ab} = 0,$$

$$\partial L / \partial N^a = 0, \partial L / \partial N = 0$$

when the gravitational field interacts with the matter field having energy-momentum tensor $T^{\mu\nu}$, there is an extra term that has to be added to the Lagrangian, namely $\int \rho\sqrt{-g}d^4x$ where ρ is the scalar density field of matter. To see this, we consider the matter to comprise of a system of N particles with rest masses $m_k, k = 1, 2, ..., N$. Suppose that m_k moves along the trajectory $x_k^\mu(\tau_k)$ where τ_k is the proper time as measured by the kth particle in his rest frame. Then the energy momentum tensor is

$$T^{\mu\nu}(x) = \sum_k m_k \frac{dx_k^\mu}{d\tau_k}(dx^\nu/dt)\delta^3(x - x_k)(-g(x_k))^{-1/2}$$

Indeed with this expression, we have

$$\int g_{\mu\nu}T^{\mu\nu}(x)\sqrt{-g}d^4x = \sum_k \int m_k g_{\mu\nu}(x_k)(dx_k^\mu/d\tau_k)(dx_k^\nu/d\tau_k)d\tau_k$$

which is the action functional for the trajectories of the N point particles. In other words, $T^{\mu\nu}$ is a tensor field and $g_{\mu\nu}T^{\mu\nu}$ is a scalar field. $T^{\mu\nu}$ can be regardes as a source field that drives the gravitational field. A simple computation shows that (using the identity $g_{\mu\nu}(dx^\mu/d\tau_k)(dx^\nu/d\tau_k = 1)$

$$g_{\mu\nu}\sqrt{-g}T^{\mu\nu} = \sum_k m_k(d\tau_k/dt)\delta^3(x - x_k) = \rho\sqrt{-g}$$

Now consider a three dimensional box at rest in the lab frame having rest frame volume V containing matter of total rest mass m_0 with its sides parallel to the cartesian axes. Suppose that this box moves with a velocity v along the x direction. Then, the side parallel to the x direction gets contracted by a factor of $1/\gamma = (1 - v^2/c^2)^{-1/2} = d\tau/dt$ while the sides parallel to the y and z directions do not undergo any change. Thus, the volume of the box as seen in the lab frame is γV. On the other hand, the mass of the matter within the box as seen in the lab frame gets increased by the factor of γ so its total mass becomes $\gamma m_0 = m_0/(1 - v^2/c^2)^{1/2}$. It follows that as seen in the lab frame, the density of matter within the box becomes $\rho_L = \rho_0\gamma^2$ where $\rho_0 = m_0/V$ is an invariant under Lorentz transformations. In the curved space-time of general relativity, the volume of an infinitesimal box d^3x becomes $\sqrt{-g}d^3x$ in a synchronous reference frame and hence the mass density is obtained by the conventional special relativistic formula after dividing by $\sqrt{-g}$. Note that the coefficient of variation of the factor $g_{\mu\nu}\sqrt{-g}$ w.r.t $g_{\mu\nu}$ is the same as the coefficient $R_{\mu\nu}$ of the variation of $g_{\mu\nu}$ appearing in the Einstein Hilbert action $\int R\sqrt{-g}d^4x$ and hence the variational principle

$$\delta_g(\int R\sqrt{-g}d^4x + \int T^{\mu\nu}g_{\mu\nu}\sqrt{-g}d^4x) = 0$$

gives the Einstein field equations in the form

$$R^{\mu\nu} == K(T^{\mu\nu} - (1/2)Tg^{\mu\nu})$$

Now we can regard the Einstein field equations as an equation for the metric tensor $g_{\mu\nu}$ with the source field being $T^{\mu\nu}$. In fact more precisely, since the

Einstein field equations imply the hydrodynamic equations $T^{\mu\nu}_{:\nu} = 0$ for the matter field, strictly speaking, we should introduce an additional source field $f^{\mu\nu}$ that can be controlled into the energy momentum tensor so that the fields controlled by this tensor are both the metric tensor and the density and velocity of the matter field in accordance with the equations

$$R^{\mu\nu} - (1/2)Rg^{\mu\nu} = K(T^{\mu\nu} + f^{\mu\nu}), T^{\mu\nu}_{:\nu} = -f^{\mu\nu}_{:\nu}$$

Thus our neural network will take as input $f^{\mu\nu}$ generate an output $\hat{g}_{\mu\nu}$ that is close to the solution $g_{\mu\nu}$ of the Einstein field equations during the training process. The control field $f^{\mu\nu}(x)$ can have a finite set of parameters $\theta \in \mathbb{R}^p$ and then we can train the neural network to output the optimal parameters θ given the solution metric, density and velocity fields to the Einstein field equations in the presence of the control field $f^{\mu\nu}(x|\theta)$. Now when presented with a new metric field, this trained neural network will output the optimal parameters θ for the control field that will to a good degree of approximation, generate the given metric tensor. In this way, the entire information contained in the metric tensor is compressed into the set of control parameters θ.

Suppose that there are other fields $q_k, k = 1, 2, ..., d$ which contribute to the energy-momentum tensor of the matter fields that affect the metric tensor of space-time. Let the Lagrangian density of these fields be given by $L_1(q_k, q_{k,\mu}, \theta)$ where θ are control fields. The energy momentum tensor of these fields is obtained by observing that

$$\partial L_1/\partial x^\mu = (\partial L_1/\partial q_k)q_{k,\mu} + (\partial L_1/\partial q_{k,\nu})q_{k,\mu\nu}$$

In the absence of other interactions, the Euler-Lagrange equations for q_k are

$$\partial L_1/\partial q_k = \partial_\nu \partial L_1/\partial q_{k,\nu}$$

and hence we get

$$(\partial/\partial x^\nu)(L_1\delta^\nu_\mu - q_{k,\mu}\partial L_1/\partial q_{k,\nu}) = 0$$

which implies that the energy-momentum tensor of the fields q_k is given by

$$T^\nu_\mu = q_{k,\mu}.\partial L_1/\partial q_{k,\nu} - L_1\delta^\nu_\mu$$

upto a constant of proportionality. If we add this component of the energy-momentum tensor to the rhs of the Einstein field equations, we get additional degrees of freedom via the control fields θ appearing in L_1 to control the metric tensor of space-time. Suppose that we know before hand that the energy-momentum tensor of the other matter/radiation fields that affect the metric tensor of space-time comes from a set of N fields having energy-momentum tensors $T^{\mu\nu}_k, k = 1, 2, ..., N$. Then assuming that all these fields are present, we may postulate that the total energy-momentum tensor of the other fields is given by

$$T^{\mu\nu}(x, \theta) = \sum_{k=1}^{N} \theta_k T^{\mu\nu}_k(x)$$

where the $\theta_k's$ are constant. This is justified because the equation $T_k^{\mu\nu}{}_{,\nu} = 0$ implies $(\theta_k T_k^{\mu\nu})_{,\nu} = 0$ and the $\theta_k's$ and the energy momentum tensor of the kth field is defined in terms of the Lagrangian only upto a multiplicative constant since if L_k is the Lagrangian of that field, then so is $\theta_k L_k$ where θ_k is any constant. For example, in the case of field having quadratic Lagrangians like the electromagnetic and Dirac fields, we can scale the fields by the factor $\sqrt{\theta_k}$ which will cause the corresponding Lagrangians to scale by the factor θ_k. Thus the Einstein field equations in the presence of these other fields can be taken as

$$G^{\mu\nu} == \sum_{k=1}^{N} \theta_k T_k^{\mu\nu}, G^{\mu\nu} = R^{\mu\nu} - (1/2)Rg^{\mu\nu}$$

We can train our neural network to generate the metric tensor $g_{\mu\nu}$ for different choices of the $\theta_k's$ provided that we also solve the Euler-Lagrange equations for the other fields interacting with the gravitational field. For that we must assume that L_k, the Lagrangian of the kth field contains the metric tensor in order that the action function $\int L_k\sqrt{-g}d^4x$ is an invariant under diffeomorphisms of space-time. For example, if the kth matter field is a tensor $q_{k\mu_1...\mu_r}^{\nu_1...\nu_p}$ of the (p,r) type, then it and its covariant derivaive w.r.t the space-time coordinates must appear in L_k in a contracted form using the metric so that L_k is a diffeomorphism invariant. For example, a term such as

$$g^{\rho\sigma} q_{k\mu_1...\mu_r:\rho}^{\nu_1...\nu_p} q_k^{\mu_1...\mu_r}{}_{\nu_1...\nu_p:\sigma\}}$$

would be one such invariant. Here $q_k^{\mu_1...\mu_r}{}_{\nu_1...\nu_p}$ is the (r,p) type tensor obtained by raising the r covariant indices and lowering the p contravariant indices of $q_{k\mu_1...\mu_r}^{\nu_1...\nu_p}$ using the metric tensor. Now given such a Lagrangian $L(q, q_{mu} g_{\mu\nu})\sqrt{-g}$, we can equivalently define its energy-momentum tensor by taking its variational derivative w.r.t $g_{\mu\nu}$:

$$T^{\mu\nu}\sqrt{-g} = \partial(L\sqrt{-g})/\partial g_{\mu\nu} - \partial_\rho \partial(L\sqrt{-g})/\partial g_{\mu\nu,\rho})$$

The change in the matter action $\delta \int L\sqrt{-g}d^4x$ under metric variations $\delta g_{\mu\nu}$ is therefore

$$\int T^{\mu\nu}\sqrt{-g}\delta g_{\mu\nu} d^4x$$

Now we know that under an infinitestimal diffeomorphism of space-time, $x^\mu \to x^\mu + \xi^\mu(x)$, the metric tensor changes by

$$\delta g_{\mu\nu} = \xi_{\mu:\nu} + \xi_{\nu:\mu}$$

It follows then that under such an infinitesmial diffeomorphism of space-time, the change in the matter action is

$$2\int T^{\mu\nu}\sqrt{-g}\xi_{\mu:\nu}d^4x = 2\int (T^{\mu\nu}\xi_\mu)_{:\nu}\sqrt{-g}d^4x - 2\int T^{\mu\nu}_{:\nu}\sqrt{-g}\xi_\mu d^4x$$

$$= \int (T^{\mu\nu}\xi_\mu\sqrt{-g})_{,\nu}d^4x - 2\int T^{\mu\nu}_{:\nu}\sqrt{-g}\xi_\mu d^4x$$

$$= -2\int T^{\mu\nu}_{:\nu}\sqrt{-g}\xi_\mu d^4x$$

and hence in the absence of other fields, we obtain the following conservation law in curved space-time

$$T^{\mu\nu}_{:\nu} = 0$$

Now coming to the original neural network problem, we generate the metric $g_{\mu\nu}$ for different choices of the parameters θ and train our neural network using the inverse of this relation, ie, input the metric tensor into the network so as to generate parameter estimates that are close in some distance measure to the parameter choice θ. Then given any other metric that falls within this class, the trained network would be able to generate the appropriate parameter estimates. In other words, if we know before hand that the metric of space-time is generated by a given set of fields like the Maxwell and Dirac fields but we do not know the scalings of these fields, then we can determine the scaling parameters using our neural network and hence compress all the information present in the metric tensor in terms of the parameters generated by the network. Is there a dynamical way of doing the same estimation using a recurrent neural network ? Specifically, we cast the Einstein field equations for a time varying metric driven by the parametric linear combination of the other field energy-momentum tensors in state variable form w.r.t the time variable and then given noisy measurements of the geodesic trajectories of particles/fluid velocity fields/electromagnetic fields we use an extended Kalman filter to dynamically estimate the parameters on a real time basis. We would have then successfully encoded/compressed the time varying metric data over the entire spatial manifold into a finite set of time varying parameters.

3.12 On modeling EEG signals using the motion of electric dipoles in the cortical region of the brain

Let the brain consist of infinitesimal electric dipoles $p_1(t), ..., p_N(t)$ located at positions $r_1(t), ..., r_N(t)$. Without taking into account relativistic effects, the polarization vector field at time t is given by

$$P(t, r) = \sum_{k=1}^{N} p_k(t)\delta(r - r_k(t))$$

The polarization charge density field is given by the well known formula

$$\rho_p(t, r) = -divP(t, r)$$

The polarization current density $J_p(t, r)$ then satisfies the equation of continuity

$$divJ_p = -\partial_t \rho_p$$

and hence we can take

$$J_p(t,r) = \partial_t P(t,r)$$

Another way to justify this formula for the polarization current density is to consider a charge q moving with a velocity v. If there are N such charges per unit volume, then the current density generated by the motion of these charges is $J_p = Nqv$ and since $P = Nqr$ represents the dipole moment of the se charges per unit volume and $v = dr/dt$, we get the relation $J_p = \partial_t P$. Now the magnetic vector potential and the electric scalar potential generated by these dipole charges are respectively given by the usual retarded potential formula

$$A(t,r) = (\mu/4\pi)\int J_p(t-|r-r'|/c, r')d^3r'/|r-r_{'}|,$$

$$V(t,r) = (1/4\pi\epsilon)\int \rho_p(t-|r-r'|/c, r')d^3r'/|r-r'|$$

If we ignore relativistic effects on the ground that the motion of the dipoles is not too rapid, then we can use the Coulomb and Biot-Savart formulae

$$A(t,r) = (\mu_0/4\pi)\int J_p(t,r')d^3r'/|r-r'|,$$

$$V(t,r) = (1/4\pi\epsilon_0)\int \rho_p(t,r')d^3r'/|r-r'|$$

to obtain

$$V(t,r) = \sum_k p_k(t).(r-r_k(t))/4\pi\epsilon_0|r-r_k(t)|^3$$

and

$$A(t,r) = (\mu_0/4\pi)\sum_k p'_k(t)/r-r_k(t)|$$

The electric field is then given by

$$E(t,r) = -\nabla V(t,r) - \partial_t A(t,r)$$

$$= -\sum_k \nabla_r(p_k(t).(r-r_k(t)/|r-r_k(t)|)$$

$$-(\mu_0/4\pi)\sum_k \partial_t(p'_k(t)/|r-r_k(t)|)$$

and the magnetic field by

$$B(t,r) = \nabla_r \times A(t,r) = (\mu_0/4\pi)\sum_k(\nabla_r(1/|r-r_k(t)|) \times p_k(t)$$

We can combine these equations in the form of a six component equation for the electromagnetic field vector $\psi(t,r) = [E(t,r), B(t,r)]$ expressible in the form

$$\psi(t,r) = F(p_k(t), r_k(t), v_k(t), k = 1,2,...,N)$$

Estimating the structure of the brain from electromagnetic field measurements then amount to estimating the dipole strengths $\{p_k(t)\}$, their positions $\{r_k(t)\}$ and their velocities $\{v_k(t)\}$ from the field measurements on a dynamic, ie, real time basis. For doing this, we must first derive the stochastic differential equations satisfied by $p_k(t), r_k(t), v_k(t)$. Assume that the motion of these dipoles takes place in an external electromagnetic field $E_e(t,r), B_e(t,r)$. To derive the Lagrangian for the interaction between the dipoles and the external fields, assume that the external fields are derivable from the electromagnetic potentials $V_e(t,r)$ and $A_e(t,r)$. A given dipole $p_k(t)$ located at $r_k(t)$ can be regarded as consisting of a pair of charges $-q_k$ and q_k located at $r_k(t)$ and $r_k(t)+d_k(t)$ where $p_k(t) = q_k d_k(t)$ with the separation vector $d_k(t)$ being very small. The contribution to the Lagrangian due to the interaction is then given by

$$U = q_k(V(t,r_k) - V(t,r_k+d_k) - q_k((v_k, A(t,r_k)) - (v_k+d'_k, A(t,r_k+d_k))$$
$$= -q_k(d_k, \nabla V(t,r_k)) - q_k(v_k, (d_k,\nabla)A(t,r_k)) - q_k(d'_k, A(t,r_k))$$
$$= -(p_k, \nabla V(t,r_k)) - (v_k, (p_k,\nabla)A(t,r_k)) - (p'_k, A(t,r_k))$$

upto linear orders in d_k, d'_k. The kinetic energy K of the dipoles is given by the sum of their translational kinetic energy and their rotational kinetic energy. Letting J_k denote the moment of inertia tensor of the kth dipole and $R_k(t)$ the 3-D rotation matrix applied to this dipole after time t around its centre of mass, we can express

$$K = \sum_k (1/2)m_k|v_k|^2 + (1/2)Tr(R'_k(t))J_k R'_k(t)^T)$$

Here, $d_k(t) = R_k(t)d_k(0)$ so that $d'_k(t) = R'_k(t)d_k(0)$ and hence, $p_k(t) = R_k(t)p_k(0)$ and $p'_k(t) = R'_k(t)p_k(0)$. Thus parametrizing the phase coordinates of the kth dipole by (r_k, v_k, p_k, p'_k) or equivalently by (r_k, v_k, R_k, R'_k) where by R_k, we mean the three Euler angles of R_k and by R'_k, we mean the rates of change of these three Euler angles, we can set up the Lagrangian of the system of dipoles as

$$L(r_k, r'_k, \phi_k, \theta_k, \psi_k, \phi'_k, \theta'_k, \psi'_k, k=1,2,...,N)$$
$$= K - U$$

from which the required equations of motion the dipoles can be derived.

3.13 On the measurement of quantum averaged densities of observables in quantum field theory

Suppose $\psi(t,r)$ is a quantum wave field whose dynamics is obtained from the Heisenberg equations of motion:

$$i\partial\psi(t,r)/\partial t = [H(t), \psi(t,r)]$$

where the Hamiltonian H is a functional of the quantum field $\psi(t,.)$. In case that the wave field ψ is a Fermionic field, its components corresponding to the different spin components will satisfy the Fermionic anticommutation relations $[\psi_s(t,r),\psi_{s'}(t,r')^*]_+ = \delta_{ss'}\delta^3(r-r'), [\psi_s(t,r),\psi_{s'}(t,r')]_+$

$$= 0, [\psi_s(t,r)^*, \psi_{s'}(t,r')^*]_+ = 0$$

An example of such a field Hamiltonian H is the Dirac field Hamiltonian with the Hartree-Fock correction terms:

$$H(t) = \int \psi(t,r)^*[(\alpha, -i\nabla+eA(t,r))+\beta m-e\Phi(t,r)]\psi(t,r)d^3r$$

$$+\int V_{s_1s_2}(t,r_1,r_2\psi_{s_1}(t,r_1)^*\psi_{s_1}(t,r_1)\psi_{s_2}(t,r_2)^*\psi_{s_2}(t,r_2)d^3r_1d^3r_2---(1)$$

It is easily verified using the Fermionic anticommutation relations that the evolution $\psi(0,r)\rightarrow\psi(t,r)$ is unitary in the sense that

$$\psi_s(t,r)^*\psi_s(t,r) = \psi_s(0,r)^*\psi_s(0,r)\forall t \geq 0$$

where summation over the repeated spin index s is implied. From this fact, it immediately follows that if the above anticommutation relations between the field operators are satisfied at time $t = 0$, then they will be satisfied at all times $t \geq 0$. Consider now a state ρ of the quantum field at time $t = 0$. Assume that this state can be expressed as a function of the field operators $\psi(0,r), \psi(0,r)^*, r \in \mathbb{R}^3$ at time $t = 0$. Note that we can describe the Heisenberg evolution of the field operator observables as

$$\psi(t,r) = U(t)^*\psi(0,r)U(t)$$

where $U(t)$ is a unitary operator in the Fermion Fock space. From this also, immediately follows the preservation of the canonical anticommutation relations at all times provided that they are valid at time $t = 0$:

$$[\psi_s(t,r),\psi_{s'}(t,r')^*]_+ = [U(t)^*\psi_s(0,r)U(t), U(t)^*\psi_{s'}(0,r')U(t)]_+ =$$

$$= U(t)^*[\psi_s(0,r),\psi_{s'}(0,r')^*]U(t) = U(t)^*\delta_{ss'}\delta^3(r-r')U(t)$$

$$= \delta_{ss'}\delta^3(r-r')U(t)^*U(t) = \delta_{ss'}\delta^3(r-r')$$

In the Heisenberg field picture, the state ρ of the system does not vary with time while the Heisenberg observables $\psi(t,r), \psi(t,r)^*$ keep evolving with time. It follows that if $X(t)$ is any evolving Heisenberg observable that can be expressed as a function of the instantaneous wave fields, then its quantum average at time t will be given by

$$Tr(\rho.X(t)) = Tr(\rho.U(t)^*X(0)U(t)) = Tr(\rho(t)X(0)), \rho(t) = U(t)\rho.U(t)^*$$

$\rho(t)$ equals the Schrodinger picture state evoution as a function of time. Suppose for example that the magnetic vector potential $A(r)$ and the electric scalar potential $\Phi(r)$ do not explicitly depend upon time. Then it is easy to see that the field Hamiltonian H will also not vary with time, ie, it will be a constant of the motion. This can be proved by noting that the anticommutation relations between the field observables imply that $H(t) = H(0)$ is a solution to the dynamical equations

$$\partial H(t)/\partial t + i[H(t), H(t)] = 0$$

It can also be seen explicitly by substituting the expression

$$\psi(t,r) = U(t)^*\psi(0,r)U(t)$$

into the expression (1) for the field Hamiltonian, using the fact that $U(t), U(t)^*$, commute with classical quantities (ie first quantized) such as $(\alpha, -i\nabla+eA(r)))+\beta m-e\Phi(r)$ and making use of the unitarity of evolution in field space (Fermionic Fock space): $U(t)^*U(t) = I$. Note that it is very important that A, Φ should not depend explicitly on time for the Hamiltonian to be a constant of the motion. Otherwise, the term $\partial H/\partial t$ would receive contributions from the terms $A(t,r), \Phi(t,r)$. We could also consider a Bosonic field like the complex Klein-Gordon scalar field interacting with the electromagnetic field after second quantization. The appropriate second quantized Hamiltonian field operator would then be

$$H(t) = \int \phi(t,r)^*((-i\nabla+eA(t,r))^2+m^2)^{1/2}-e\Phi(t,r))\phi(t,r)d^3r$$
$$+\int V(r_1,r_2)\phi(r_1)^*\phi(r_1)\phi(r_2)^*\phi(r_2)d^3r_1d^3r_2$$

However, much meaning cannot be attached to this expression because the square root of the operator $((-i\nabla + eA(t,r))^2 + m^2)$ cannot be defined. However, we can consider expansion of the KG operator Suppose $A(r), \Phi(r)$ do not depend explicitly upon time. Then, we diagonalize the Klein-Gordon operator as follows:

$$(E_n + e\Phi(r))^2u_n(r) = ((-i\nabla + eA(r))^2 + m^2)u_n(r), n = 1,2,...$$

The $u_n(r), n = 1,2,...$ form an onb for $L^2(\mathbb{R}^3)$. Thus, we can write

$$\sum_n u_n(r)\bar{u}_n(r') = \delta^3(r-r'),$$

so that

$$[(-i\nabla+eA(r))^2+m^2]\delta^3(r-r') = \sum_n(E_n+e\Phi(r))^2u_n(r)\bar{u}_n(r') = (H+e\Phi(r))^2\delta^3(r-r')$$

since

$$Hu_n(r) = E_nu_n(r), n = 1,2,...$$

The kernel of the Hamiltonian H is given by

$$H(r,r') = \sum_n E_nu_n(r)\bar{u}_n(r')$$

Note that $E_n = E_n(A,\Phi)$ are functionals of A, Φ. From the functionals $E_n(A,\phi), n \geq 1$, we form the field Hamiltonian operator

$$H_F = \int \phi(t,r)^*(\sum_n E_n(A,\Phi)u_n(r)\bar{u}_n(r'))\phi(t,r')d^3rd^3r'$$
$$+\int V(r,r')\phi(r)^*\phi(r)\phi^*(r')\phi(r')d^3rd^3r'$$

or equivalently in terms of the operator coefficients

$$a_n = \int \bar{u}_n(r)\phi(r)d^3r, a_n^* = \int u_n(r)\phi(r)^* d^3r, \phi(r) = \phi(0, r)$$

we get

$$H_F = \sum_n E_n(A, \Phi)a_n^* a_n + \int V(r, r')\phi(r)^*\phi(r)\phi^*(r')\phi(r')d^3rd^3r'$$

The CCR

$$\phi(r), \phi(r')^*] = \delta^3(r - r')$$

and the orthogonality relations

$$\int \bar{u}_n(r)u_m(r)d^3r = \delta_{nm}$$

imply

$$[a_n, a_m^*] = \delta_{n,m}, [a_n, a_m] = 0$$

It should be noted however that the eigenfunctions $u_n(r)$ are functionals of the classical fields A, Φ just in the same way are the eigenvalues E_n.

Now if $X(t) = F(t, r, -i\nabla)$ is a first quantized observable like for example position r, momentum $-i\nabla$, angular momentum $-ir \times \nabla$, free energy $(\alpha, -i\nabla)+\beta m$, total energy in the presence of em fields $(\alpha, -i\nabla + eA(t, r)) - e\Phi(t, r)$, then the position space density operator field of $X(t)$ is given by $\psi(t, r)^* X(t)\psi(t, r)$ and its quantum average

$$f_X(t, r) = Tr(\rho.\psi(t, r)^* X(t)\psi(t, r))$$

gives the classical position space density of the observable X at time t.

Chapter 4

Radon Transform Applications in Computer Vision and Image Reconstruction

4.1 Radon transform applied to the recovery of a 3-D object from a finite number of 2-=D projections

Let $f(r), r = (x, y, z)$ be a three dimensional object field. It projection onto the 2-D plane $n.r = d$ where n is a unit vector or equivalently, $n_x x + n_y y + n_z z = d$ is given by the radon transform

$$Rf(d, n) = \int f(r)\delta(d - n.r)d^3 r = \int f(x, y, z)\delta(d - n_x x - n_y y - n_z z)dxdydz$$

The radon transform is a function of three independent real variable, two variables specify the unit vector n that is normal to the plane and the third variable is d, the distance of the plane from the origin. More generally, we can consider projections of the object on a family of two dimensional curved surface $g(r|\theta) = 0$ as the parameter θ varies over a set E. We can call the resultant transform as a generalized Radon transform $R_g f(\theta)$:

$$R_g f(\theta) = \int f(r)\delta(g(r|\theta)d^3 r) = \int f(x, y, z)\delta(g(x, y, z|\theta))dxdydz$$

The problem is to recover the object field function $f(r)$ from $R_g f(\theta)$ as θ varies over a finite subset of E. In the special case of the Radon transform, we allow the distance parameter d to assume negative values too and then forming the 1-D Fourier transform of $Rf(d, n)$ w.r.t d gives us

$$F_1 Rf(\omega, n) = \int_{\mathbb{R}} Rf(d, n)exp(-j\omega d)dd = \int f(r)exp(-j\omega n.r)d^3 r = F_3 f(\omega n)$$

where F_3 denotes the three dimensional Fourier transform of the object intensity function f. Inverting this 3-D Fourier transform then yields the desired algorithm for recovering f from Rf:

$$f(r) = (2\pi)^{-3} \int F_1 Rf(\omega, n)exp(j\omega n.r)\omega^2 d\omega.d\Omega(n)$$

where the integral w.r.t ω is taken over $[0, \infty)$ while the integral w.r.t $d\Omega(n)$ is taken over all solid angles of the surface of the unit sphere in 3 dimensions. The above inversion formula can be cast as

$$f(r) = (2\pi)^{-3} \int Rf(d, n)exp(j\omega(n.r - d))\omega^2 d\omega d\Omega(n)dd$$

DOI: 10.1201/9781003728627-4

Consider the integral

$$h(t) = \int_0^\infty exp(j\omega t)d\omega = lim_{N\to\infty}(exp(jNt)-1)/jt$$

$$= lim_{N\to\infty)((cos(Nt)-1)/jt+sin(Nt)/t)}$$

We shall evaluate the weak limit in the two terms above. let $\phi(t)$ be a smooth function. Then,

$$\int_{\mathbb{R}} \phi(t)(cos(Nt)/t)dt = \int_{\mathbb{R}} \phi(t/N)cos(t)/t \to \phi(0) \int cos(t)dt/t = 0$$

provided that we use the principal value in evaluating the above integral, noting that the principal value converges to zero because $cos(t)/t$ is an odd function. Likewise,

$$\int_{\mathbb{R}} \phi(t)sin(Nt)dt/t = \int \phi(t/N)()sin(t)/t)dt \to \phi(0) \int sin(t)dt/t = \pi\phi(0)$$

which amounts to saying that the weak limit of $sin(Nt)/t$ as $N \to \infty$ equals $\pi\delta(t)$. Thus we obtain the formula

$$h(t) = -1/jt + \pi\delta(t)$$

and we have the following inversion formula for the Radon transform:

$$f(r) = -(2\pi)^{-3} \int Rf(d,n)h''(n.r-d)dd.d\Omega(n)$$

Remark: The Dirac δ function as well as its derivatives are defined as generalized functions or more precisely as distributions acting on a space of test functions with an approrpriate topology so as to make these distributions continuous functionals. For example if $\phi(t)$ is a rapidly decreasing function, then and $\delta_n = \delta^{(n)}(t)$, then

$$\delta_n(\phi) = \int \delta_n(t)\phi(t) = (-1)^n\phi^{(n)}(0)$$

and if we introduce the Sobolev norm

$$\| \phi \|_N = sup_{\in\mathbb{R},0\le n\le N}|\phi^{(n)}(t)|$$

then δ_n becomes a continuous linear functional on the test function space w.r.t this Sobolev norm:

$$|\delta_n(\phi)| \le \| \phi \|_N, 0 \le n \le N$$

4.2 Radon transform of stochastic fields under rotation, recovery of 3-D object field from 2-D projections

Let G be the rotation group $SO(3)$ acting on three dimensional space. For $g \in G$, the object field on rotation by g gets transformed to the object field $f_g(r) = f(g^{-1}r)$ and its Radon transform is

$$Rf_g(d,n) = \int f(g^{-1}r)\delta(d - n.r)d^3r = \int f(r)\delta(d - n.g^{-1}r)d^3r =$$

$$\int f(r)(d - gn.r)d^3r = Rf(d, gn)$$

Let π be a unitary representation of G. Then choose and fix a point r_0 in the object field. We get

$$\int f_g(hr_0)\pi(h)dh = \int f(g^{-1}hr_0)\pi(h)^*dh = \int f(h.r_0)\pi(gh)^*dh$$

$$= \int f(hr_0)\pi(h)^*\pi(g)^*dh = (\int f(hr_0)\pi(h)^*dh)\pi(g)^*$$

This equation is expressed as

$$\pi(f_g) = \pi(f)\pi(g)^* - - - (1)$$

where

$$\pi(f) = \int f(hr_0)(h)^*dh$$

Let

$$H = \{h \in G : hr_0 = r_0\}$$

Then, H is a subgroup of G and we have

$$f(ghr_0) = f(gr_0), g \in G, h \in H$$

It follows then that

$$\pi(f_g) = \int_{G\times H} f_g(hh_1r_0)\pi(h)^*dhdh_1$$

$$= \int_{G\times H} f_g(hr_0)\pi(hh_1^{-1})^*dhdh_1 = (\int_H \pi(h_1)dh_1)\pi(f_g) = P_H\pi(f_g)$$

and of course likewise,

$$\pi(f) = P_H\pi(f)$$

where P_H is the orthogonal projection of $\mathbb{C}^d$ ($d \times d$ being the size of π) onto the subspace

$$V = \{x \in \mathbb{C}^d : \pi(h)x = x\forall h \in H\}$$

Thus, equation (1) is equivalent to the reduced equation

$$P_H\pi(f_g) = P_H\pi(f)\pi(g)^* --- (2)$$

We now fix the distance d of the plane on which the Radon transform is being carried out and consider the equation

$$Rf_g(d,n) = Rf(d,gn), n \in S^2 --- (3)$$

This equation is more convenient than the equation $f_g(r) = f(g^{-1}r), r \in \mathbb{R}^3)$ because in (3), g acts on the unit vector n and n varies over the surface of the entire unit sphere S^2 and G acts transitively on S^2. This means that we record the radon transform of the original and rotated object fields after these fields are projected using the Radon transform onto all the planes that have a fixed distance d from the origin but whose unit normal varies over all possible directions. From these pairs of planar projections, we seek to determine the group transformation element g. This can be done by application of the Peter-Weyl theorem to (3) or equivalently, by expanding $Rf(d,n)$ and $Rf_g(d,n)$ as linear combinations of the spherical harmonic functions of n:

$$Rf(d,n) = \sum_{l,m} Rf(d,l,m)Y_{lm}(n), Rf_g(d,n) = \sum_{l,m} Rf_g(d,l,m)Y_{lm}(n)$$

so that

$$Rf(d,l,m) = \int Rf(d,n)\bar{Y}_{lm}(n)d\Omega(n), Rf_g(d,l,m) = \int Rf_g(d,n)\bar{Y}_{lm}(n)d\Omega(n)$$

$$Rf(d,gn) = \sum_{l,m} Rf(d,l,m)Y_{lm}(gn) = \sum_{l,m,m'} Rf(d,l,m)[\bar{\pi}_l(g)]_{mm'}Y_{lm'}(n)$$

and hence we get the identity

$$Rf_g(d,l,m) = \sum_{m'} [\bar{\pi}_l(g)]_{m'm} Rf(d,l,m')$$

or equivalently, in matrix-vector notation,

$$Rf_g(d,l) = \pi_l(g)^* Rf(d,l)$$

In the presence of noise, this identity will only approximately be valid and hence we use a least squares method to estimate g:

$$\hat{g} = argmin_{g_1 \in G} W(l) \parallel Rf_g(d,l) - \pi_l(g_1)^* Rf(d,l) \parallel^2$$

where $W(l)$ are all positive weights.

Remark: In the presence of noise, the transformation equation of the object field is

$$f_g(r) = f(g^{-1}r) + w(r)$$

and after taking the Radon transform,

$$Rf_g(d,n) = Rf(d,gn) + Rw(d,n)$$

so in the spherical harmonic domain, with d fixed,

$$Rf_g(d,l) = \pi_l(g)^*)Rf(d,l) + Rw(d,l)$$

Now suppose that w has a G-invariant correlation function, ie,

$$R_w(gr, gr') = R_w(r, r'), \forall r, r' \in \mathbb{R}^3, g \in G, Rw(r, r') = \mathbb{E}(w(r)w(r'))$$

Then

$$\mathbb{E}(Rw(d, gn)Rw(d, gn')) = \int R_w(r, r')\delta(gn.r-d)\delta(gn'.r'-d)d^3rd^3r'$$

$$= \int R_w(r, r')\delta(n.g^T r-d)\delta(n'.g^T r'-d)d^3rd^3r'$$

$$= \int R_w(gr, gr')\delta(n.r-d)\delta(n'.r'-d)d^3rd^3r'$$

$$= \int R_w(r, r')\delta(n.r-d)\delta(n'.r'-d)d^3rd^3r' = \mathbb{E}(Rw(d, n)Rw(d, n'))$$

This equation shows that the radon transform of the noise also has G-invariant correlation function. In fact, the same argument shows that more generally for all d, d', G-invariance of the noise correlation function implies

$$\mathbb{E}(Rw(d, gn)Rw(d', gn')) = \mathbb{E}(Rw(d, n)Rw(d', n'))$$

We write down the spherical harmonic expansion of $Rw(d, n)$ as

$$Rw(d, n) = \sum_{lm} a(l, m)Y_{lm}(n)$$

and show that G-invariance of the correlation function of Rw implies that

$$\mathbb{E}(a(l, m)\bar{a}(l', m')) = b(l)\delta(ll')\delta(mm')$$

In fact, writing

$$\mathbb{E}(a(l, m)\bar{a}(l', m')) = R_a(lm, l'm')$$

we get

$$\mathbb{E}(Rw(d, n)Rw(d, n')) = \sum R_a(lml'm')Y_{lm}(n)\bar{Y}_{l'm'}(n')$$

Applying the G-invariance property of this function gives us

$$\sum R_a(lml'm')Y_{lm}(n)\bar{Y}_{l'm'}(n') = \sum R_a(lml'm')Y_{lm}(g^{-1}n)\bar{Y}_{l'm'}(g^{-1}n')$$

$$= \sum R_a(lml'm')[\pi_l(g)]_{m_1m}\bar{\pi}_{l'}(g)]_{m_1'm'}Y_{lm_1}(n)\bar{Y}_{l'm_1'}(n')$$

from which we deduce by linear independence of the functions $Y_{lm} \otimes \bar{Y}_{l'm'}$ (in fact orthnormality) in $L^2(S^2 \times S^2)$ that

$$R_a(lm_1l'm_1') = \sum_{mm'} R_a(lml'm')[\pi_l(g)]_{m_1m}\bar{\pi}_{l'}(g)]_{m_1'm'}$$

Integrating both sides w.r.t g over G using the Schur orthogonality relations gives us

$$R_a(lm_1l'm'_1) = \sum_{mm'}(2l+1)\delta(ll')\delta(m_1m'_1)\delta(mm')R_a(lml'm')$$
$$= (2l+1)\delta(ll')\delta(m_1m'_1)\sum_m R_a(lmlm)$$

which is precisely the desired result.

4.3 Characterization and analysis of G-invariant nonlinear systems using the theory of decompostion of the representation of a group into irreducible representations

Problem: Consider a Volterra system of the form

$$f_2(x) = \int K(x, x_1, ..., x_N)f_1(x_1)...f_n(x_N)dx_1...dx_N$$

where $x, x_1, ..., x_N$ vary over a manifold M on which a Lie group G of transformations acts transitively. Assume that this system is G-invariant, ie

$$K(gx, gx_1, ..., gx_N) = K(x, x_1, ..., x_N), x, x_1, ..., x_N \in M, g \in G$$

Then, the response of the above Volterra system to $f_1(gx)$ is given by

$$\int K(x, x_1, ..., x_N)f_1(gx_1)...f_N(gx_N)dx_1...dx_N$$
$$= \int K(x, g^{-1}x_1, ..., g^{-1}x_N)f_1(x_1)...f_1(x_N)dx_1...dx_N$$
$$= \int K(gx, x_1, ..., x_N)f_1(x_1)...f_1(x_N) = f_2(gx)$$

where we are assuming that the measure dx on the manifold M is G-invariant. Note that we have used in the above derivation, the G-invariance of the kernel K. In othr words, if the Volterra kernel is G-invariant, then the response of this system to a G-shifted input is the corresponding G-shifted output. Now, consider the standard representation U of G acting in $L^2(M)$, with the G-invariant measure being used to define $L^2(M)$. We can decompose

$$U =_{n=1}^{\infty} m_n\pi_n, L^2(M) = \bigoplus_n \bigoplus_{i=1}^{m_n} \mathcal{H}_{n,i}$$

where $\pi_n, n = 1, 2, ...$ are irreducible unitary representations of G and the m'_ns are positive integers that determine the multiplicity with which each irreducible representation of G appears in the decomposition of U. The restriction of U to the invariant subspace $\mathcal{H}_{n,i}$ in an appropriate onb $B_{n,i}$ equals π_n. Note that U is a unitary representation since

$$< U(g)f_1, U(g)f_2 >= \int \bar{f}_1(g^{-1}x).f_2(g^{-1}x)dx = \int \bar{f}_1(x)f_2(x)dx$$

because we are using a G-invariant measure on M. Writing

$$B_{n,i} = \{e(nik) : k = 1, 2, ..., d_n\}$$

we get

$$f_1(x) = \sum_{nik} < f_1, e(nik) > e(nik)(x), f_2(x) = \sum_{nik} < f_2, e(nik) > e(nik)(x)$$

Writing

$$K(x, x_1, ..., x_N) =$$

$$\sum_{nikn_1i_1k_1...n_Ni_Nk_N} < K, e(nik) \otimes e(n_1i_1k_1) \otimes e(n_2i_2k_2) \otimes ... \otimes e(n_Ni_Nk_N) >$$

$$\times e(nik)(x)e(n_1i_1k_1)(x_1)...e(n_Ni_Nk_N)(x_N)$$

we see that the condition for G-invariance of the system K can be expressed in terms of its coefficients

$$K(nik|n_{11}k_1, ..., n_Ni_Nk_N) =< K, e(nik)\otimes e(n_1i_1k_1)\otimes e(n_2i_2k_2)\otimes ...\otimes e(n_Ni_Nk_N) >$$

$$= \int K(x, x_1, ..., x_N)\bar{e}(nik)(x)\bar{e}(n_1i_1k_1)(x_1)...\bar{e}(n_Ni_Nk_N)(x_N)dx_1...dx_N$$

as

$$K(nik|n_1i_1k_1...n_Ni_Nk_N)$$

$$= \sum_{ll_1...l_N} K(nil|n_1i_1l_1...n_Ni_Nl_N)[\bar{\pi}_n(g)]_{lk}[\bar{\pi}_{n_1}(g)]_{l_1k_1}...[\bar{\pi}_{n_N}(g)]_{l_Nk_N}$$

Another way to write the same thing is as follows: Write

$$K(g^{-1}x, g^{-1}x_1, ..., g^{-1}x_N) = (U(g)^{\otimes(N+1)}K)(x, x_1, ..., x_N)$$

Then the condition becomes

$$U(g)^{\otimes(N+1)}K = K\forall g \in G$$

Now we can write

$$U(g) = T.diag[\pi_1(g)], ..., \pi_1(g), \pi_2(g), ..., \pi_2(g), ...]T^{-1}, g \in G = TV(g)T^{-1}$$

where π_j occurs m_j times as a block diagonal entry and T is a non-singular operator. Then

$$U(g)^{\otimes(N+1)} = T^{\otimes(N+1)}V(g)^{\otimes(N+1)}(T^{\otimes(N+1)}\})^{-1}$$

and hence the condition becomes

$$V(g)^{\otimes(N+1)}L = L\forall g \in G$$

where

$$L = T^{\otimes(N+1)}K$$

In this formalism, we are regarding the kernel K as an element in the Hilbert space $L^2(M)^{\otimes N+1}$. However, it is more natural to regard K as a linear operator from $L^2(M)^{\otimes N}$ into $L^2(M)$ and then the condition of G-invariance can be expressed as

$$KU(g)^{\otimes N}f_1^{\otimes N} = U(g)Kf_1^{\otimes N}, f_1 \in L^2(M)$$

so that

$$KU(g)^{\otimes N} = U(g)K, g \in G$$

Writing

$$U(g) = TV(g)T^{-1}, V(g) = diag[\pi_1(g), ..., \pi_1(g), \pi_2(g), ..., \pi_2(g), ...]$$

as above, we then get

$$M.V(g)^{\otimes N} = V(g)M, g \in G$$

where now, M is the infinite matrix

$$M = T^{-1}KT^{\otimes N}$$

Now using the Clebsch-Gordon formula for decomposing a tensor product of several irreducible representations into a direct sum of irreducibles, we have on observing that $V(g)^{\otimes N}$ is block diagonal with diagonal blocks of the form

$$\pi_{i_1}(g) \otimes ...\pi_{i_N}(g) - - - (a)$$

with the $\pi_j's$ all irreducible representations, that each of the terms in (a) can be expressed as a direct sum of the $\pi_j's$ upto an isomorphism and hence

$$V(g)^{\otimes N} = SW(g)S^{-1}$$

where S is a non-singular matrix and

$$W(g) = diag[\pi_1(g), ..., \pi_1(g), \pi_2(g), ..., \pi_2(g), ...] = \bigoplus_{j \geq 1} n_j \pi_j(g)$$

where the $n_j's$ are non-negative integers possibly different from the $m_j's$ and hence the condition for G-invariance can be expressed as

$$MSW(g)S^{-1} = V(g)M$$

or equivalently,

$$MSW(g) = V(g)MS$$

or

$$QW(g) = V(g)Q, g \in G - - - (2)$$

where Q is an infinite matrix. Note that $W(g)$ and $V(g)$ are both block diagonal with the irreducibles of G appearing on their block diagonals. From (2), the general structure of Q can be derived by using Schur's lemma.

4.4 Information pumped into the state of quantum particles by a blackhole

Consider the quantum mechanical equation satisfied by an electron in the vicinity of a Schwarzchild blackhole. The metric of space-time is given by

$g_{00} = 1-2m/r = \alpha(r), g_{11} = -(1-2m/r)^{-}1 = \alpha(r)^{-}1, g_{22} = -r2, g_{33} = -r2sin2(\theta)$

A tetrad basis for this metric is e_μ^a where we require that

$$\eta_{ab} e_\mu^a e_\nu^b = g_{\mu\nu}$$

This is achieved by setting

$$e_\mu^0 dx^\mu = \sqrt{\alpha(r)}dt, e_\mu^1 dx^\mu = \sqrt{\alpha(r)}dr, e_\mu^2 dx^\mu = rd\theta, e_\mu^3 = rsin(\theta)d\phi$$

Noting that

$$x^0 = t, x^1 = r, x^2 = \theta, x^3 = \phi,$$

we thus get

$$e_0^0 = \beta(r), e_m^0 = 0, m = 1,2,3,$$
$$e_0^1 = 0 = e_2^1 = e_3^1, e_1^1 = \beta(r),$$
$$e^2 = 0 = e_1^2 = e_3^2 = 0, e_2^2 = r,$$
$$e_0^2 = e_1^2 = e_2^2 = 0, e_3^2 = rsin(\theta)$$

where

$$\beta(r) = \sqrt{\alpha(r)}$$

and then we can compute the spinor connection corresponding to this metric:

$$\Gamma_\nu = (1/8)[\gamma^a, \gamma^b] e_{a\mu:\nu} e_b^\mu$$

Check that this is the right spinor connection: Let ω_μ^{ab} be the spinor connection. Then we require the covariant derivative of e_μ^a to vanish:

$$e_{\mu:\nu}^a + \omega_\nu^{ab} e_{b\mu} = 0$$

This yields the required formula for the spinor connection Note that while taking the covariant derivative of e_μ^a, we must use the Riemann-Christoffel connection $\Gamma_{\nu\rho}^\mu$ for the space-time index μ and the spinor connectioo for the spinor/tetrad index a. The total covariant derivative is obtained by summing these two contributions. Another way to check that this is the right spinor connection is to show that the spinor covariant derivative

$$\nabla_\mu = \partial_\mu + \Gamma_\mu, \Gamma_\mu = \omega_\mu^{ab}[\gamma^a, \gamma^b]/4$$

transforms under a local Lorentz transformation $\Lambda_b^a(x)$ to

$$\nabla'_\mu = D(\Lambda(x))(\partial_\mu + \Gamma_\mu)D(\Lambda(x))^{-1}$$

where $\Lambda \rightarrow D(\Lambda)$ is the spinor representation of the Lorentz group. It satisfies

$$D(\Lambda)\gamma^a D(\Lambda)^{-1} = \Lambda_b^a \gamma^b$$

or equivalently when

$$\Lambda(x) = I + \theta(x)$$

is an infinitesimal local Lorentz transformation,

$$\nabla'_\mu = \partial_\mu + \Gamma'_\mu, \Gamma'_\mu - \Gamma_\mu = \delta\Gamma_\mu = -\partial_\mu D(\theta(x)) + [D(\theta(x)), \Gamma_\mu(x)]$$

where $\theta \to D(\theta)$ is to be regarded as the differential of the spinor representation $\Lambda \to D(\Lambda)$ of the Lorentz group. It is well known that if $\epsilon_{\mu\nu}$ are the standard generators of the Lie algebra of the Lorentz group, ie

$$\epsilon_{\mu\nu} + \epsilon_{\nu\mu} = 0$$

then the map

$$\epsilon^{\mu\nu} \to (1/4)[\gamma^\mu, \gamma^\nu] = J^{\mu\nu}$$

defines the spinor representation of the Lie algebra of the Lorentz group. This is proved by showing the commutation relations

$$[J^{\mu\nu}, J^{\rho\sigma}] = \eta^{\nu\sigma} J^{\mu\rho} + \eta^{\mu\rho} J^{\nu\sigma} - \eta^{\nu\rho} J^{\mu\sigma} - \eta^{\mu\sigma} J^{\nu\rho}$$

It now becomes clear that

$$D(\theta(x) = \theta_{ab}(x) J^{ab}$$

where

$$\theta_{ab} = \eta_{ac}\theta^c_b$$

Now under the infinitesimal local Lorentz transformation $I + \theta(x)$, e^a_μ changes by

$$\delta e^a_\mu = \theta^a_b(x) e^\mu_b(x)$$

and hence $e^a_{\mu:\nu} e^{b\mu} J_{ab}$ changes by

$$J_{ab}(\delta e^a_{\mu:\nu} e^{b\mu} + e^a_{\mu:\nu} \delta e^{b\mu})$$

$$= J_{ab}(\theta^a_{c,\nu} e^c_\mu e^{b\mu} + \theta^a_c e^c_{\mu:\nu} + \theta^b_c e^a_{\mu:\nu} e^{c\mu})$$

Note that $\theta^a_b(x)$ should be regarded as scalars w.r.t diffeomorphisms of space-time and hence its covariant derivative w.r.t a space time index should coincide with an ordinary partial derivative. Note also that

$$J_{ab} e^a_{\mu:\nu} e^{b\mu} = J_{ab} e^{b\mu}(e^a_{\mu,\nu} - \Gamma^\rho_{\mu\nu} e^a_\rho)$$

and further,

$$J_{ab} e^{b\mu} \Gamma^\rho_{\mu\nu} e^a_\rho = J_{ab} \Gamma_{\rho\mu\nu} e^{b\mu} e^{a\rho} =$$

$$(1/2) J_{ab}(g_{\rho\nu,\mu} - g_{\mu\nu,\rho}) e^{b\mu} e^{a\rho} = J_{ab} g_{\rho\nu,\mu} e^{b\mu} e^{a\rho}$$

since $J_{ab} = -J_{ba}$. Thus, we get

$$\Gamma_\nu = (1/2 J_{ab} e^a_{\mu:\nu} e^{b\mu} = (1/2) J_{ab} e^a_{\mu,\nu} e^{b\mu} - (1/2) J_{ab} g_{\rho\nu,\mu} e^{b\mu} e^{a\rho}$$

or in other words, the spinor connection cannot in general be defined by using the ordinary space-time partial derivatives of the tetrad in place of the covariant derivative. That the stated spinor connection satisfies the right transformation law under infinitesimal local Lorentz transformations can now be easily verified. It should be noted that when the metric tensor is a small perturbation of the flat Minkowski metric, we can write

$$e^a_\mu = \delta^a + h^a_\mu, e_{a\mu} = \eta_{a\mu} + h_{a\mu}$$

where the $h's$ are of the first order of smallness. Then, in this first order approximation, we have

$$g_{\mu\nu} = \eta_{\mu\nu} + 2h_{\mu\nu},$$

$$g_{\rho\nu,\mu}e^{b\mu}e^{a\rho} = 2\eta^{b\mu}\eta^{a\rho}h_{\rho\nu,\mu} = 2h^{a,b}_\nu$$

and

$$e^a_{\mu,\nu}e^{b\mu} = \eta^{b\mu}h^a_{\mu,\nu} = h^{ab}_{,\nu}$$

so upto first order smallness, we get

$$\Gamma_\nu = (1/2)J_{ab}(h^{ab}_{,\nu} - 2h^{a,b}_\nu)$$

The Dirac equation in the Schwarzchild space-time can now be easily formulated as

$$[i\gamma^a e^\mu_a(\partial_\mu + \Gamma_\mu) - m]\psi(x) = 0$$

When there are random fluctuations in the gravitational field of the blackhole, it becomes clear that such perturbations will be manifest in the form of random perturbations of the tetrad e^a_μ and hence if $\psi(x), x = (t, r)$ is the wave function at time t, then the mixed state of the electron after averaging out over the gravitational field flucutations will be given by

$$\rho(t, r, r') = \mathbb{E}(\psi(t, r)\psi(t, r'^*))$$

and this mixed state will satisfy a Lindblad like master equation.

4.5 Curvature of the spinor connection

The spinor covariant derivative is

$$D_\mu = \partial_\mu + (1/2)J_{ab}\omega^{ab}_\mu, J_{ab} = [\gamma_a, \gamma_b]/4$$

Then the curvature of this connection $[D_\mu, D_\nu]$ satisfies

$$R_{\mu\nu} = [D_\mu, D_\nu] = (1/2)(\omega^{ab}_{\nu,\mu} - \omega^{ab}_{\mu,\nu})J_{ab} + (1/4)\omega^{ab}_\mu\omega^{cd}_\nu[J_{ab}, J_{cd}]$$

Now, using the Lorentz algebra commutation relations

$$[J_{ab}, J_{cd}] = \eta_{bc}J_{ad} + \eta_{ad}J_{bc} - \eta_{ac}J_{bd} - \eta_{bd}J_{ac}$$

we can write

$$R_{\mu\nu} = R^{ab}_{\mu\nu}J_{ab}$$

where $R^{ab}_{\mu\nu}$, the spinor representation of the Riemann curvature tensor, is given by

$$4.R^{ab}_{\mu\nu}J_{ab} = (2.\omega^{ab}_{\nu,\mu} - 2.\omega^{ab}_{\mu,\nu} + \omega^{ad}_{\mu}\omega^{cb}_{\nu}\eta_{cd} - \omega^{cb}_{\mu}\omega^{ad}_{\nu}\eta_{cd} + \omega^{db}_{\mu}\omega^{ca}_{\nu}\eta_{dc}$$
$$-\omega^{ac}_{\mu}\omega^{bd}_{\nu}\eta_{cd})J_{ab}$$

Using the antisymmetry of J_{ab} and ω^{ab}_{μ} w.r.t interchange of a and b, this expression simplifies to

$$4.R^{ab}_{\mu\nu}J_{ab} = (2.\omega^{ab}_{\nu,\mu} - 2.\omega^{ab}_{\mu,\nu} + 2\omega^{ad}_{\mu}\omega^{cb}_{\nu}\eta_{cd} - 2\omega^{cb}_{\mu}\omega^{ad}_{\nu}\eta_{cd})J_{ab}$$

so we can take

$$R^{ab}_{\mu\nu} = (1/2)[\omega^{ab}_{\nu,\mu} - \omega^{ab}_{\mu,\nu} + \omega^{ad}_{\mu}\omega^{cb}_{\nu}\eta_{cd} + \omega^{ca}_{\mu}\omega^{bd}_{\nu}\eta_{cd}]$$

4.6 Some remarks on quantum field theory with respect to the Dirac equation for an electron in the vicinity of a quantum blackhole

Consider the metric field perturbations to be quantized according to a linearized Einsteinian gravity. Thus the tetrad can be expressed as

$$e^{a}_{\mu} = \delta^{a}_{\mu} + h^{a}_{\mu}$$

where $h^{a}_{\mu}(x)$ is of the first order of smallness. It is related to the metric perturbation upto first order of smallness terms by

$$\eta_{ab}(\delta^{a}_{\mu} + h^{a}_{\mu})(\delta^{b}_{\nu} + h^{b}_{\nu}) = g_{\mu\nu} = \eta_{\mu\nu} + \delta g_{\mu\nu}$$

and therefore,

$$\eta_{a\nu}h^{a}_{\mu} + \eta_{a\mu}h^{a}_{\nu} = \delta g_{\mu\nu}$$

or equivalently, if we lower the indices of h using the Minkowski metric η, we can express this as

$$\delta g_{\mu\nu} = 2h_{\mu\nu}$$

Now in the linearized Einsteinian gravity, $\delta g_{\mu\nu}$ after adopting an appropriate system of coordinates, satisfies the standard flat space-time wave equation and hence can be expanded as a superposition of plane waves with coefficients being the graviton creation and annihilation operators in momentum/wave vector space. Thus, the tetrad perturbations h^{a}_{μ} are also linear superpositions of plane waves with graviton creation and annihilation operators in momentum space. The gravitational energy-momentum pseudo-tensor is thus in upto quadratic terms in the metric perturbations, a quadratic functional of the graviton creation and annihilation operators. The Dirac equation in this background lineraized gravitational field will thus be of the form

$$[\gamma^{a}(\delta^{\mu}_{a} + h^{\mu}_{a}(x))(i\partial_{\mu} + i\Gamma_{\mu}) - m]\psi(x) = 0$$

or after linearization,

$$[i\gamma^\mu\partial_\mu + i\gamma^\mu\Gamma_\mu + i\gamma^a h_a^\mu\partial_\mu - m]\psi(x) = 0$$

In this equation, according to linearized quantum gravity, Γ_μ and h_a^μ are linear superposition of plane waves with coefficients being graviton creation and annihilation operators. Thus, assuming the gravitons to be in a coherent state, we can in principle calculate the Schrodinger wave function $\psi(t,r) = \psi(x)$ at any time t as a functional of the intial state and the graviton creation and annihilation operators. After partial averaging over the graviton coherent state, we then obtain the following mixed state for the electron at time t:

$$\rho(t,r,r') =< \phi(u)|\psi(t,r)\psi(t,r')^*|\phi(u) >$$

where $|\phi(u) >$ is the graviton coherent state. Thus, even though we start with a pure state for the electrons and positrons, it evolves to a mixed state in finite time owing to the background gravitational field being quantized in a coherent state. This phenomenon can be interpreted by saying that the quantum gravitational field pumps in entropy into the electron-positron field.

4.7 Corrections to the electron photon and graviton propagators due to mutual interactions and due to interaction with the quantum gravitational field

We write in perturbative form

$$e_a^\mu = \delta_a^\mu - h_a^\mu(x), e_\mu^a = \delta_\mu^a + h_\mu^a$$

for the tetrad of space-time. Here $h_a^\mu(x)$ satisfies the linearized Einstein field equation which can be derived as follows:

$$2h_{\mu\nu} = \delta g_{\mu\nu}, h_{\mu\nu} = \eta_{\nu a}h_\mu^a$$

so that

$$\delta g_{\mu\nu} = 2\eta_{\mu a}h_\nu^a$$

Note that

$$e_a^\mu = g^{\mu\nu}\eta_{ab}e_\nu^b$$

and thus,

$$\delta e_a^\mu = \eta_{ab}(\delta g^{\mu\nu}\delta_\nu^b + \eta^{\mu\nu}\delta e_\nu^b)$$

$$\delta g^{\mu\nu} = -\eta^{\mu\rho}\eta^{\nu\sigma}\delta g_{\rho\sigma}$$

and hence

$$\delta e_a^\mu = -\eta^{\mu\nu}h_{a\nu} = -\eta^{\mu\nu}\eta_{\nu\rho}h_a^\rho = -h_a^\mu$$

which is consistent with the above notation. To see what the linearized Einstein field equations are, we observe that upto linear orders in $\delta g_{\mu\nu}$,

$$R_{\mu\nu} = \Gamma^\alpha_{\mu\alpha,\nu} - \Gamma^\alpha_{\mu\nu,\alpha}$$

$$= \eta_{\alpha\beta}(\Gamma_{\beta\mu\alpha,\nu} - \Gamma_{\beta\mu\nu,\alpha})$$

$$= (h_{,\mu\nu} - h^\alpha_{\mu,\alpha\nu} - h^\alpha_{\nu,\alpha\mu} + \quad h_{\mu\nu})$$

If we choose coordinates so that

$$h^\alpha_{\mu,\alpha} - (1/2)h_{,\mu=0}$$

then the above expression simplifies to

$$R_{\mu\nu} = \Box h_{\mu\nu}$$

Also upto linear orders

$$R = \eta_{\mu\nu} R_{\mu\nu} = \Box h$$

so again upto linear orders, the Einstein tensor is given by

$$G_{\mu\nu} = R_{\mu\nu} - (1/2)Rg_{\mu\nu} = \Box(h_{\mu\nu} - (1/2)h\eta_{\mu\nu})$$

and upto linear orders in the metric perturbations, the Einstein-Maxwell equations are

$$\Box h'_{\mu\nu} = KT_{\mu\nu}$$

where $T_{\mu\nu}$ is the energy-momentum tensor of the electromagnetic field, $K = 8\pi G/c^2$,

$$h'_{\mu\nu} = h_{\mu\nu} - (1/2)h\eta_{\mu\nu}$$

We next require to compute the energy-momentum tensor of the Dirac field and add this to the energy-momentum tensor of the electromagnetic field in the rhs of the Einstein field equations. The Lagrangian density of the Dirac field is

$$L_D = \psi^*[\alpha^a e_a^\mu (i\partial_\mu + eA_\mu + i\Gamma_\mu) - m]\psi.\sqrt{-g}$$

We calculate the variation of $e_a^\mu, \Gamma_\mu, \sqrt{-g}$ in terms of $\delta g_{\mu\nu}$ as follows:

$$e_a^\mu = \delta_a^\mu - h_a^\mu = \delta_a^\mu - \eta^{\mu\nu} h_{\nu a}$$

upto linear orders in the metric perturbation, so that upto the same linear orders, we have

$$\delta e_a^\mu = -\eta_{\mu\nu}\delta h_{\nu a} = -\eta_{\mu\nu}\delta g_{\nu a}/2$$

Likewise,

$$\Gamma_\mu = (1/2)J_{ab}(e^a_{\nu,\mu} - \Gamma_{\rho\nu\mu}e^{a\rho})e^b_\nu = (1/2)J^{ab}(e^a_{\nu,\mu} - g_{\rho\mu,\nu}e^{a\rho})e^{b\nu}$$

$$= (1/2)J^{ab}(h^a_{\nu,\mu} - 2h_{\rho\mu,\nu}\eta_{a\rho})\eta_{b\nu}$$

$$= (1/2)J^{ab}(h^{ab}_{,\mu} - 2h^{a,b}_\mu)$$

upto first order. Now,

$$\delta\Gamma_\mu = (1/2)J_{ab}(e^b_\nu \delta e^{a\nu}_{,\mu} + e^{a\nu}_{,\mu}\delta e^b_\nu)$$

where

$$\delta e^a_\nu = (1/2)\eta_{ac}\delta g_{c\nu}, \delta e^\mu_a = (-1/2)\eta_{\mu\nu}\delta g_{a\nu},$$

$$\delta e^{a\mu} = \eta_{ab}\delta e^\mu_b = (-1/2)\eta_{\mu\nu}\eta_{ab}\delta g_{b\nu}$$

Thus, writing

$$\Gamma_\mu = \Gamma^1_\mu + \Gamma^2_\mu,$$

where

$$\Gamma^1_\mu = (1/2)J_{ab}e^{a\nu}_{,\mu}e^b_\nu, \Gamma^2_\mu = (1/2)J_{ab}g_{\mu\nu,\rho}e^\rho_a e^\nu_b$$

we have upto first order,

$$\delta\Gamma^1_\mu = (1/4)J_{ab}(e^{a\nu}_{,\mu}\eta_{bc}\delta g_{c\nu} - e^b_\nu\eta_{ac}\eta_{\nu\rho}\delta g_{c\rho,\mu})$$

$$= (1/4)J_{ab}(e^{a\nu}_{,\mu}\eta_{b\rho}\delta g_{\nu\rho} - e^{b\rho}\eta_{a\nu}\delta g_{\nu\rho,\mu})$$

$$\delta\Gamma^2_\mu =$$

Also,

$$\delta e^\mu_a = (-1/2)\eta_{\mu\nu}\delta g_{\nu a}$$

Finally,

$$\delta\sqrt{-g} = -\delta g/2\sqrt{-g} = -gg^{\nu\rho}\delta g_{\nu\rho}/2\sqrt{-g} = (1/2)\sqrt{-g}g^{\nu\rho}\delta g_{\nu\rho}$$

It follows then that with neglect of a total space-time divergence (which will not contribute to the energy-momentum tensor), the variation in the Dirac-Lagrangian w.r.t the metric tensor around a Minkowskian background space-time is given by

$$\delta L_D =$$

$$\delta(\psi^*[\alpha^a e^\mu_a(i\partial_\mu + eA_\mu + i\Gamma_\mu) - m]\psi.\sqrt{-g}) =$$

$$(1/2)\psi^*\alpha^\rho\eta_{\mu\nu}(i\partial_\mu+eA_\mu+i\Gamma_\mu)\psi\sqrt{-g}\delta g_{\nu\rho}$$

$$+(i/4)\psi^*\alpha^c J_{ab}\psi.e^\mu_c e^a_{\nu,\mu}\eta_{b\rho}\sqrt{-g}\delta g_{\nu\rho}-(i/4)\psi^*\alpha^c J_{ab}\psi.e^\mu_c e^{b\nu}\eta_{a\rho})\psi.\sqrt{-g}\delta g_{\nu\rho,\mu})$$

$$+(1/2)(\psi^*[\alpha^a e^\mu_a(i\partial_\mu + eA_\mu + i\Gamma_\mu) - m]\psi.\eta_{\nu\rho}\delta g_{\nu\rho}$$

$$= [(1/2)\psi^*\alpha^\rho\eta_{\mu\nu}(i\partial_\mu + eA_\mu + i\Gamma_\mu)\psi + (i/4)\psi^*\alpha^\mu J_{ab}\psi e^a_{\nu,\mu}\eta_{b\rho}$$

$$+(i/4)(\psi^*\alpha^c J_{ab}\psi.e^\mu_c e^{b\nu}\eta_{a\rho})\psi.\sqrt{-g})_{,\mu}(-g)^{-1/2}$$

$$+(1/2)(\psi^*[\alpha^a e^\mu_a(i\partial_\mu + eA_\mu + i\Gamma_\mu) - m]\psi.\eta_{\nu\rho}]\sqrt{-g}\delta g_{\nu\rho}$$

This gives the energy-momentum tensor of the Dirac field upto linear orders in the metric perturbations. Note: In this expression, we must replace e^μ_a by $\delta^\mu_a - h^\mu_a$, Γ_μ by $(1/2)J_{ab}h^{ab}_{,\mu}$ and retain only terms upto linear orders in the h_{ab}. Note that $h^{ab} = \eta_{a\mu}\eta_{b\nu}h_{\mu\nu}$. Now the linearized Einstein field equations are

$$h'_{\mu\nu}(x) = (\partial_0^2 - \nabla^2)h'_{\mu\nu}(x) = KT_{\mu\nu}(x)$$

where

$$T_{\mu\nu} = T_{M\mu\nu} + T_{D\mu\nu}$$

with T_M the energy momentum tensor of the Maxwell field and T_D the energy momentum tensor of the Dirac field, both expressed upto linear orders in the metric perturbations $h_{\mu\nu}$ and its partial derivatives. We can write

$$T_{M\mu\nu} = T^{(0)}_{M\mu\nu} + T^{1ab)}_{M\mu\nu}h_{ab} + T^{2abc}_{M\mu\nu}h_{ab,c}$$

where $T^{(k)}_{M\mu\nu}, k = 0,1,2$ do not contain any metric pertrubation term, they are only functions of the electromagnetic four vector potential A_μ and its first order partial derivative $A_{\mu,\nu}$. Likewise for the Dirac field, we can write

$$T_{D\mu\nu} = T^{(0}_{D\mu\nu}) + T^{1ab}_{D\mu\nu}h_{ab} + T^{(2abc}_{D\mu\nu}h_{ab,c}$$

Now we can compute the equations for the exact corrected propagator of the gravitational field in the linearized Einsteinian gravity model: Let

$$D_{\mu\nu ab}(x,x') =< T(h'_{\mu\nu}(x)h'_{ab}(x')) >$$

With T_M the energy momentum tensor of the Maxwell field and T_D the energy momentum tensor of the Dirac field, both expressed upto linear orders in the metric perturbations $h_{\mu\nu}$ and its partial derivatives. We can write

$$T_{M\mu\nu} = T^{(0)}_{M\mu\nu} + T^{1ab)}_{M\mu\nu}h_{ab} + T^{2abc}_{M\mu\nu}h_{ab,c}$$

where $T^{(k)}_{M\mu\nu}, k = 0,1,2$ do not contain any metric pertrubation term, they are only functions of the electromagnetic four vector potential A_μ and its first order partial derivative $A_{\mu,\nu}$. Likewise for the Dirac field, we can write

$$T_{D\mu\nu} = T^{(0}_{D\mu\nu}) + T^{1ab}_{D\mu\nu}h_{ab} + T^{(2abc}_{D\mu\nu}h_{ab,c}$$

Now we can compute the equations for the exact corrected propagator of the gravitational field in the linearized Einsteinian gravity model: Let

$$D_{\mu\nu ab}(x,x') =< T(h'_{\mu\nu}(x)h'_{ab}(x')) >= \theta(x^0 - x^{'0}) < h'_{\mu\nu}(x)h_{ab}(x') >$$

$$+\theta(x^{'0} - x^0) < h'_{ab}(x')h_{\mu\nu}(x) >$$

It is clear that in the linearized theory of gravity, we can express the Lagrangian density of the gravitational field as a quadratic form in $h'_{\mu\nu,a}(x)$ and hence by adding a term that is a quadratic form in $h'_{\mu\nu,a}(x)$ to the Lagrangian density (which can be regarded as a coordinate constraint), the position fields in linearized gravity can be taken as $h'_{\mu\nu 98x)}(x)$ and the momentum fields $P_{\mu\nu}(x) = \partial L_D/\partial h'_{\mu\nu,0}(x)$ will then become non-singular linear functionals of $h'_{\mu\nu,0}(x) and h'_{\mu\nu,r}(x), r =$ $1,2,3$ with $h'_{\mu\nu,0}(x)$ appearing in each $P_{\mu\nu}$. Solving this equation gives us $h'_{\mu\nu,0}(x)$ as linear combination of $P_{\mu\nu}$ and $h'_{\mu\nu,r}$ as

$$h_{\mu\nu}, 0'(x) = C_1(\mu\nu|\alpha\beta)P_{\alpha\beta}(x) + C_2(\mu\nu|\alpha\beta r)h_{\alpha\beta,r'}(x)$$

where the sum in r ranges over the spatial indices $1,2,3$ and $C_1(\mu\nu|\alpha\beta)$ is a non-singular 16×16 matrix. Thus on introducting the Bosonic equal time commutation relations

$$[h'_{\mu\nu}(t,r), P_{\alpha\beta}(t,r')] = i\eta_{\mu\alpha}\eta_{\nu\beta}\delta^3(r-r'),$$

$$[h'_{\mu\nu}(t,r), h'_{\alpha\beta}(t,r')] = 0, [P_{\mu\nu}(t,r), P_{\alpha\beta}(t,r'] = 0$$

we get the following equal time Bosonic commutation relations between the canonical position and canonical velocity fields for gravity:

$$[h'_{\mu\nu}(t,r), h'_{\alpha\beta,0}(t,r')]$$

$$= -C(\mu\nu|\alpha\beta)\delta^3(r-r'), [h'_{\mu\nu}(t,r), h'_{\alpha\beta}(t,r')] = 0, [h'_{\mu\nu,0}(t,r), h'_{\alpha\beta,0}(t,r')] = 0$$

where $((C(\mu\nu|\alpha\beta)))$ is a skew-Hermitian matrix. We then get using these commutation relations,

$$D_{\mu\nu ab}(x,x') = \delta(x^0 - x'^0) < [h'_{\mu\nu,0}(x), h'_{ab}(x')] > + < T(\Box h'_{\mu\nu}(x), h'_{ab}(x')) >$$

$$= C(ab|\mu\nu)\delta^4(x-x') + K. < T_{\mu\nu}(x)h'_{ab}(x') >$$

This equation will remain true even in the exact nonlinear Einsteinian gravity provided that we include in $T_{\mu\nu}$ the higher order terms in the Einstein tensor. More precisely, we observe that the exact Einstein field equations can be expressed as

$$h'_{\mu\nu}(x) = -G^{(2)}_{\mu\nu}(x) + T_{\mu\nu}(x), T_{\mu\nu} = T_{M\mu\nu} + T_{D\mu\nu}$$

where $G^{(2)}_{\mu\nu}$ contains the quadratic and higher order terms of the Einstein tensor and $T_{M\mu\nu}$ and $T_{D\mu\nu}$ also contain all the higher order terms in the metric tensor. Then, the propagator equation for the metric tensor assumes the form

$$D_{\mu\nu ab}(x,x') = C(ab|\mu\nu)\delta^4(x-x') + K. < T_{\mu\nu}(x)h'_{ab}(x') > - < G^{(2)}_{\mu\nu}(x)h'_{ab}(x') >$$

This equation can be used to perturbatively calculate the gravitational propagator upto any given order.

Remark: Formally solving the Einstein-Maxwell-Dirac equations

$$R_{\mu\nu} - (1/2)Rg_{\mu\nu} = K.(T_{M\mu\nu} + T_{D\mu\nu}),$$

$$(F^{\mu\nu}\sqrt{-g}))_{,\nu} = J^\mu\sqrt{-g}, F^{\mu\nu} = g^{\mu\alpha}g^{\nu\beta}F_{\alpha\beta}, F_{\alpha\beta} = A_{\beta,\alpha} - A_{\alpha,\beta}$$

$$J^\mu = -e\psi(x)^*\gamma^0\gamma^a e^\mu_a(x)\psi(x), g_{\mu\nu}(x) = \eta_{ab}e^a_\mu e^b_\nu, e^\mu_a = g^{\mu\rho}\eta_{ab}e^b_\rho,$$

$$[\gamma^a e^\mu_a(i\partial_\mu + eA_\mu + i\Gamma_{\mu)} - m]\psi = 0, \Gamma_\mu = (1/2)J^{ab}e^a_{\nu,\mu}e^{b\nu}, J^{ab} = [\gamma^a, \gamma^b]/4$$

where

$$T_{M\mu\nu} = -(1/4)F_{\alpha\beta}F^\alpha g_{\mu\nu} + F_{\mu\alpha}F^\alpha_\nu,$$

The exact energy-momentum tensor of the Dirac field in curved space-time must be calculated by evaluating δe^a_μ in terms of $\delta g_{\mu\nu}$. This is carried out as follows:

$$\delta g_{\mu\nu} = \eta_{ab}(e^a_\mu \delta e^b_\nu + e^b_\nu \delta e^a_\mu) = e_{a\mu}\delta e^a_\nu + e_{a\nu}\delta e^a_\mu$$

and hence, we can assume without loss of generality that

$$e_{a\mu}\delta e^a_\nu = (1/2)\delta g_{\mu\nu}$$

because $\delta g_{\mu\nu}$ is symmetric w.r.t interchange of μ and ν. This gives us

$$\delta e^a_\nu = (1/2)e^{a\mu}\delta g_{\mu\nu}$$

Then, noting that

$$\delta\Gamma_\mu = (1/2)J^{ab}\delta(e_{a\nu,\mu}e^\nu_b) = (1/2)J^{ab}(e_{a\nu,\mu}\delta e^\nu_b + e^\nu_b\delta e_{a\nu,\mu})$$

$$= (1/2)J^{ab}e_{a\nu,\mu}\delta e_b^\nu - (1/2)J^{ab}e_{b,\mu}^\nu\delta e_{a\nu} + (1/2)J^{ab}(e_b^\nu\delta e_{a\nu})_{,\mu}$$

Note that δe_a^μ can be evaluated as follows.

$$e_a^\mu e_\nu^a = \delta_\nu^\mu$$

so

$$\delta e_a^\mu . e_\nu^a + e_a^\mu \delta e_\nu^a = 0$$

and hence

$$\delta e_a^\mu = -e_a^\nu e_b^\mu \delta e_\nu^b = (-1/2)e_a^\nu e_b^\mu e^{b\rho}\delta g_{\rho\nu} = (-1/2)g^{\mu\rho}e_a^\nu\delta g_{\nu\rho} --- (1)$$

So we get writing

$$\Gamma_\mu = \Gamma_\mu^1 + \Gamma_\mu^2$$

where

$$\Gamma_\mu^1 = (1/2)J^{ab}e_{a\nu,\mu}e_b^\nu, \Gamma_\mu^2 = (1/2)J^{ab}g_{\rho\nu,\mu}e_a^\rho e_b^\nu,$$

that

$$\delta\Gamma_\mu^1 =$$

$$(1/2)J^{ab}\delta(e_{a\nu,\mu}e_b^\nu) = (-1/4)J^{ab}e_{a\nu,\mu}g^{\nu\rho}e_b^\alpha\delta g_{\alpha\rho}$$
$$+(1/4)J^{ab}e_b^\nu e_a^\rho\delta g_{\nu\rho,\mu} --- (2)$$

and

$$\delta\Gamma_\mu^2 = (1/2)J^{ab}(e_a^\rho e_b^\nu\delta g_{\rho\nu,\mu} + g_{\rho\nu,\mu}(e_b^\nu\delta e_a^\rho + e_a^\rho\delta e_b^\nu)$$
$$= (-1/2)g_{\nu\rho,\mu}(g^{\rho\alpha}e_a^\beta e_b^\nu + g^{\nu\alpha}e_b^\beta e_a^\rho)\delta g_{\alpha\beta} --- (3)$$

Equns (1), (2) and (3) are exact expressions. These can be used in calculating the variation of the Dirac Lagrangian density w.r.t the metric variations and hence obtain the exact expression for the energy-momentum tensor of the Dirac field in background space-time.

4.8 Lagrangian density for the Dirac field in a curved background space-time metric and also taking into account its interaction with the electromagnetic field

Remark: Consider the following Lagrangian density for the Dirac field in a curved background space-time metric and also taking into account its interaction with the electromagnetic field:

$$L(\psi, \chi, \partial_\mu\psi)\sqrt{-g}$$

$$= Re[\chi(x)^*[e_a^\mu(x)\gamma^0\gamma^a(i\partial_\mu + eA_\mu(x) + \Gamma_\mu(x)) - m\gamma^0]\psi(x)]\sqrt{-g(x)} --- (a)$$

where

χ is another bisipinor field, called the dual of the Dirac bispinor field ψ. The variational equation $\delta(L\sqrt{-g})/\delta\chi^*(x) = 0$ gives the Dirac equation for ψ in curved space-time:

$$[e_a^\mu(x)\gamma^0\gamma^a(i\partial_\mu + eA_\mu(x) + \Gamma_\mu(x)) - m\gamma^0]\psi(x)] = 0$$

or equivalently in terms of the anticommuting Hermtian matrices(for $\mu = 1, 2, 3$, note that $\alpha^0 = I]$) $\alpha^\mu =^0 \gamma^\mu, \beta\gamma^0$, this equation can be expressed as

$$[e_a^\mu\alpha^a(i\partial_\mu + eA_\mu + i\Gamma_\mu) - m\beta]\psi = 0 - - - (1)$$

The field equation satisfied by the auxiliary field χ is obtatined from the variational equation $\delta(L\sqrt{-g})/\delta\psi(x) = 0$ and is given by

$$-i\partial_\mu(e_a^\mu\sqrt{-g}\chi^*\alpha^a) + \chi^*\alpha^a(eA_\mu + i\Gamma_\mu)e_a^\mu\sqrt{-g} - m\chi^*\beta\sqrt{-g} = 0 - - - (2)$$

Note that the $\alpha^{a's}$ are constant matrices. (1) and (2) together imply (on premultiplying (1) by $\sqrt{-g}\chi^*$, postmultiplying (2) by ψ and subtracting)

$$\partial_\mu(\chi^*\alpha^a e_a^\mu\psi\sqrt{-g}) = \partial_\mu(\chi^*\alpha^a e_a^\mu\sqrt{-g})\psi + \chi^*\alpha^a e_a^\mu\partial_\mu\psi\sqrt{-g} = 0$$

Thus, defining the real curved space-time Dirac four current density J^μ by

$$J^\mu = Re(\chi^*\alpha^a e_a^\mu\psi)$$

we obtain the conservation law

$$\partial_\mu(J^\mu\sqrt{-g}) = 0$$

So in fact, we should correctly speaking use the Lagrangian density (a) in order to derive the energy-momentum tensor of the Dirac field. In the special case of flat Minkowskian space-time, we have $e_a^\mu = \delta_a^\mu$ and it follows that χ and ψ satisfy the same flat space-time Dirac equation.

4.9 Interaction of light with matter in a quantum antenna modeled as a resonator cavity in a background curved space-time metric of the gravitational field

Consider a resonator cavity of arbitrary shape containing electrons, positrons and photons interacting with each other. We insert into this cavity probes containing classical current and classical electromagnetic field sources. Denote these classical sources by J_c^μ and A_c^μ. The Maxwell and Dirac equations that govern the dynamics of the two fields interacting with each other and with the classical sources are

$$(F^{\mu\nu}\sqrt{-g})_{,\nu} = eRe(\chi^*\alpha^a\psi).e_a^\mu\sqrt{-g} - J_{c\mu}\sqrt{-g}$$

$$\gamma^a e_a^\mu(i\partial_\mu + eA_\mu] + eA_{c\mu} + i\Gamma_\mu)\psi = 0$$

Here, $\alpha^a = \gamma^0\gamma^a$ are the Dirac α-matrices and $g_{\mu\nu}$ is the metric tensor of the background gravitational field. We are assuming this background gravitational field to be classical. e_a^μ is a tetrad for the gravitational field metric tensor and $\Gamma_\mu = (1/2)J^{ab}e_a^\nu e_{b\nu,\mu}$ is the spinor connection of the gravitational field where $J^{ab} = [\gamma^a, \gamma^b]/4$ are the standard Lie algebra generators of the spinor representation of the Lorentz group. Note that if Λ is a Lorentz transformation with ω a Lie algebra element of the Lorentz group so that $\Lambda = exp(\omega)$, and if D is the spinor representation of the Lorentz group, then

$$D(\Lambda) = exp(dD(\omega)), dD(\omega) = \omega.J = \omega_{ab}J^{ab}$$

Note that since ω is an infinitesimal Lorentz transformation,

$$\omega_{ab} + \omega_{ba} = 0$$

This is a consequence of the fact that

$$\Lambda_b^a x^b = (\Lambda x)^a$$

satisfies

$$(\Lambda x)^a(\Lambda x)_a = x^a x_a$$

and hence

$$(\omega.x)^a x_a + x^a(\omega.x)_a = 0$$

or equivalently,

$$\omega_b^a x^b x_a + \omega_{ab}x^b x^a = 0$$

or equivalently,

$$\omega_{ab}x^a x^b = 0 \forall x \in \mathbb{R}^4$$

implying that

$$\omega_{ab} + \omega_{ba} = 0$$

We impose the Loretnz gauge condition on the electromagnetic field:

$$(A^\mu\sqrt{-g})_{,\mu} = 0$$

and linearize our equations upto linear orders in the metric perturbation so that

$$e_a^\mu = \delta_a^\mu + h_a^\mu$$

where h is of the first order of smallness. The gauge condition can then be approximately expressed as

$$((\eta^{\mu\nu} - 2h^{\mu\nu})(1+h)A_\nu)_{,\mu} = 0, h = h_\mu^\mu = \eta_{\mu\nu}h_{\mu\nu}$$

or equivalently,

$$\eta_{\mu\nu}A_{\nu,\mu} + (\rho_{\mu\nu}A_\nu)_{,\mu} = 0$$

where

$$\rho_{\mu\nu} = h\eta_{\mu\nu} - 2h^{\mu\nu}, h^{\mu\nu} = \eta_{\mu\alpha}\eta_{\nu\beta}h_{\alpha\beta}$$

This gauge condition is valid upto linear orders in the metric perturbations. Further, we may also write down the Maxwell equations upto linear orders in the metric perturbations:

$$F^{\mu\nu}\sqrt{-g} = g^{\mu\alpha}g^{\nu\beta}\sqrt{-g}F_{\alpha\beta} =$$

$$(\eta_{\mu\alpha} - 2h^{\mu\alpha})(\eta_{\nu\beta} - 2h^{\nu\beta)})(1+h)F_{\alpha\beta} =$$

$$\eta_{\mu\alpha}\eta_{\nu\beta}F_{\alpha\beta} + \tau_{\mu\nu\alpha\beta}F_{\alpha\beta}F_{\alpha\beta}$$

where

$$\tau_{\mu\nu\alpha\beta} = h\eta_{\mu\alpha}\eta_{\nu\beta} - 2\eta_{\mu\alpha}h_{\nu\beta} - 2\eta_{\nu\beta}h_{\mu\alpha}$$

upto linear orders in the metric perturbations or equivalently upto linear orders in the tetrad perturbations. Using the gauge condition,

$$(\eta_{\mu\alpha}\eta_{\nu\beta}F_{\alpha\beta})_{,\nu} =$$

$$\eta_{\mu\alpha}\eta_{\nu\beta}F_{\alpha\beta,\nu} = \eta_{\mu\alpha}\eta_{\nu\beta}(A_{\beta,\alpha\nu} - A_{\alpha,\beta\nu})$$

$$= -\eta_{\mu\alpha} \quad A_\alpha - \eta_{\mu\alpha}(\rho_{\beta\nu}A_\nu)_{,\beta\alpha}$$

So the Maxwell equations upto linear orders in the metric perturbations are given by

$$-\eta_{\mu\alpha} \quad A_\alpha - \eta_{\mu\alpha}(\rho_{\beta\nu}A_\nu)_{,\beta\alpha} = -J^\mu(1+h) - - - (1)$$

where

$$h = \eta_{\mu\nu}h_{\mu\nu} = h^\mu_\mu$$

In this expression, the Dirac current J^μ is given by

$$J^\mu = -e.Re(\chi^*\alpha^a\psi)e^\mu_a$$

so that

$$J^\mu(1+h) = -eRe(\chi^*\alpha^a\psi)e^\mu_a(1+h) = -eRe(\chi^*\alpha^a\psi)(\delta^\mu_a - h^\mu_a)(1+h)$$

$$= -eRe(\chi^*\alpha^\mu\psi) - \delta.e(h\delta^\mu_a - h^\mu_a).Re(\chi^*\alpha^a\psi)$$

Here ψ is the Dirac wave field and χ is its dual. The Maxwell equations upto linear orders in the metric perturbations can thus be expressed as

$$-\eta_{\mu\alpha} \quad A_\alpha + eRe(\chi^*\alpha^\mu\psi) = \delta.[\eta_{\mu\alpha}(\rho_{\beta\nu}A_\nu)_{,\beta\alpha} - e(h\delta^\mu_a - h^\mu_a).Re(\chi^*\alpha^a\psi)] - - - (2)$$

Upto first order perturbations in tbe metric (We shall be assuming that the spinor connection Γ_μ has been expressed upto linear orders in the metric/tetrad perturbation as discussed above), the Dirac equation is

$$[\alpha^\mu(i\partial_\mu + eA_\mu) - \beta m]\psi(x) = \delta.[i\alpha^a h^\mu_a(i\partial_\mu + eA_\mu) - i\alpha^\mu\Gamma_\mu]\psi - - - (3)$$

where δ is a perturbation tag used to signify the fact that the metric/tetrad perturbations are of the first order of smallness. Likewise, the dual of the Dirac equation is

$$i\partial_\mu(\chi^*\alpha^a e(\delta^\mu_a - h^\mu_a)(1+h))(1-h) + \chi^*\alpha^a(\delta^\mu_a - h^\mu_a)(eA_\mu + i\Gamma_\mu) - \chi^*\beta.m = 0$$

which simplifies using first order perturbation theory in the metric pertubations to

$$(\alpha^\mu(-i\partial_\mu + eA_\mu) - \beta m)\chi = \delta.[ih.\alpha^\mu(\partial_\mu\chi) + i\alpha^a\partial_\mu((h^\mu_a - h\delta^\mu_a)\chi) - - - (4)$$

$$+i\Gamma^*_\mu\alpha^\mu\chi + eA_\mu.h^\mu_a\alpha^a\chi]$$

Finally, the Einstein field equations upto linear orders in the metric/tetrad perturbations are

$$h'_{\mu\nu} = K(T^0_{\mu\nu} + \delta.T^1_{\mu\nu\alpha\beta}h_{\alpha\beta} + \delta.T^2_{\mu\nu\alpha\beta\rho}h_{\alpha\beta\rho}) - - - (5)$$

where $T_{\mu\nu}$ is the sum of the Dirac and Maxwell energy-momentum tensor in a background curved metric expressed upto linear orders in the metric perturbations as given in the expression above within the bracket on the rhs. $T^k_{\mu\nu}, k = 0, 1, 2$ depend only on the Maxwell field A_μ and the Dirac fields ψ, χ and do not depend explicitly on the metric perturbations $h_{\mu\nu}$. Note that

$$h'_{\mu\nu} = h_{\mu\nu} - (h/2)\eta_{\mu\nu}$$

which implies $h' = -h$ and hence

$$h_{\mu\nu} = h'_{\mu\nu} - h'\eta_{\mu\nu}/2$$

The idea is to first solve perturbatively the coupled Maxwell Dirac equations in flat space-time perturbatively in powers of the electron charge which acts as the first perturbation parameter. These equations are obtained from (2),(3),(4) by setting the perturbation parameter $\delta = 0$. After obtaining these unperturbed solutions for ψ, χ, A_μ, we denote these by ψ^0, χ^0, A^0_μ and we solve (5) with $\delta = 0$ and $T^0_{\mu\nu}$ computed using ψ^0, χ^0, A^0_μ in the flat space-time metric. We denote the solution to the resultant metric by $h^{'0}_{\mu\nu}$. Following this, we solve perturbatively in powers of δ for $h'_{\mu\nu}$. However, if we want to develop a full perturbation theory for simultaneously solving for the Maxwell, Dirac and metric fields taking all the nonlinearities and interactions into account, we must introduce a perturbation parameter for the nonlinear terms in the Einstein field tensor as well as the electronic charge perturbation parameter in the interactions between the different fields and write down a two dimensional perturbation expansion, ie, a bivariate power series in the two parameters.

4.10 A remark on the solution of the Einstein-Maxwell-Dirac equations within a cavity resonator using modal expansions using an alternate method of perturbation theory

In this method of perturbation, we assume the gravitational perturbation parameter δ to be the same as the electron charge e. In other words, the perturbation of Minkowski metric is assumed to be of the same order of smallness as the electronic charge. We develop an expansion of the metric field and the Maxwell and Dirac fields in powers of the electronic charge. First, we make a computation upto linear orders in the electronic charge.

We choose a set of orthonormal basis functions $u_{\mu,n}(r), w_{nl}(r), v_{nl}(r), f_{\mu\nu,n}(r) n = 1, 2, \ldots$ for the cavity satisfying the Dirichlet boundary conditions and expand the unperturbed electromagnetic four vector potential $A^0_\mu(t, r)$, the Dirac fields $\psi^0(t, r), \chi^0(t, r)$ and the unperturbed metric/tetrad perturbations $h^0_{\mu\nu}(t, r)$ in terms of these basis functions so that the coefficients in these expansions will be functions of time only:

$$A^0_\mu(t, r) = \sum_n (c(n, s, t)u_{\mu,n,s}(r) + c(n, s, , t)^* u_{\mu,n,s}(r)^*),$$

$$\psi^0(t, r) = \chi^0(t, r) = \sum_n (a(n, s, t)w_{n,s}(r) + b(n, s, t)^* v_{n,s}(r)),$$

$$h^0_{\mu\nu}(t, r) = \sum_n (e(n, s, t)f_{\mu\nu,n,s}(r) + e(n, s, t)^* f_{\mu\nu,n,s,}(r)^*) + K \int G(x - x')T^0_{\mu\nu}(x')d^4x'$$

The basis functions will satisfy the unperturbed equations

$$\nabla^2 u_{m,n,s}(r) + \omega_1(n)^2 u_{m,n,s}(r) = 0, \partial_m u_{m,n,s}(r) = 0, u_{0,n,s}(r) = 0$$

where we are adopting the Coulomb gauge for the unperturbed em potential, ie, $A^0_0 = 0, div(A^0) == A^0_{m,m} = 0$

$$(\nabla^2 + \omega_2(n)^2)f_{mn,k,s}(r) = 0, f_{0\mu,n,s}(r) = 0$$

$$(((\alpha, -i\nabla) + \beta m)w_{n,s}(r) = E(n)w_{n,s}(r), ((\alpha, -i\nabla) + \beta m)v_{n,s}(r) = -E(n)v_{n,s}(r)$$

Thus,

$$c(n, s, t) = c(n, s)exp(-i\omega_1(n)t), c(n, s, t)^* = c(n, s)^* exp(i\omega_2(n)t)$$

$$e(n, s, t) = e(n, s)exp(-i\omega_2(n)t), e(n, s, t)^* = e(n, s)^* exp(i\omega_2(n)t)$$

$$a(n, s, t) = a(n, s)exp(-iE(n)t), b(n, s, t) = b(n, s)exp(-iE(n)t)$$

$e(n, s)(a(n, s)^*)$ creates (annihilates) a graviton of modal value n and spin s with $s = -2, -1, 0, 1, 2$, $a(n, s)$ $(a(n, s)^*)$ annihilates (creates) an electron of modal value n ie energy $E(n)$ with $s = \pm 1/2$. $b(n, s)$ $(b(n, s)^*)$ annihilates (creates) a positron of modal value n, ie, energy $E(n)$. The first order perturbations $A^1, \psi^1, h^1_{\mu\nu}$ to the Maxwell, Dirac and metric fields satisfy the equations

$$i\partial_t\psi^1 - ((\alpha, -i\nabla) + \beta m)\psi^1 - e(\alpha, A^0)\psi^0 = [i\alpha^a h^{0\mu}_a(i\partial_\mu + eA^0_\mu) - i\alpha^\mu\Gamma_\mu]\psi^0$$

or equivalently,

$$(\alpha^\mu(i\partial_\mu) - \beta m)\psi^1 - e(\alpha, A^0)\psi^0 = [i\alpha^a h^{0\mu}_a(i\partial_\mu + eA^0_\mu) - i\alpha^\mu\Gamma_\mu]\psi^0,$$

$$(\alpha^\mu(-i\partial_\mu) - \beta m)\chi^1 + e(\alpha, A^0)\chi^0 = [ih^0.\alpha^\mu(\partial_\mu\chi^0) + i\alpha^a\partial_\mu((h^{0\mu}_a - h^0\delta^\mu_a)\chi^0)$$
$$+i\Gamma^*_\mu\alpha^\mu\chi + eA^0_\mu.h^{0\mu}_a\alpha^a\chi^0]$$

and

$$-\eta_{\mu\alpha}A^1_\alpha + eRe(\chi^{0*}\alpha^\mu\psi^0) = \delta.[\eta_{\mu\alpha}(\rho^0_{\beta\nu}A^0_\nu)_{,\beta\alpha} - e(h^0\delta^\mu_a - h^{0\mu}_a).Re(\chi^{0*}\alpha^a\psi^0)]$$

$$h^{'1}_{\mu\nu} = K(T^1_{\mu\nu\alpha\beta}h^0_{\alpha\beta} + T^2_{\mu\nu\alpha\beta\rho}h^0_{\alpha\beta\rho})$$

By using the Green's functions for the free Dirac operator and its dual and for the D'Alembertian wave operator, we can solve these equations to obtain the first order perturbations $\psi^1, \chi^1, A^1, h^1_{\mu\nu}$ in terms of the Dirac electron-positron creation and annihilation operators and the graviton and photon creation and annihilation operators. These expressions show that ψ^1 is the sum of two terms, the first is bilinear in the photon creation-annihilation operators and the Dirac electron annihilation and positron creation operators and the second is bilinear in in the graviton creation-annihilation operators and the Dirac electron annihilation and positron creation operators. Likewise for χ^1. Further, $h^1_{\mu\nu}$ is the sum of two terms. The first is bilinear in the graviton creation and annihilation operators and a quadratic combination of the photon creation and annihilation operators and the second is bilinear in the graviton creation and annihilation operators and a quadratic combination of the Dirac electron and positron creation and annihilation operators. It should be noted that the quadratic combinations of the Dirac creation and annihilation operators exclude terms that are quadratic in the electron creation operators alone, or in the electron annihilation operators alone or in the positron creation operators alone or in the positron annihilation operators alone or terms bilinear in the electron annihilation and positron annihilation operators or terms bilinear in the electron creation and positron creation operators alone.

4.11 Work with Vijay: Designing a quantum neural network using a two dimensional oscillator in a magnetic field. MATLAB program for simulating the qnn

Designing a quantum neural network using a two dimensional oscillator in a magnetic field. MATLAB program for simulating the qnn.

The Hamiltonian of the oscillator is

$$H(t) = ((p_x + eA_x)^2 + (p_y + eA_y)^2)/2 + (\omega_0^2/2)(x^2 + y^2)$$

where

$$A_x(t, x, y) = -B(t)y/2, A_y(t, x, y) = B(t)x/2$$

so that the magnetic field becomes

$$curlA(t, x, y) = B(t)\hat{z}$$

Thus

$$H(t) = H_0 + e^2B(t)^2(x^2 + y^2)/4 + eB(t)(xp_y - yp_x) + (\omega_0^2/2)(x^2 + y^2)$$

$$= H_0 + (e^2B(t)^2/4)r^2 + eB(t)L_z$$

where

$$r^2 = x^2 + y^2, L_z = xp_y - yp_x = -i(x\partial_y - y\partial_x)$$

and

$$H_0 = (p_x^2+p_y^2)/2+(\omega_0^2/2)r^2$$

$$= a_x^*a_x+a_y^*a_y+\omega_0, a_x = (p_x-i\omega_0 x)/\sqrt{2}, a_y = (p_y-i\omega_0 y)/\sqrt{2}$$

We have the obvious Bosonic commutation relations

$$[a_x, a_x^*] = [a_y, a_y^*] = \omega_0, [a_x, a_y] = 0$$

This Hamiltonian can be expressed as

$$H(t) = H_0 + eB(t)V_1 + e^2B(t)^2V_2$$

where V_1, V_2 are two Hermitian operators. Specifically,

$$V_1 = L_z, V_2 = r^2/4$$

Discretizing this Hamiltonian w.r.t the spatial arguments x, y, ie writing $x = n_x\Delta, y = n_y\Delta, n_x, n_y = -M, -M+1, ..., M-1, M$ where M is a large positive integer, we get an approximation to the Hamiltonian as a $(2M+1)^2 \times (2M+1)^2$ time varying Hermitian matrix and the time dependent wave function $\psi(t, x, y)$ on discretization becomes a $(2M+1)^2 \times 1$ complex vector $\psi(t)$ having unit length. The time varying Schrodinger equation satisfied by $\psi(t, x, y)$ is

$$i\partial_t\psi(t, x, y) = H(t)\psi(t, x, y)$$

which on spatial discretization becomes

$$i\partial_t\psi(t, n_x, n_y) =$$

$$H_0(t)psi(t, n_x, n_y) + B(t)V_1\psi(t, n_x, n_y) + B(t)^2V_2\psi(t, n_x, n_y)$$

where

$$H_0\psi(t,n_x,n_y) = [(-1/2)(P_x^2+P_y^2)+\omega_0^2(X^2+Y^2)/2]\psi(t,n_x,n_y)$$

where P_x is the $(2M+1)^2\times(2M+1)^2$ matrix defined by

$$P_x = -i\Delta^{-1}\sum_{n_x,n_y} e(n_x,n_y)(e(n_x{+}1,n_y)-e(n_x,n_y))^T, e(n_x,n_y)$$

$$= e(n_x)\otimes e(n_y), e(n) = [\delta(n,1),...,\delta(n,2M{+}1)]^T$$

Note that we can write

$$P_x = P_0 \otimes I_{2M+1}$$

where

$$P_0 = -i\Delta^{-1}\sum_n e(n)(e(n+1)-e(n))^T \in \mathbb{C}^{2M+1\times 2M+1}$$

and likewise

$$P_y = I_{2M+1}\otimes P_0$$

Again, we observe that V_1 gets replaced by a $(2M+1)^2\times(2M+1)^2$ Hermitian matrix defined by

$$V_1 = -i(X\otimes P_0 - P_0\otimes X)$$

where X is the $(2M+1)\times(2M+1)$ matrices defined by

$$X = \Delta\sum_n ne(n)e(n)^T$$

Finally V_2 is the $(2M+1)^2\times(2M+1)^2$ Hermitian matrix defined by

$$V_2 = (X^2\otimes I_{2M+1} + I_{2M+1}\otimes X^2)/4$$

Now Schrodinger's equation can be approximated by the following $(2M+1)^2$-finite state equation

$$i\psi'(t) = (H_0 + eB(t)V_1 + e^2B(t)^2V_2)\psi(t)$$

and it remains to set up the adaptive update equation for the control magnetic field $B(t)$.

4.12 Lecture plan for analog and digital communication

Block diagram of an analog communication system. Analog signal is generated and is modulated either using AM and its modifications like DSB, DSBSC, SSB, PMm, FM and the modulated signal is transmitted over the channel and at the reciever end, the signal is demodulated to recover an estimate of the original analog signal that was transmitted.

Statistical models of a noisy channel. The channel introduces linear/nonlinear distortions into the signal and couples it either additively or non-additively with noise which may be Gaussian or non-Gaussian. If $x(t)$ is the input message transmitted and $w(t)$ is channel noise, then the channel output/the received signal can be expressed as

$$y(t) = F(t, x(s), s \leq t, w(t))$$

Examples of special cases include (a) LTI channels with additive noise:

$$y(t) = h(t) * x(t) + w(t)$$

(b) linear time varying channels with additive noise:

$$y(t) = \int h(t, s)x(s)ds + w(t)$$

(b) Nonlinear TI channels with additive noise:

$$y(t) = \int F(x(t - \tau), \tau)d\tau + w(t)$$

and more generally,

$$y(t) = \sum_{r \geq 1} \int F_r(x(t - \tau_1, ..., x(t - \tau_r), \tau_1, ..., \tau_r)d\tau_1...d\tau_r + w(t)$$

(c) channels with nonlinearly coupled noise:

$$y(t) = F(x(t-\tau_1), ..., x(t-\tau_r), w(t-s_1), ..., \\ w(t-s_m), \tau_1, ..., \tau_r, s_1, ..., s_m)d\tau_1...d\tau_r ds_1...ds_m$$

The role of a transmitter and receiver antenna transmitting the modulated message in the form of an electromagnetic wave over the channel and in recieving the em wave and then demodulating it.

Channels which take into account the effects of the gravitational field based on using the covariant form of the Maxwell equations in curved space-time. The generalized wave equation for the propagation of the four vector potential in curved space-time with the metric being a small perturbation of the Minkowskian flat space-time metric is derived by applyiing the covariant form of the Lorentz gauge condition in curved space-time. This wave equation contains a perturbation term that is bilinear in the vector potential derivatives and the metric perturbations.

Basic components of a digital communication system: Generation of an analog signal followed by a sampler, quantizer, encoder, modulator that converts the one and zero bits into signal waveforms using ASK, PSK, FSK, QPSK, DPSK etc, followed by the transmitter antenna that transmits em waves over the noisy channel, and at the reeeiver end we have a receiver antenna, a demoulator/detector that converts the received waveforms into bits, decoder that converts binary strings into discrete amplitude levels, D to A convertor that involves using a version of a holde circuit/interpolator that converts a discrete time signal into a continuous time analog signal.

How to compute the transition probability matrix for a channel that involves m-ary shift keying using likelihood ratio based detection theory.

The notion of information content in a discrete memoryless source, information rate of an ergodic source. The Shannon-Mcmillan-Breiman theorem for an ergodic source:

$$limn^{-1}log(\mu[x_1,...,x_n]) \to \bar{H}(x) = H(x_0|x_{-1},x_{-2},...) = -\mathbb{E}(log(\mu(x_0|x_{-1},x_{-2},...)))$$

Basic block diagram of a quantum communication system involving transmission of classical alphabets by encoding them into a quantum state transmitted over a quantum noisy channel described by the Stinespring/Choi-Kraus representation.

Noise analysis of an AM system, bandpass noise represented using the complex lowpass envelope based on the Hilbert transform, coherent AM detector, square law AM detector, envelope detector.

Estimating the parameters of a quantum state using successive POVM measurements on the state that evolves in accordance with the Schrodinger-Lindblad master equation for open quantum systems.

Examples of Cq channels using the Schrodinger-Lindblad equation. The information bearing sequence is modulated using pulses to get an analog current waveform. This current is fed into an antenna that generates and electromagnetic field at the receiver end expressed in terms of the retarded Green's function acting on the current. This electromagnetic field at the receiver end is incident upon an electron bound to its nucleus causing the electron to make time dependent transitions between stationary states. By measuring these time dependent transition probabilities of the electron using relative population of the electron in excited states, we infer the original information bearing sequence. More generally, by making POV measurements on the evolving mixed state at the receiver end, we infer about the information bearing sequence using the maximum likelihood method coupled to the state collapse principle.

Transmission of the quantum electromagnetic field over a channel comprising photons, electrons, positrons, gravitons and Yang-Mills matter and gauge particles. The transmitted photon em field interacts with these other fields and using perturbation theory, the received em field can be computed as the sum of the unperturbed transmitted component convolved with the linear time invariant impulse response component of the channel and interaction components expressible as nonlinear functionals of the transmitted em field and the channel fields. The former component is expressible as a linear superposition of the creation and annihilation operators of the transmitted em photon field while the latter component is expressible as nonlinear combinations of the creation and annihilation operators of the transmitted photon field and those of the channel photon, Dirac and graviton field. Then given a joint coherent state of the tramsmitted photons, the channel photons and the channel electrons, positrons and gravitons, we can evaluate the quantum expectations of functionals of the received em field in such a state.

In order to see the role played by graviton interactions with the photons and electrons and positrons in communication of signals using quantum electromagnetic fields over a quantum channel comprising gravitons, electrons and positrons, we have to first be able to expand the linearized solution to the Einstein field equations for the metric perturbations in the absence of other interactions, as plane waves with coefficients being the graviton creation and annihilation operators. For each propagation wave vector, there are five degrees of polarization of the graviton wave, ie, at each wave vector/four momentum of the field, there are associated five creation and five annihilation operators for gravitons. This follows from the fact that gravitons are spin two particles. A derivation of this fact is presented below:

4.13 Proof that gravitons are spin two particles using the plane wave solution to the linearized Einstein field equations

The solution to the linearized Einstein field equations for the metric perturbations can be expressed as a superposition of plane waves:

$$h_{\mu\nu}(x) = e_{\mu\nu}.exp(-ik.x), k.x = k_0 t - K.r$$

The coordinate condition chosen is

$$h^{\nu}_{\mu,\nu} - h_{,\mu}/2 = 0$$

which gives

$$e^{\nu}_{\mu}k_{\nu} - (1/2)e^{\nu}_{\nu}k_{\mu} = 0$$

where the indices of $e_{\mu\nu}$ are raised using the Minkowski metric. Note that $e_{\mu\nu}$ represents a small amplitude corresponding to the metric perturbations. Assuming for simplicity the gravitational waves to be propagating along the z directions so that $k^1 = k^2 = 0, k^3 = k^0 = \omega$ say, the above coordinate constraints give

$$\omega(e^0_{\mu} - e^3_{\mu}) - (1/2)e.k_{\mu} = 0$$

which evaluates for $\mu = 0, 1, 2, 3$ to

$$e^0_0 - e^3_0 - e/2 = 0, e^0_1 - e^3_1 = 0, e^0_2 - e^3_2 = 0, e^0_3 - e^3_3 + e/2 = 0$$

where

$$e = e^0_0 + e^1_1 + e^2_2 + e^3_3 = e_{00} - e_{11} - e_{22} - e_{33}$$

These equations can equivalently be expressed as

$2(e_{00}+e_{30})-(e_{00}-e_{11}-e_{22}-e_{33}) = 0, e_{01}+e_{13} = 0,$

$e_{02}+e_{23} = 0, 2(e_{03}+e_{33})+(e_{00}-e_{11}-e_{22}-e_{33}) = 0$

or

$$e_{00}+e_{33}+2e_{03}+e_{11}+e_{22}=0, e_{13}=-e_{01}, e_{23}=-e_{02}, e_{00}+e_{33}+2e_{03}-(e_{11}+e_{22})=0$$

or

$$e_{00}+e_{33}=-2e_{03}, e_{11}+e_{22}=0, e_{13}=-e_{01}, e_{23}=-e_{02}$$

It follows from these linear relations that a set of linearly independent linear combinations of the metric perturbations that will completely determine the metric may be chosen as

$$e_{00}, e_{03}, e_{11}-e_{22}, 2e_{12}, e_{01}, e_{02}$$

Now under a rotation of the coordinate axes around the x^3 axis by an angle θ, the space-time coordinates transform as $x \to x'$, where

$$x^{0'}=x^0, x^{3'}=x^3, x^{1'}=x^1.cos(\theta)+x^2.sin(\theta), x^{2'}=x^1.sin(\theta)-x^2.cos(\theta)$$

and under such a coordinate transformation, the above metric components transform as

$$e_{00}'=e_{00}, e_{03}'=e_{03}, e_{11}'-e_{22}'=(e_{11}-e_{22}).cos(2\theta)-2e_{12}.sin(2\theta), 2e_{12}'$$
$$=(e_{11}-e_{22}).sin(2\theta)+2e_{12}.cos(2\theta),$$

$$e_{01}'=e_{01}.cos(\theta)-e_{02}.sin(\theta), e_{02}'=e_{01}.sin(\theta)+e_{02}.cos(\theta)$$

These transformation equations follow from the relations

$$\partial_0'=\partial_0, \partial_3'=\partial_3,$$

$$\partial_1'=cos(\theta)\partial_1-sin(\theta)\partial_2, \partial_2'=sin(\theta)\partial_1+cos(\theta)\partial_2$$

so that

$$\partial_1'2=(cos(\theta)\partial_1-sin(\theta)\partial_2)^2=cos^2(\theta)\partial_1^2+sin^2(\theta)\partial_2^2-sin(2\theta)\partial_1\partial_2,$$

$$\partial_2'2=(sin(\theta)\partial_1+cos(\theta)\partial_2)^2=cos^2(\theta)\partial_2^2+sin^2(\theta)\partial_1^2+sin(2\theta)\partial_1\partial_2,$$

$$\partial_1'\partial_2'=(cos(\theta)\partial_1-sin(\theta)\partial_2)(sin(\theta)\partial_1+sin(\theta)\partial_2)$$
$$=cos^2(\theta)\partial_1^2-sin^2(\theta)\partial_2^2-sin(2\theta)\partial_1\partial_2$$

4.14 Non-Abelian gauge fields interacting with the photon, electron-positron and graviton fields

Let G be the gauge group. It is a subgroup of $U(n)$, If $n=1$ the gauge group is $U(1)$ and the associated gauge field is the electromagnetic field which taken along with the Dirac matter field has local U(1)-invariance. Consider now another matter field ψ interacting with the gauge field A_μ which is now a field on the space-time manifold but with values in the Lie algebra $\mathfrak{g} \subset \int u(n)$ of G). More precisely if $i\tau_a, a = 1, 2, ..., N$ are skew-Hermitian generators of the Lie algebra $\mathfrak{g}$ with structure constants $c(abc)$, ie,

$$[\tau_a, \tau_b] = -i.c(abc)\tau_c$$

then we can write with the usual summation convention of Einstein,

$$A_\mu(x) = A_\mu^a(x)\tau_a$$

Note that $A_\mu(x)dx^\mu$ defines a $\mathfrak{g}$ valued one form on the space-time manifold. Let B_μ be the photon field. Then, the total matter and gauge field Lagrangian in curved space-time is given by

$$L = Re(\chi^*(e_a^\mu \alpha^a(i\partial_\mu + eA_\mu + eB_\mu + i\Gamma_\mu) - \beta m)\psi\sqrt{-g}) - (1/4)(Tr(F_{\mu\nu}F^{\mu\nu})$$

$$-(1/4)H_{\mu\nu}H^{\mu\nu})\sqrt{-g}$$

where

$$F_{\mu\nu} = A_{\nu,\mu} - A_{\mu,\nu} + ie[A_\mu, A_\nu] = (ie)^{-1}[\partial_\mu + ieA_\mu, \partial_\nu + ieA_\nu]$$

$$= (A^a_{\nu,\mu} - A^a_{\mu,\nu})\tau_a + c(bca)A^b_\mu A^c_\nu \tau_a = F^a_{\mu\nu}\tau_a$$

$$H_{\mu\nu} = B_{\nu,\mu} - B_{\mu,\nu}$$

It should be noted that a sum such as $eA_\mu + eB_\mu + i\Gamma_\mu$ is to be interpreted as a direct sum of Lie algebra elements. $-iA_\mu$ is an element of the gauge Lie algebra $\mathfrak{g}$, $-iB_\mu$ is an element of the $U(1)$ Lie algebra $\mathfrak{u}(1)$ which is $i\mathbb{R}$ and Γ_μ is an element of the Lie algebra of the spinor representation of the Lorentz group and $-i(eA_\mu + eB_\mu + i\Gamma_\mu$ represents an element in the direct sum of these three Lie algebras. Correspondingly the matter fields ψ, χ are elments in the tensor product of the vector spaces $\mathbb{C}^N$, $\mathbb{C}$ and $C^4, orequivalentlyof C^N$ and $\mathbb{C}^4$. ψ, χ represent joint wave functions of the electrons, positrons and G-gauge field particles.

4.15 The Yang Mills field equations in the background gravitational field derived from the variational principle

$$\int d^4xTr[(\delta A_{\nu,\mu} - \delta A_{\mu,\nu})\sqrt{-g}F^{\mu\nu} + ie(\delta A_\mu.A_\nu + A_\mu.\delta A_\nu - A_\nu\delta A_\mu - \delta A_\nu A_\mu)\sqrt{-g}F^{\mu\nu}) = 0$$

which results on integration by parts and equating coefficients of δA_ν in

$$[F^{\mu\nu}\sqrt{-g}]_{,\nu} + ie[A_\nu, F^{\mu\nu}]\sqrt{-g} = 0$$

or equivalently, in the language of covariant derivatives,

$$F^{\mu\nu}_{:\nu} + ie[A_\nu, F^{\mu\nu}] = 0$$

Let us now quantize the above Yang-Mills field equations using the Hamiltonian formalism. The Lagrangian density is

$$L_Y = Tr(F^{\mu\nu}F_{\mu\nu}\sqrt{-g})$$

We adopt the gauge condition

$$\partial_\mu(A^{\mu N}\sqrt{-g}) = 0$$

Note that this gauge condition is a tensor equation, if it is valid in one coordinate system, it will be valid then in all coordinate systems. In order to incorporate this gauge condition, we modify our Lagrangian to

$$L_Y = (-1/4)Tr(F^{\mu\nu}F_{\mu\nu}\sqrt{-g}) + a.(\partial_\mu(A^{\mu N}\sqrt{-g}))^2$$

Note that a gauge transformation of the YM field is defined by the formula

$$\partial_\mu + ieA'_\mu(x) = g(x)(\partial_\mu + ieA_\mu(x))g(x)^{-1}$$

or equivalently,

$$A'_\mu(x) = g(x)A_\mu(x)g(x)^{-1} - (i/e)g(x)\partial_\mu(g(x)^{-1})$$
$$= g(x)A_\mu(x)g(x)^{-1} + (i/e)\partial_\mu(g(x)).g(x)^{-1}$$

In case the local gauge group transformation $g(x)$ is infinitesimal, we can write

$$g(x) = I + i\epsilon(x) = I + \epsilon^a(x)i\tau_a$$

where $\epsilon^a(x), a = 1, 2, ..., N$ are real valued functions of space-time and then we get upto first order of smallness,

$$g(x)A_\mu(x)g(x)^{-1} = A_\mu(x) + [i\epsilon(x), A_\mu(x)] = A^a_\mu(x)\tau_a + i[\epsilon^b(x)\tau_b, A^c_\mu(x)\tau_c]$$
$$= [A^a_\mu(x) + c(bca)\epsilon^b(x)A^c_\mu(x)]\tau_a$$

and

$$i(\partial_\mu g(x)).g(x)^{-1} = -\partial_\mu\epsilon(x) = -\partial_\mu\epsilon^a(x)\tau_a$$

so the infinitesimal gauge transformation of the YM gauge potential field is given by

$$A'_\mu(x) - A_\mu(x) = \delta A_\mu(x) = [i\epsilon(x), A_{mu}(x)] - e^{-1}\partial_\mu\epsilon(x)$$
$$= \delta A^a_\mu(x)\tau - a = [c(bca)\epsilon^b(x)A^c_\mu(x)] - \partial_\mu\epsilon^a(x)]\tau_a$$

so that

$$\delta A^a_\mu(x) = -\partial_\mu\epsilon^a(x) + c(bca)\epsilon^b(x)A_\mu +^c (x)$$

Note that a gauge transfomation has N degrees of freedom where N is the dimension of the gauge group, ie the dimension of the Lie algebra of the gauge group. Hence, by choosing an appropriate gauge transformation, we can ensure that the N conditions $\partial_\mu(A^{\mu a}\sqrt{-g}) = 0, a = 1, 2, ..., N$ are met. This suggests that we can add the term $C.\sum_{a=1}^N(\partial_\mu(A^{\mu a}\sqrt{-g}))^2$ to the Lagrangian density of the YM field.

4.16 The relationship between gauge anomalies for Yang-Mills field theories and the Atiyah-Singer index theorem, computing the approximate propagator for Yang-Mills gauge theories, Evaluating using the path integral, the exact propagator for both the matter and gauge fields in a non-Abelian gauge theory in the absence of gravity using properties of Gaussian Bosonic Fermionic Berezin integrals

First consider the case of flat space-time. The gauge field Lagrangian is

$$L_Y = (-1/4)F^a_{\mu\nu}F^{a\mu\nu}+c_0.\partial_\mu A^{a\mu})^2$$

$$= (-1/4)(A^a_{\nu,\mu}-A^a_{\mu,\nu}+eC(abc)A^b_\mu A^c_\nu)(A^{a\nu,\mu}-A^{a\mu,\nu}+eC(aef)A^{e\mu}A^{f\nu})$$

$$+c_0(\partial_\mu A^{a\mu})^2$$

$$= L_2 + eL_3 + e^2 L_4$$

where

$$L_2 = (-1/4)F_0^{a\mu\nu}F^a_{mu\nu} + c_0(\partial_\mu A^{a\mu})^2$$

is the quadratic term,

$$L_3 = (-1/2)C(aef)(A^a_{\nu,\mu} - A^a_{\mu,\nu})A^{e\mu}A^{f\nu} = -eC(aef)A^{a\nu,\mu}A^{e\mu}A^{f\nu}$$

is the cubic term, and

$$L_4 = (-1/4)C(abc)C(aef)A^b_\mu A^c_\nu A^{e\mu}A^{f\nu}$$

is the fourth degree term. The propagator is

$$D^{ab}_{\mu\nu}(x,y) =< 0|T(A^a_\mu(x)A^b_\nu(y))|0 >= \int exp(i\int L_Y(A^a_\mu(z), A^a_{\mu,\nu}(z))d^4z)A^a_\mu(x)A^b_\nu(y)DA$$

We approximate the propagator upto $O(e^4)$ as

$$\int exp(i\int L_2(z)d^4z)(1+e\int L_3(z)d^4z+e^2.((1/2)\int L_3(z_1)L_3(z_2)d^4z_1d^4z_2+\int L_4(z)d^4z)$$

$$+e^3((1/6)\int L_3(z_1)L_3(z_2)L_3(z_3)d^4z_1d^4z_2d^4z_3 + \int L_3(z_1)L_4(z_2)d^4z_1d^4z_2)$$

$$+e^4((1/24)\int L_3(z_1)L_3(z_2)L_3(z_3)L_3(z_4)d^4z_1d^4z_2d^4z_3d^4z_4)+(1/2)$$

$$\int L_3(z_1)L_3(z_2)L_4(z_3)d^4z_1d^4z_2d^4z_3$$

$$+(1/2)\int L_4(z_1)L_4(z_2)d^4z_1d^4z_2))A^a_\mu(x)A^b_\nu(y)DA$$

Evaluating this path integral involves using standard formulas for upto the fourteenth order moments of a multivariate Gaussian random vector in standard probability theory.

Evaluating using the path integral, the propagator for both the matter and gauge fields in a non-Abelian gauge theory in the absence of gravity. The Lagrangian density is

$$L = \psi^*[\alpha^\mu(i\partial_\mu + eA^a_\mu\tau_a) - m]\psi - (1/4)F^a_{\mu\nu}F^{a\mu\nu} = L_M + L_G + L_I$$

where L_M is the matter action:

$$L_M = \psi^*(\alpha^\mu]i\partial_\mu - m)\psi$$

L_G is the gauge Lagrangian:

$$L_G = (-1/4)F^a_{\mu\nu}F^{a\mu\nu}$$

and L_I is the interaction Largrangian between the matter and gauge fields:

$$L_I = e\psi^*\alpha^\mu\tau_a\psi.A^a_\mu$$

where by $\alpha^\mu\tau_a$, we actually mean $\alpha^\mu \otimes \tau_a$. Note that when the gauge group $G \subset U(n)$, then L is invariant under gauge transormations $g(x)$

$$\psi(x) \to g(x)\psi(x),$$

$$A_\mu(x) \to g(x)A_\mu(x)g(x)^{-1} + (i/e)g(x)(\partial_\mu g(x)^{-1})$$

or equivalently under infinitesimal gauge transformations $g(x) = I + i\epsilon(x)$ according to

$$\psi(x) \to \psi(x) + i\epsilon(x)\psi(x),$$

$$A_\mu(x) \to A_\mu(x) + i[\epsilon(x), A_\mu(x)] + (1/e)\partial_\mu\epsilon(x)$$

The exact matter field (Fermion) propagator is

$$S(x,y) =< T(\psi(x)\psi(y)^*) >= \int exp(i\int L(z)d^4z)\psi(x)\psi(y)^*D\psi.D\psi^*.DA$$

amd the exact gauge field propagator is

$$D^{a,b}_{\mu\nu}(x,y) = \int exp(i\int L(z)d^4z)A^a_\mu(x)A^b_\nu(y)^*D\psi.D\psi^*.DA$$

Note that since ψ is a Fermionic field, the path integral w.r.t it is a Berezin integral. Thus, if $\theta_1, ..., \theta_n$ are Fermionic/Grassmannian/Berezin parameters, they anticommute with each other and then we define

$$\int f(\theta_1, ..., \theta_n)d\theta_1...d\theta_n = f_n$$

where f_n is the coefficient of $\theta_1...\theta_n$ in the expansion of f. Note that any such f can be expressed as

$$f(\theta_1,...,\theta_n) = f_0 + \sum_{k=1}^{n} \sum_{1\leq i_1<i_2<...<i_k\leq n} f_k[i_1,...,i_k]\theta_{i_1}...\theta_{i_k}$$

Then we find for example the following evaluation of the Gaussian Berezin integral: Let A be a skewsymmetric matrix of size $n\times n$ where $n=2k$. Then,

$$\int exp(\theta^T A\theta)d\theta_1...d\theta_n = (n!)^{-1}\sum sgn(i_1,j_1,...,i_k,j_k)A(i_1,j_1...A(i_k,j_k) = det(A)$$

where the sum is over all partitions $\{(i_1,j_1),...,(i_k,j_k)\}$ of $\{1,2,...,n\}$ into k pairs taking the ordering of these pairs as well as the ordering within each pair into account. Equivalently, we can say that the above sum is over all permutations $i_1,j_1,...,i_k,j_k$ of $\{1,2,...,n\}$ with $sgn(i_1,j_1,...,i_k,j_k)$ denoting the signature of this permutation. Applying this to the above matter field path integral, we obtain

$$\int exp(i\int\int \psi(z)^*(\alpha^\mu(i\partial_\mu+eA_\mu)(z)-m)\psi(z)d^4z)D\psi.D\psi^*$$
$$= det(\alpha^\mu(i\partial_\mu+eA_\mu)-m)$$

where by this determinant, we mean the determinant of the kernel

$$K(x,y) = \alpha^\mu\otimes(i\partial_\mu\delta^4(x-y)I_N + eA_\mu(x)\delta^4(x-y)) - m.\delta^4(x-y)I_{4N}$$

Likewise, we have the following evaluation of the second moment of a Gaussian function in the Berezin context:

$$\int exp(\theta^T A\theta)\theta_1\theta_2 d\theta_1...d\theta_n = adj(A)(1,2) = det(A)A^{-1}(1,2)$$

Equivalently,

$$\frac{\int exp(\theta^T A\theta)\theta_1\theta_2 d^n\theta}{\int exp(\theta^T A\theta)d^n\theta} d^n\theta = A^{-1}(1,2)$$

which is analogous to the standard Bosonic Gaussian integral. Applying this to our above Lagrangian gives the exact matter field propagator as

$$S(x,y) = \frac{\int exp(i\int L(z)d^4z)\psi(x)\psi(y)^*D\psi.D\psi^*DA}{\int exp(i\int L(z)d^4z)\psi(x)\psi(y)^*D\psi.D\psi^*DA}$$

$$= \frac{\int det(\alpha^\mu(i\partial_\mu+eA_\mu)-m))[(\alpha^\mu(i\partial_\mu+eA_\mu)-m)^{-1}(x,y)]DA}{\int det(\alpha^\mu(i\partial_\mu+eA_\mu)-m))DA}$$

or equivalently, in the four momentum domain:

$$S(p,p') = \frac{\int det((\alpha^\mu q_\mu - m)\delta^4(q-q') + eA_\mu(q-q')))((\alpha^\mu p_\mu - m) + e\alpha^\mu\hat{A}_\mu)^{-1}(p,p')DA}{\int det((\alpha^\mu q_\mu - m)\delta^4(q-q') + eA_\mu(q-q'))DA}$$

where by $\hat{A}_\mu$, we denote the momentum space Kernel $A_\mu(p-p')$. Note that when $A=0$,

$$(\alpha^\mu p_\mu - m + e\alpha^\mu \hat{A}_\mu)^{-1}(p,p') = (\alpha^\mu p_\mu - m)^{-1}\delta^4(p-p')$$

and hence, in the special case when there is no gauge field A, the result reduces to the standard Dirac Fermion propagator:

$$S(p,p') = (\alpha^\mu p_\mu - m)^{-1}\delta^4(p-p')$$

4.17 A remark concerning the design of a quantum unitary gate to match a sequence of pure state pairs

Let $|\phi_k>, |\psi_k>, k=1,2,...,N$ be pure states in a finite dimensional Hilbert space $\mathcal{H}$. The aim is to design a unitary operator U in $\mathcal{H}$ so that

$$E(U) = \sum_{k=1}^{N} \| U|\psi_k> - |\phi_k>\|^2$$

is a minimum. To do so we can parametrize U as $exp(i(\theta_1 H_1 + ... + \theta_k H_k))$ where $H_1, ..., H_k$ is a basis for the space of all Hermitian operators in $\mathcal{H}$ and $\theta_1, ..., \theta_k$ are real numbers. Alternatively, we can parametrize U as $exp(i\theta_{H1})...exp(i\theta_k H_k)$. If we adopt the former parametrization and optimize using the gradient search algorithm, things become complicated as to differentiate $exp(i(\theta_1 H_1 + ... + \theta_k H_k))$ w.r.t θ_j we require to use the familiar formula for the differential of the exponential map in Lie algebra theory which is a rather complicated formula. On the other hand, if we use the second parametrization, things become less complicated. For example, writing down the spectral decomposition of H_m as

$$H_m = \sum_r c_m(r) P_m(r), m = 1,2,...,k$$

where $\{P_m(r), r = 1,2,...,n_m\}$ is an orthogonal spectral resolution of the identity, we get

$$exp(i\theta_m H_m) = \sum_r exp(ic_m(r)\theta_m) P_m(r)$$

and then

$$< v|U|u >= Re(< v|exp(i\theta_1 H_1)...exp(i\theta_k H_k)|u >)$$

$$= \sum_{r_1,...,r_k} Re(< v|P_1(r_1)...P_k(r_k)|u > exp(i(c_1(r_1)\theta_1 + ... + c_k(r_k)\theta_k)))$$

and this formula can be substituted into

$$E(U) = 2N - 2.\sum_k Re(< \phi_k|U|\psi_k >$$

to express the required optimization problem as

$$argmax_{\theta_1,...,\theta_k} \sum_{s,r_1,...,r_k} Re(< \phi_k|P_1(r_1)...P_k(r_k)|\psi_k > exp(i(c_1(r_1)\theta_1 + ... + c_k(r_k)\theta_k))))$$

Chapter 5

Chiral Anomalies and Atiyah-Singer Index Theorem Applications

5.1 Problems in communication systems:Calculating the spectral and cross spectral densities of the in-phase and quadrature components in the complex lowpass representation of a stationary process with bandpass power spectral density

Consider a real noise signa $n(t)$ that is stationary with correlation $R_n(\tau) = \mathbb{E}(n(t+\tau)n(t))$ having spectral density $S_n(\omega) = \int_{\mathbb{R}} R_n(\tau)exp(-j\omega\tau)d\tau$ such that $S_n(\omega)$ is concentrated on the band $[\omega_c - \sigma, \omega_c + \sigma] \cup [-\omega_c - \sigma, -\omega_c + \sigma]$ where $\omega_c > 2\sigma$. Let $\hat{n}(t)$ denote the Hilbert transform of $n(t)$:

$$\hat{n}(t) = (1/\pi t) * n(t)$$

Note that the Fourier transform of $1/\pi t$ is $-j.sgn(\omega)$. Show that

$$S_{\hat{n}n}(\omega) = \mathcal{F}(R_{\hat{n}n})(\omega = -jsgn(\omega).S_n(\omega),$$

$$S_{n\hat{n}}(\omega = jsgn(\omega))S_n(\omega), S_{\hat{n}}(\omega) = S_n(\omega)$$

Deduce that if $x(t) = n(t) + j\hat{n}(t)$, then $x(t)$ can be obtained by passing $n(t)$ through an LTI system with transfer function $2U(\omega)$ where $U(\omega)$ is the unit step function.

$$Hence, with$$

$\mathrm{R}_x(\tau) = \mathbb{E}(x(t+\tau)x(t)^*)$ and $S_x(\omega)$ its Fourier transform, we have

$$S_x(\omega) = \mathcal{F}(R_x(\tau)) = 4U(\omega)S_n(\omega)$$

is concentrated over the band $[\omega_c - \sigma, \omega_c + \sigma]$ having no negative frequency components. Define the complex process

$$s(t) = (n(t) + j\hat{n}(t))exp(-j\omega_c t) = x(t).exp(-j\omega_c t)$$

Show that

$$R_s(\tau) = \mathbb{E}(s(t+\tau).s(t)^*) = R_x(\tau).exp(-j\omega_c\tau)$$

-

$$S_s(\omega) = S_x(\omega + \omega_c) = 4U(\omega + \omega_c)S_n(\omega + \omega_c)$$

is concentrated over the lowpass band $[-\sigma, \sigma]$. Show that

$$n(t) = Re(s(t).exp(j\omega_c t)$$

This is called the complex lowpass representation of a stationary bandpass stochastic process. Show that

DOI: 10.1201/9781003728627-5

$$n(t) = s_1(t).cos(\omega_c t) - s_2(t).sin(\omega_c t)$$

where

$s_1(t) = Re(s(t)) = m(t).cos(\omega_c t + \hat{m}(t).sin(\omega_c t), s_2(t) = -m(t)sin(\omega_c t + \hat{m}(t)cos($

$\omega_c t) Expressthespectraoftherealllowpasssignals s_1(t), s_2(t)$ as well as their cross spectra in terms of $S_n(\omega)$. Evaluate the total power in these signals by integrating their power spectra over their concentration band.

5.2 Gauge fixing for path integrals of matter and gauge fields. To show that the path integral is essentially independent of the gauge fixing functional

Let ϕ be a set of fields and Λ an element of the gauge group. Under the gauge transformation $\phi \to \phi_\Lambda$, we assume that the action $I(\phi)$ for the fields is invariant along with the path measure $D\phi$. In particular, we assume that $exp(iI(\phi))D\phi$ is invariant, ie

$$exp(iI(\phi_{)D\phi_\Lambda = exp(iI(\phi))D\phi}$$

Consider the path integral for these fields given by

$$P = \int exp(iI(\phi))B[f(\phi)]F(\phi)D\phi$$

where $f(\phi)$ is a gauge fixing functional and

$$F(\phi) = (df(\phi_\lambda)/d\lambda)|_{\lambda=e}$$

We wish to prove the independence of the path integral P w.r.t the choice of the gauge fixing functional f. To this end, consider changing the path integration variable from ϕ to ϕ_Λ where Λ is an elenent of the gauge group. Since the path integral should not change under a change of the integqation variable, it must follow that

$$P = \int exp(iI(\phi_\Lambda))B[f(\phi_\Lambda)]F(\phi_\Lambda)D\phi_\Lambda$$

$$= \int exp(iI(\phi))B[f(\phi_\Lambda)]F(\phi_\Lambda)D\phi$$

where in the second step, we've used gauge invariance. Now,

$$F(\phi_\Lambda) = det[(df(\phi_{\Lambda o \lambda})/d\lambda)|_{\lambda=e}]$$

$$= det[(df(\phi_\Lambda)/d\Lambda)]det[(d\Lambda o\lambda/d\lambda)|_{\lambda=e}]$$

as infinite dimensional Jacobian determinants. Thus, denoting the left invariant Haar density on the gauge group by $\rho(\Lambda)$, we have

$$\rho(\Lambda) = [det[d\Lambda o\lambda/d\lambda]|_{\lambda=e}]^{-1}$$

and hence

$$F(\phi_\Lambda) = \rho(\Lambda)^{-1}.det[(df(\phi_\Lambda)/d\Lambda)]$$

Then, since P is independent of Λ and the Haar measure is assumed to be normalized, we get

$$P = \int P\rho(\Lambda)D\Lambda = \int exp(iI(\phi))B[f(\phi_\Lambda)]det[(df(\phi_\Lambda)/d\Lambda)]D\Lambda.D\phi$$

Noting that we can write

$$\int B[f(\phi_\Lambda)]det[(df(\phi_\Lambda)/d\Lambda)]D\Lambda = \int B[f]Df = B_0$$

a constant assuming that for any fixed field ϕ, $\Lambda \rightarrow f(\phi_\Lambda)$ is surjective in the sense that its range covers possible values gauge fixing functionals evaluated all field configurations, it follows that

$$P = B_0 \int exp(i\phi))D\phi$$

proving that P does not depend upon the gauge fixing functional $f(\phi)$ although it depends upon the function B via the multiplicative constant $B_0 = \int B[f]Df$.

5.3 The relationship of the Atiyah-Singer index theorem to gauge anomalies in non-Abelian quantum field theory and to the existence of a non-zero value of the difference between the number of left Chiral and right Chiral Fermions

Consider the action $I(\psi, \bar{\psi})$ for a non-Abelian Fermionic matter field ψ. The associated path integral is

$$P = \int exp(iI(\psi, \bar{\psi}))D\psi.D\bar{\psi}$$

where

$$\bar{\psi} = \psi^*\gamma^0$$

Here, we are assuming I to be a gauge invariant action functional of the matter fields, ie, I also contains the non-Abelian gauge field $A_\mu(x)$ and under a gauge transformation of the form $\psi(x) \rightarrow exp(i\alpha(x)T)$ where iT is an element of the Lie algebra of the gauge group, while simultaneously transforming the gauge field A_μ according to the corresponding adjoint representation of the gauge group, or more precisely, $\partial_\mu + ieA_\mu(x) \rightarrow g(x)(\partial_\mu + ieA_\mu(x))g(x)^{-1}$ where $g(x) = exp(i\alpha(x)T)$, the action functional I remains invariant. However, we shall see that the path integral P does not remain invariant under a Chiral gauge transformation even though I remains owing to the fact that the path measure $D\psi.D\bar{\psi}$ does not remain invariant under such a transformation. Therefore, an extra term has to be added to the Gauge Lagrangian to cancel out this anomaly resulting for lack of Chiral gauge invariance of the path integral.

Under an infinitesimal Chiral gauge transformation of the matter field defined by the matrix $exp(i\alpha(x)T\gamma^5) = I + i\alpha(x)T\gamma^5$ where T is a Hermitian matrix, so that iT belongs to the Lie algebra of the gauge group $G \subset U(N)$ and $\alpha(x)$ is a real valued function of the space-time coordinate x and $T\gamma^5$ is to be interpreted as $T \otimes \gamma^5$, the Chiral matter fields transform as $\psi(x) \to \psi(x) + i\alpha(x)T\gamma^5\psi(x)$ and $\bar{\psi}(x) \to \bar{\psi}(x) - i\psi(x)^*\alpha(x)T\gamma^5\gamma^0 = \bar{\psi}(x) + i\bar{\psi}(x)T\gamma^5$ because γ^5 anticommutes with γ^0. Thus, the path measure $D\psi.D\bar{\psi}$ transforms to $(1 + 2iTr(T\gamma^5K))D\psi.D\bar{\psi}$ where K is the kernel $K(x,y) = \delta^4(x-y)$. The infinitesimal change in P resulting from such an infinitesimal Chiral gauge transformation is therefore

$$\delta P = 2i \int \alpha(x).Tr(T\gamma^5 K(x,x))d^4 5x$$

which does not make senst because $K(x,x) = \delta^4(x-x) = \delta^4(0)$ is infinity. However, we can modify this computation of the change in the path integral in a gauge invariant way by interpreting the kernel $K(x,y)$ appearing in the integral $\int \alpha(x)K(x,x)d^4x$ as the limit as $M \to \infty$ of $\int f((\gamma^\mu D_\mu)^2/M^2)exp(ik.(x-y))d^4k/(2\pi)^4$ where $f(w)$ is a twice differentiable function satisfying $f(0) = 1$ and $D_\mu = \partial_\mu + ieA_\mu$ is the gauge covariant derivative. Noting that

$$D^2 = (\gamma^\mu D_\mu)^2 = \gamma^\mu\gamma^\nu D_\mu D_\nu = (1/4)[\gamma^\mu,\gamma^\nu][D_\mu,D_\nu]+(1/4)[\gamma^\mu,\gamma^\nu]_+[D_\mu,D_\nu]_+ =$$

$$= (1/4)[\gamma^\mu,\gamma^\nu](ieF_{\mu\nu}) + (1/2)\eta_{\mu\nu}[D_\mu,D_\nu]_+(x)$$

$$= (1/4)[\gamma^\mu,\gamma^\nu](ieF_{\mu\nu}) + \eta_{\mu\nu}D_\mu D_\nu$$

$$= (1/4)[\gamma^\mu,\gamma^\nu](ieF_{\mu\nu}) + \Delta$$

where $\Delta = D_\mu D^\mu$ is the Laplacian of the gauge covariant derivative. Thus $\int \alpha(x)Tr(T\gamma^5K(x,x))d^4x$ gets replaced by

$$\int (f(0)Tr(\gamma^5 T)\alpha(x)exp(ik.(x-x)))d^4kd^4x+$$

$$(f'(0)/M^2)\int \alpha(x)Tr(\gamma^5)Tr(T\Delta)exp(ik.(x-x))d^4kd^4x$$

$$-(e^2f''(0)/16M^4)\int \alpha(x)Tr(\gamma^5[\gamma^\mu,\gamma^\nu].[\gamma^\rho,\gamma^\sigma]).Tr(TT_aT_b)(F^a_{\mu\nu}(x)F^b_{\rho\sigma)}(x)d^4x$$

The first two terms vanish because $Tr(\gamma^5T) = Tr(\gamma^5)Tr(T)$ and $Tr(\gamma^5) = 0$. Multiplying by M^4 and taking the limit $M \to \infty$ gives the result that this integral is proportional to

$$\int \alpha(x)\epsilon(\mu\nu\rho\sigma)Tr(F_{\mu\nu}(x)F_{\rho\sigma(x)})d^4x = \int \alpha(x)F(x)\wedge F(x)d^4x$$

Taking $\alpha(x) = \alpha$ as an infinitesimal real constant, this expression is simply proportional to $\int Tr(F\wedge F)d^4x$ which is proportional to the index of the Dirac-Yang-Mills operator according to the Atiyah-Singer index theorem. This index is precisely equal to the difference between the dimensions of the zero eigenvalue subspace of the Dirac operator $D = \gamma^\mu D_\mu$ having eigenvalue $+1$ for γ^5 and that having eigenvalue -1 for γ^5. In other words, the gauge anomaly for the path integral is closely related to the Chiral Fermion asymmetry, ie, the difference between the number of zero mass Dirac Fermions having eigenvalues $+1$ and -1 for γ^5 or equivalently the difference between the number of zero mass Dirac Fermions that have right handed and left handed chirality. Note that since $\gamma^\mu, \mu = 0, 1, 2, 3$ anticommute with γ^5, it follows that D anitcommutes with γ^5 and hence $D\psi = 0$ implies $D\gamma^5\psi = 0$ which implies that $D[(1 \pm \gamma^5)/2)\psi] = 0$. Noting that $(1 + \gamma^5)/2$ is the orthogonal projection onto the eigensubspace of γ^5 with eigenvalue $+1$ while $(1 - \gamma^5)/2$ is the orthogonal projection onto the eigensubspace of γ^5 with eigenvalue -1. Thus, the zero mass Dirac operator has a complete set of solutions each of which is either right or left Chiral.

Let now ϕ be a solution to $D\phi = \lambda\phi$. Then

$$D\gamma^5\phi = -\gamma^5 D\phi = -\lambda\gamma^5\phi$$

$$ie, if$$

ϕ is an eigen-function of the Dirac operator D with eigenvalue λ, then $\gamma^5\phi$ is an eigenfunction of the Dirac operator D with eigenvalue $-\lambda$. We can therfore write down the spectral decomposition of the identity as

$$\delta(x-y)I = \sum_k \phi_k(x)\phi_k(y)^* + \sum_k \phi_{-k}(x)\phi_{-k}(y)^* + \sum_m \phi_m(x)\phi_m(y)^*$$

where ϕ_k ranges over the eigenfunctions of D with non-zero eigenvalue λ_k and ϕ_{-k} ranges over the eigenfunctions of D with eigenvalue $-\lambda_k$, more specifically, $\phi_{-k} = \gamma^5\phi_k.\phi_m$ ranges over an orthogonomal basis of eigenfunctions of D with zero eigenvalue. Now if $D\phi = 0$, then $D((1 + \gamma^5)\phi/2 = 0$ and also $D((1 - \gamma^5)\phi/2) = 0$ and since $(1 + \gamma^5)\phi/2$ has eigenvalue $+1$ for γ^5 while $(1 - \gamma^5)\phi/2$ has eigenvalue -1 for γ^5, it follows that without loss of generality we can write

$$\sum_m \phi_m(x)\phi_m(y)^* = \sum_p \phi_p(x)\phi_p(y)^* + \sum_p \phi_{-p}(x)\phi_{-p}(y)^*$$

where ϕ_p has eigenvalue $+1$ for γ^5 while ϕ_{-p} has eigenvalue -1 for γ^5. Note that if $D\phi = \lambda\phi$ with $\lambda \neq 0$, then $D((1 + \gamma^5)\phi/2) = \lambda(1 - \gamma^5)\phi/2$ and $D((1 - \gamma^5)\phi/2) = \lambda(1+\gamma^5)\phi/2$, ie, we do not have Chiral solutions to the Dirac operator for non-zero eigenvalues.

Thus we obtain the following expression for the gauge anomaly when $T = 1, \alpha(x) = \alpha$ as

$$\alpha.Tr(\gamma^5.I) = \alpha.Tr(\gamma^5(\sum_k(|\phi_k><\phi_k|+|\phi_{-k}><\phi_{-k}|)+\sum_p(|\phi_p><\phi_p|)+\sum_p|\phi_{-p}><\phi_{-p}|))$$

$$= \sum_k <\phi_k|\gamma^5|\phi_k> + <\phi_{-k}|\gamma^5|\phi_{-k}>)+\sum_p(<\phi_p|\gamma^5|\phi_p>)+\sum_p(<\phi_{-p}|\gamma^5|\phi_{-p}>)$$

$$= n_+ - n_-$$

where

$$n_+ = \sum_p < \phi_p|\phi_p >, n_- = \sum_p < \phi_{-p}|\phi_{-p} >$$

are respectively the number of zero modes of the Dirac operator having eigenvalue +1 and −1. Note that herein we have used the fact that $< \phi_k|\gamma^5|\phi_k >= 0$ since $|\phi_k >$ and $\gamma^5|\phi_k >$ are respectively zero modes of the Dirac operator with different eigenvalues, namely λ_k and $-\lambda_k$ with $\lambda_k = 0$ and likewise $< \phi_{-k}|\gamma^5|\phi_{-k} >= 0$. Note that the zero modes of the Dirac operator need not be paired. For example, if $D\phi = 0, wehave$D$(1+\gamma^5)\phi/2 = 0$ and $D(1-\gamma^5)\phi/2 = 0$ but it is possible that one of $(1+\gamma^5)\phi/2$ and $(1-\gamma^5)\phi/2$ can be zero. This is not true for the non-zero modes since $D\phi = \lambda\phi$ with $\lambda \neq 0$ implies $D(1+\gamma^5)\phi/2 = \lambda(1-\gamma^5)\phi/2)$ and hence if $(1+\gamma^5)\phi/2 = 0$, then it would follow that $(1-\gamma^5)\phi/2 = 0$ and vice versa. Thus we have established a remarkable relationship between the index of the Dirac-Yang-Mills operator, Chiral asymmetry for Fermions and gauge anomalies while simultaneously establshing the Atiyah-Singer index theorem for the Dirac-Yang-Mills operator in flat space-time.

Under an infinitesimal gauge transformation defined by $\alpha(x), T, \gamma^5$, the action $I(\psi, \bar{\psi})$ must change by a quantity of the form $\int J^\mu(x)\partial_\mu\alpha(x)d^4x$ because we are assuming that I is invariant under a global gauge transfomation, ie, when $\alpha(x) = \alpha$ is a constant. Now the local infinitesimal Chiral gauge transformation described above changes the path measure $D\psi.D\bar{\psi}$ by $2iTr(TT_aT_b)\int \alpha(x)\epsilon(\mu\nu\rho\sigma)F^a_{\mu\nu}F^b_{\rho\sigma}d^4x$ which evaluates when

$T = I$ to $2i\int \alpha(x)\epsilon(\mu\nu\rho\sigma)F^a_{\mu\nu}F^b_{\rho\sigma}d^4x$.

Under such a transformation, as remarked above, I changes by $\int J^\mu(x)\partial_\mu\alpha(x)d^4x$. Since the total path integral P must remain invariant under any transformation of the wavefunction, it must therefore follow that

$$i\int J^\mu(x)\partial_\mu\alpha(x)d^4x + 2iTr(TT_aT_b)\int \alpha(x)\epsilon(\mu\nu\rho\sigma)F^a_{\mu\nu}F^b_{\rho\sigma}d^4x = 0$$

and hence the anomalous current J^μ is not conserved but rather satisfies

$$\partial_\mu J^\mu(x) = 2.Tr(TT_aT_b)\epsilon(\mu\nu\rho\sigma)F^a_{\mu\nu}(x)F^b_{\rho\sigma}(x)$$

which in the special case when $T = I$ and $Tr(T_aT_b) = N\delta_{ab}$ for the Lie algebra generators, reduces to

$$\partial_\mu J^\mu(x) = 2N.\epsilon(\mu\nu\rho\sigma)F^a_{\mu\nu}(x)F^b_{\rho\sigma}(x) = 2N.Tr(F(x)\wedge F(x))$$

As a product of operators, we have with $D_\mu = \partial_\mu + ieA_\mu$

$$\gamma^\mu D_{mu}exp(ik.(x-y)) = exp(ik.(x-y))[i\gamma.k + \gamma^\mu D_\mu]$$

and hence as operators,

$$f((\gamma^\mu D_\mu)^2/M^2)exp(ik.(x-y)) = exp(ik.(x-y))f((i\gamma.k + \gamma^\mu D_\mu)^2/M^2)$$

Thus, as functions,

$$f((\gamma.D)^2/M^2)(exp(ik.(x-y))) = exp(ik.(x-y))f((i\gamma.k+\gamma.D)^2/M^2)(1)$$

Now, it follows that

$$Tr(\alpha(x)\gamma^5 Tf((\gamma.D)^2/M^2))$$

$$= \int \alpha(x)Tr(\gamma^5 Tf((\gamma/D)^2/M^2).(exp(ik.(x-y)))|_{y=x}d^4kd^4x$$

$$= \int \alpha(x)Tr(\gamma^5 Tf((i\gamma.k+\gamma^\mu D_\mu)^2/M^2)(1))d^4kd^4x$$

$$= M^4 \int \alpha(x))Tr(\gamma^5 T.f((i\gamma.k+\gamma.D/M)^2))(1))d^4kd^4x$$

But

$$f((i\gamma.k+\gamma.D/M)^2)) = f(-k^2+i[\gamma.k,\gamma.D]_+/M+(\gamma.D)^2/M^2)$$

$$= f(-k^2)+f'(-k^2)(i[\gamma.k,\gamma.D]_+/M$$

$$+(\gamma.D)^2/M^2)+(f''(-k^2)/2)(i[\gamma.k,\gamma.D]_+/M+(\gamma.D)^2/M^2)^2$$

$$= f(-k^2)+(f''(-k^2)/2)(\gamma.D)^4/M^4 = (f''(-k^2)/2)(\gamma.D)^4/M^4$$

where we have neglected the higher terms in the Taylor series expansion which are either $O(1/M^5)$ or else do not contribute to the trace over product of the γ's after operating on 1 and then multiplying by $\gamma^5 T$. Note that

$$((\gamma.k)(\gamma.D))^2 = (\gamma^\mu\gamma^\nu k_\mu D_\nu)^2 = (1/16)([\gamma^\mu,\gamma^\nu](k_\mu D_\nu - k_\nu D_\mu))^2$$

where we neglect the terms involving symmetric combinations $[\gamma^\mu,\gamma^\nu]_+ = 2\eta_{\mu\nu}$ because they do not contribute to the trace after multiplying by $\gamma^5 T$ since $Tr(\gamma^5)=0$ and the trace of the product of three Dirac matrices is zero. Thus

$$Tr(\gamma^5.((\gamma.k)(gamma.D))^2)$$

$$= (1/16)Tr(\gamma^5[\gamma^\mu,\gamma^\nu][\gamma^\rho,\gamma^\sigma])Tr(T.(k_\mu D_\nu - k_\nu D_\mu).(k_\rho D_\sigma - k_\sigma D_\rho))$$

$$= \epsilon(\mu\nu\rho\sigma).Tr(T.(k_\mu D_\nu - k_\nu D_\mu).(k_\rho D_\sigma - k_\sigma D_\rho))$$

Consider a typical term in this expansion for example

$$\epsilon(\mu\nu\rho\sigma)k_\mu k_\rho D_\nu D_\sigma)$$

This clearly evaluates to zero since $k_\mu k_\rho$ is symmetric in (μ,ρ) while $\epsilon(\mu\nu\rho\sigma)$ is antisymmetric. Proceeding in this way, we get that

$$Tr([\gamma.k,\gamma.D]_+)^2(1) = 0$$

Thus finally the anomaly in the path integral evaluates to

$$\int \alpha(x)(f''(-k^2)/2)Tr(\gamma^5 T(\gamma.D)^4(1))d^4kd^4x$$

Now, writingo

$$C = \int f''(-k^2)d^4k/2$$

this becomes

$$C.\int \alpha(x)Tr(\gamma^5 T(\gamma.D)^4(1))d^4x = C.Tr(\gamma^5\gamma^\mu\gamma^\nu\gamma^\rho\gamma^\sigma)Tr(D_\mu D_\nu E_\rho D_\sigma(1))$$
$$= C.\epsilon(\mu\nu\rho\sigma)Tr(D_\mu D_\nu D_\rho D_\sigma(1))$$
$$= (C/16)\epsilon(\mu\nu\rho\sigma)Tr([D_\mu, D_\nu][D_\rho, D_\sigma])) = (C/16)\epsilon(\mu\nu\rho\sigma)Tr(F_{\mu\nu}F_{\rho\sigma})$$

Defining the two form fields

$$F = F_{\mu\nu}dx^\mu \wedge dx^\nu$$

we can thus express the gauge anomaly as

$$i\int \alpha(x)\epsilon(\mu\nu\rho\sigma))Tr(F_{\mu\nu}(x)F_{\rho\sigma}(x)d^4x$$
$$= i\int \alpha(x)Tr(F \wedge F)$$

because,

$$dx^\mu \wedge dx^\nu \wedge dx^\rho \wedge dx^\sigma = \epsilon(\mu\nu\rho\sigma)dx^0 \wedge dx^1 \wedge dx^2 \wedge dx^3$$

It is clear that when $\alpha(x) = 1$, we obtain that this gauge anomaly is precisely the index of the Dirac operator:

$$n_+ - n_- = C\int Tr(F \wedge F)$$

5.4 The Atiyah-Singer theorem in a curved space-time. Computing the square of the Dirac operator

In this case, the relevant Dirac-Yang-Mills operator is

$$\gamma^\mu(x)\nabla_\mu, \nabla_\mu = \partial_\mu + \Gamma_\mu + ieA_\mu, \Gamma_\mu = \Gamma_{\mu ab}[\gamma^a, \gamma^b]/4, A_\mu = A_\mu^a\tau_a$$

The curvature of the covariant derivative ∇_μ is given by the sum of the curvatures of the gravitational part and the Yang-Mills part because the Dirac Gamma matrices that define the spinor representation of the Lorentz group commute with the Yang-Mills gauge Lie algebra generators ($[\Gamma_\mu, \gamma^a] = 0$) . Note that the gravitational curvature tensor is the Riemann curvature tensor expressed in the spinor representation:

$$[\nabla_\mu, \nabla_\nu] = [\Gamma_{\nu,\mu} - \Gamma_{\mu,\nu} + [\Gamma_\mu, \Gamma_\nu]] + ie[A_{\nu,\mu} - A_{\mu,\nu} + ie[A_\mu, A_\nu]]$$
$$= R_{\mu\nu} + K_{\mu\nu}$$

Note that

$$\gamma^\mu(x) = \gamma^a e_a^\mu(x)$$

so that

$$[\gamma^\mu(x), \gamma^\nu(x)]_+ = [\gamma^a, \gamma^b]_+ e_a^\mu(x)e_b^\nu(x) == 2\eta^{ab}e_a^\mu(x)e_b^\nu(x) = 2g^{\mu\nu}(x)$$

Then, the square of the Dirac operator evaluates to

$$D^2 = (\gamma^\mu(x)\nabla_\mu)^2 = \gamma^\mu(x)\nabla_\mu\gamma^\nu(x)\nabla_\nu$$

$$= \gamma^\mu(x)([\nabla_\mu, \gamma^\nu(x)] + \gamma^\nu(x)\nabla_\mu)\nabla_\nu$$
$$= \gamma^\mu(x)\gamma^\nu(x)\nabla_\mu\nabla_\nu + \gamma^\mu(x)(\nabla_\mu\gamma^\nu(x))\nabla_\nu$$

This is the first step in the proof of the Atiyah-Singer index theorem. To proceed further, we must consider the two block diagonal entries of the square of the Dirac operator. These are of the form $D_1^* D_1$ and $D_1 D_1^*$. Using the Clifford algebra identities

$$\{\gamma^\mu, \gamma^\nu\} = 2\eta^{\mu\nu}$$

we can express

$$\gamma^\mu(x)\gamma^\nu(x)\nabla_\mu\nabla_\nu = \eta^{\mu\nu}\nabla_\mu\nabla_\nu + (1/2)[\gamma^\mu, \gamma^\nu]R_{\mu\nu}$$

where

$$R_{\mu\nu} = [\nabla_\mu, \nabla_\nu]$$

is the sum of the curvature tensors of the gravitational spinor connection and that of the Yang-Mills connection. The index of the Dirac operator equals the difference of the nullities of these two operators $D_1^* D_1$ and $D_1 D_1^*$. Clearly, this is also the same as the difference between the dimensions of the zeros of the Dirac operator corresponding to eigenvalues ± 1 for γ^5. Since the non-zero eigenvalues of $D_1^* D_1$ and $D_1 D_1^*$ are the same inclusive of multiplicities, it follows that the above index is also the same as

$$Ind(D) = Tr(exp(-tD_1^* D_1) - exp(-tD_1 D_1^*))$$

for any $t > 0$. This is called the supertrace of $exp(-tD^2)$. It can be evaluated using (a) Mehler's formula for the evolution kernel of a multidimensional quantum harmonic oscillator, and using Getzler's scaling method which transforms the Dirac matrices into a scaled sum of Fermionic creation and annihilation operators acting on the space of skewsymmetric tensors. The above supertrace in the limit of $t \to 0$ then becomes the integral of a differential form on the space-time manifold with the differential forms being constructed out of the Riemann and Yang-Mills curvature forms. These forms are called the Todd class and the Chern class.

5.5 The index of an elliptic complexes, especially the de-Rham complex. The isomorphism between harmonic forms and closed forms modulo exact forms. Finds application to the Chiral asymmetry problem

Let E be a vector bundle with base manifold M and fibre space V. Consider the Dirac operator $D = \gamma.\nabla$ where $\nabla = d + A$ is a covariant derivative A being a V valued function on M. We assume that V is a Lie algebra.

For example, V can be the direct sum of the Lie algebra of a Lie subgroup $G \subset U(n)$ and the Lie algebra of the spinor representation of the Lorentz group. Such a covariant derivative describes Dirac Fermions interacting with a gravitational field and a Yang-Mills non-Abelian gauge field. In terms of coordinates $\nabla_\mu = \partial_\mu + A_\mu(x)$. Let $\Lambda^k(M,V)$ denote the space of all alternating k-forms on M with values in V. Let $H^k(M,V)$ denote the space of all harmonic elements of $\Lambda^k(M,V)$, ie all elements $\omega \in H^k(M,V)$ for which $(d^*d + dd^*)\omega = 0$. This equation is equivalent to saying that $d\omega = 0, d^*\omega = 0$. Suppose that $\omega \in \Omega^k$ is arbitrary. Then, we can find a $\theta \in \Omega^{k-1}$ such that $< \omega - d\theta, \omega - d\theta >$ is a minimum. Clearly, this minimum is attained iff $d^*(\omega - d\theta) = 0$ or equivalently, iff $d^*d\theta = d\omega$. Further, if $\omega \in \Omega^k$ is closed, ie, $d\omega = 0$, then $d(\omega - d\theta = 0)$. Thus, we have a one-one correspondence between all the closed elements $\omega \in \Omega^k$ and all closed elements $\phi = \omega - d\theta \in \Omega^k$ that satisfy $d^*\phi = 0$. Now $d\phi = 0, d^*\phi = 0$ hold iff $(d^*d + dd^*)\phi = 0$, ie We also note that $\omega - \phi = d\theta$ is exact, ie, a total differential. Thus, we obtain a one-one correspondence between the closed k forms on M modulo the exact k-forms and Harmonic k-forms on M with values in V. Note that $d\Lambda^{k-1}(M,V) subset \Lambda^k(M,V)$. Thus, we have a vector space-isomorphism between $\Lambda^k(M,V)/d\Lambda^{k-1}(M,V)$ and $H^k(M,V)/EH^k(M,V)$ where $EH^k(M,V)$ is the vector space of exact harmonic k-forms. Note that the equation $d^*d\theta = d^*\omega$ implies $d^*d\Delta\theta = d^*\Delta\omega$ where $\Delta = d^*d + dd^*$ is the Laplacian on $\Lambda(T,V) = \bigoplus_{k\geq 0} \Lambda^k(TM,V)$. In particular, $\Delta\omega = 0$ implies $d^*d\Delta\theta = 0$ implies $d\Delta\theta = 0$ implies $\Delta d\theta = 0$ implies $d\theta$ is an exact harmonic form. Now consider the sequence of mappings $d_{k-1} = d : H^{k-1}(M,V) \to H^k(M,V), k = 1,2,....$ Clearly the nullspace of $d_k : H^k(M,V) \to H^{k+1}(M,V)$ is the space of all closed harmonic forms of order k which is the whole of $H^k(M,V)$, ie, $d_k = 0$ for all k. On the other hand, consider $d_{k-1} : \Lambda^{k-1}(M,V) \to \Lambda^k(M,V)$. The null-space of d_k is the space of all closed forms in $\Lambda^k(M,V)$ while the range of d_{k-1} is the space of all the exact forms in $\Lambda^k(M,V)$. Since there can exist closed forms that are not exact, it follows that the sequence $d_{k-1} : \Lambda^{k-1}(M,V) \to \Lambda^k(M,V), k = 1,2,...$ is not exact. From this discussion, it is clear that the degree of non-exactness of this sequence is thus measured by the dimensions of $H^k(M,V)/EH^k(M,V), k = 1,2,....$ Note that we have seen that the equivalence class of closed-k-forms modulo exact k-forms is the same as the equivalence class of harmonic-k-forms modulo exact harmonic k-forms. Now we shall see that $EH^k(M,V) = 0$, ie, an exact harmonic form is zero. This would establish a vector space isomorphism between the space $H^k(M,V)$ of all harmonic k-forms and the class of closed k-forms modulo the exact forms. To see this, suppose ω is an exact k-form that is harmonic, ie, $\omega = d\theta$ and $\Delta\omega = 0$. Then, it follows that $0 = (d^*d + dd^*)d\theta = dd^*d\theta$ and this implies $d^*d\theta = 0$ which in turn implies $d\theta = 0$, ie, $\omega = 0$. This proves our claim. Note that we have used the fact that if T is any linear operator, then $\mathcal{N}(T^*T) = \mathcal{N}(T)$ and $\mathcal{N}(TT^*) = \mathcal{N}(T^*)$

5.6 Representation of frequency modulated waves for stationary random processes

The baseband signal consists of a discrete or continuous superposition of sinusoids. When a carrier sinusoid is FM modulated with this signal, we express the resulting modulated wave in complex exponential form and then the representation problem amounts to writing down the Fourier series of the complex exponential of a sinusoid and then multiplying all the resulting Fourier series representation. The Fourier series coefficients for each component are Bessel functions and this representation finds application in calculation of the approximate bandwith of the FM signal when the modulating wave is an arbitrary stationary process by representing this process as a stochastic Fourier integral.

Representation of frequency modulated waves for stationary random processes. Consider a random process $m(t)$ that is stationary with spectral representation

$$m(t) = \int exp(j\omega t)dZ(\omega), \mathbb{E}(dZ(\omega)dZ(\omega)'*) = S(\omega)\delta(\omega - \omega')d\omega.d\omega'$$

Then, we have

$$\int_0^t m(s)ds = \int (exp(j\omega t) - 1)dZ(\omega)/j\omega) = c + \int exp(j\omega t)dZ(\omega)/j\omega$$

where c is a constant. Discretizing the frequency integral, we can write

$$\int_0^t m(s)ds = c + \sum_k exp(j\omega_k t)Z_k$$

where

$$Z_k = \Delta Z(\omega_k)/j\omega_k, \Delta Z(\omega_k) = Z(\omega_{k+1}) - Z(\omega_k)$$

In other words, we have approximated $\int_0^t m(s)ds$ by a discrete sum of complex sinusoids with random amplitudes. The FM wave is then

$$x(t) = A_c.cos(\omega_c t + K\int_0^t m(s)ds) = A_c.cos((\omega_c + Kc)t + K\sum_k Z_k exp(j\omega_k t)$$

$$= A_c.Re(exp(j(\omega_c + Kc)t)\Pi_k exp(jKZ_k)exp(j\omega_k t))$$

$$= Re[A_c.exp(j(\omega_c + Kc)t).\Pi_k \sum_n J_n(KZ_k)exp(jn\omega_k t)]$$

$$= Re[A_c.exp(j(\omega_c + Kc)t) \sum_{\{n_k\}} \Pi_k(J_{n_k}(KZ_k)exp(jn_k\omega_k t))]$$

$$= A_c. \sum_{\{n_k\}} [\Pi_l J_{n_l}(KZ_{n_l})]cos((\omega_c + Kc + \sum_l n_l.\omega_l)t)$$

Here, the $Z_k's$ are infinitesimal complex random variables the expectation of whose products with their complex conjugates are of linear orders in the frequency differentials and hence we must expand the Bessel's functions upto quadratic orders in these random variables and take the limit.

5.7 Phase locked loop for demodulating FM waves

Noise in slope circuit demodulation of FM waves. Formulation of the problem. When additive noise is present in the received FM wave, we express this total signal in the complex lowpass envelope representation. Applying this signal to a slope circuit involves differentiating this signal and then passing it through an envelope detector. The problem is to express this envelope in terms of the original baseband signal and the additive noise in the FM signal.

The received FM wave is

$$x(t) = A_c.cos(\omega_c t + K\int_0^t m(s)ds) + n(t)$$

where owing to the application of a bandpass filter,

$$n(t) = n_c(t).cos(\omega_c t) - n_s(t).sin(\omega_c t)$$

with $n_c(t), n_s(t)$ being real stationary lowpass noise processes with spectra a constant N_0 over the message frequency band $[-\sigma, \sigma]$. In phasor notation at frequency ω_c, the received FM wave is

$$y(t) = A_c.exp(j\phi(t)) + r(t).exp(j\psi(t))$$

where

$$\phi(t) = K\int_0^t m(s)ds, r(t) = \sqrt{n_c(t)^2 + n_s(t)^2}, \psi(t) = tan^{-1}(n_s(t)/n_c(t))$$

Thus

$$y(t) = A_c.exp(j\phi(t))(1 + (r(t)/A_c))exp(j(\psi(t) - \phi(t)))$$

The actual received FM wave is

$$x(t) = Re(y(t).exp(j\omega_c t))$$

We apply this signal to the slope circuit detector which involves differentiating the signal and calculating its envelope via an envelope detector.

Chapter 6

Anomalies in Quantum Field Theories and Quantum Neural Networks

6.1 Transformation laws of the components of the wave function, the gauge field and the gauge covariant derivative corresponding to the unbroken subgroup when the original local gauge group symmetry is broken

The new gauge field is to be expressed in terms of the original gauge field and the Goldstone fields corresponding to the coset space, ie, to the broken part of the gauge group. Basically, when we remove away the coset or Goldstone components associated with the broken part of the gauge group from the matter field part as well as from the gauge field part, we are left respectively with the corresponding unbroken part of the matter and gauge fields. These unbroken components of the matter and gauge fields transform according to the unbroken subgroup. The gauge covariant derivative of the unbroken group involves in addition to the original gauge field, extra terms coming from the Goldstone/broken part and it also transforms according to the unbroken subgroup.

Let matter fields $\psi(x)$ transform according to the vector representation of the gauge group G and gauge fields according to the adjoint representation of G acting in the Lie algebra $\mathfrak{g}$ of G. Under a local gauge transformation $g(x) \in G, x \in \mathbb{R}^4$, these these transformations are given by

$$\psi'(x) = g(x)psi(x), A'_\mu(x) = g(x)A_\mu(x)g(x)^{-1} + (i/e)(\partial_\mu g(x))g(x)^{-1}$$

These transformations imply the following transformation laws of the covariant derivative $\nabla_\mu = \partial_\mu + ieA_\mu(x)$ and its action on matter fields:

$$\nabla'_\mu = \partial_\mu + ieA'_\mu(x) = g(x)\nabla_\mu g(x)^{-1}$$

$$\nabla'_\mu\psi'(x) = g(x)\nabla_\mu\psi(x) NowsupposeH isaLiesubgroupo$$
$$fG.Wewishtoextractoutfromthematterfield\psi(x)$$

and the gauge field $A_\mu(x)$, that component that transforms according to H. This means that we must decompose $g(x)\gamma(x) = \gamma(g,x)h(g,x)$ in a unique way so that $h(g,x) \in H$ while $\gamma(g,x)$ is a representative of a particular coset ie of an element in G/H. This means that we must "subtract off" those components of $\psi(x)$ and $\psi'(x)$ that fall in the coset space G/H so that the resultant fields $\psi_1(x)$ and $\psi'_1(x)$ transform according to H:

$$\psi_1(x) = \gamma(x)^{-1}\psi(x), \psi(x) = \gamma(x)\psi_1(x)$$

DOI: 10.1201/9781003728627-6

Note that

$$D(g_2g_1) = D(g_2)D(g_1)$$

Since $iA_\mu(x) \in \mathfrak{g}$, it follows that we can expand

$$A_\mu(x) = A^a_\mu(x)T_a$$

Then,

$$A_{\mu 1}(x) = Ad(\gamma(x))^{-1}A_\mu(x) = A^a_\mu(x)Ad(\gamma(x))^{-1}(T_a)$$

$$= [D(\gamma(x))^{-1}]_{ba}T_bA^a_\mu(x) = A^a_{\mu 1}(x)T_a$$

where

$$A^a_{\mu 1}(x) = [D(\gamma(x))^{-1}]_{ab}A^b_\mu(x)$$

which is the same as

$$A^b_\mu(x) = [D(\gamma(x))]_{ab}A^b_{\mu 1}(x)$$

We wish to determine the following:

[a] What is the transformation law of $\psi_1(x)$ under H that corresponds to the transformation law $\psi(x) \to g(x)\psi(x)$? We have already seen that the answer to this question is

$$\psi'_1(x) = h(g,x)\psi_1(x), g(x)\gamma(x) = \gamma(g,x)h(g,x)$$

Note that For each $g(x) \in G$, we have chosen a unique $\gamma(g,x) \in G$ so that $g(x) \in \gamma(g,x)H$.

[b] What is the transfgormation law of $A_{\mu 1}(x)$ under H that corresponds to the transformation law of $A_\mu(x) \to A'_\mu(x)$ under $g(x)$?

[c] What does the gauge covariant derivative acting on ψ look like when acting on ψ_1 when this covariant derivative is expressed in terms of $A_{\mu 1}$ and perhaps some additional gauge fields involving the coset space G/H, ie, expressed in terms of the fields $\gamma(x)$ that represent the component of the matter field $\psi(x)$ that do not transform according to H, ie, that fall in G/H.

To answer [b], we note that

$$g(x)\gamma(x) = \gamma(g,x)h(g,x)$$

so that if we define

$$A'_{\mu 1}(x) = \gamma(g,x)^{-1}A'_\mu(x)\gamma(g,x)$$

then we have

$$A'_{\mu 1}(x) = \gamma(g,x)^{-1}g(x)A_\mu(x)g(x)^{-1}\gamma(g,x) + (i/e)\gamma(g,x)^{-1}(\partial_\mu g(x))g(x)^{-1}\gamma(g,x)$$

$$= \gamma(g,x)^{-1}g(x)\gamma(x)A_{\mu 1}(x)\gamma(x)g(x)^{-1}\gamma(g,x) + (i/e)\gamma(g,x)^{-1}(\partial_\mu g(x))g(x)^{-1}\gamma(g,x)$$

But since

$$g(x) = \gamma(g,x)h(g,x)\gamma(x)^{-1}$$

it follows that

$$\gamma(g,x)^{-1}g(x)\gamma(x) = h(g,x)$$

and

$$\partial_\mu g(x) = (\partial_\mu\gamma(g,x)).h(g,x)\gamma(x)^{-1}$$
$$+\gamma(g,x)(\partial_\mu h(g,x))\gamma(x)^{-1} - \gamma(g,x)h(g,x)\gamma(x)^{-1}(\partial_\mu\gamma(x))\gamma(x)^{-1}$$

so that

$$\gamma(g,x)^{-1}(\partial_\mu g(x))g(x)^{-1}\gamma(g,x)$$
$$= \gamma(g,x)^{-1}(\partial_\mu\gamma(g,x)) + (\partial_\mu h(g,x)))h(g,x)^{-1} - h(g,x)\gamma(x)^{-1}(\partial_\mu\gamma(x))h(g,x)^{-1}$$

so we finally get

$$A'_{\mu 1}(x) = h(g,x)[A_{\mu 1}(x) - (i/e)\gamma(x)^{-1}\partial_\mu\gamma(x)]h(g,x)^{-1}$$
$$+(i/e)(\partial_\mu h(g,x)))h(g,x)^{-1} + (i/e)\gamma(g,x)^{-1}\partial_\mu\gamma(g,x)$$

which is better expressed as

$$A'_{\mu 1}(x) - (i/e)\gamma(g,x)^{-1}\partial_\mu\gamma(g,x)$$
$$= h(g,x)[A_{\mu 1}(x) - (i/e)\gamma(x)^{-1}\partial_\mu\gamma(x)]h(g,x)^{-1} + (i/e)(\partial_\mu h(g,x)))h(g,x)^{-1}$$

We thus define the fields

$A_{\mu 2}(x) = A_{\mu 1}(x) - (i/e)\gamma(x)^{-1}\partial_\mu\gamma(x)$, $A'_{\mu 2}(x) = A'_{\mu 1}(x) - (i/e)\gamma(g,x)^{-1}\partial_\mu\gamma(g,x)$

and note that the transformation law that takes the former gauge field to the latter is simply given by an H-group gauge transformation:

$$A'_{\mu 2}(x) = h(g,x)A_{\mu 2}(x) + (i/e)(\partial_\mu h(g,x)))h(g,x)^{-1}$$

Now let us using this identity express the gauge covariant derivative w.r.t the H-subgroup matter and gauge transformation. The starting point is:

$$g(x)(\partial_\mu + ieA_\mu(x))\psi(x) = \partial_\mu + ieA'_\mu(x))\psi'(x)$$

6.2 Vestigial sideband modulation

Here, the baseband lowpass signal is modulated using the Hilbert transformer into an upper-sideband bandpass signal but with a small residual component in the spectrum seeping into the lower sideband region in order to take into account that the bandpass filter cannot have a sharp cutoff. If we use a pure USB signal without including a vestigial component then during the process of demodulation using a coherent demodulator, some portion of the lower frequency part of the demodulated output is likely to be lost. Specifically, if $m(t)$ is the original lowpass signal occupying the frequency band $[-\sigma, \sigma]$ then the USB signal is

$$x_1(t) = (m(t) + j\hat{m}(t))exp(j\omega_c t)/2 + (m(t) - j\hat{m}(t))exp(-j\omega_c t)/2$$

$$= m(t)cos(\omega_c t) - \hat{m}(t)sin(\omega_c t)$$

Likewise, the LSB signal is

$$x_2(t) = m(t)cos(\omega_c t) + \hat{m}(t)sin(\omega_c t)$$

$x_1(t)$ has its spectrum concentrated over the USB $[\omega_c, \omega_c + \sigma] \cup [-\omega_c, -\omega_c - \sigma]$ while $x_2(t)$ has its spectrum concentrated over the LSB $[\omega_c - \sigma, \omega_c] \cup [-\omega_c, -\omega_c - \sigma]$. The sum of these two signals is the DSBSC signal

$$x(t) = x_1(t) + x_2(t) = 2m(t)cos(\omega_c t)$$

Let $h(t)$ be a filter whose frequency response is concentrated over the band $[\omega_c - \delta, \omega_c] \cup [-\omega_c, -\omega_c + \delta]$ where $0 < \delta << \sigma$. The VSB signal is then

$$x_3(t) = x_1(t) + h(t) * x_2(t)$$

It is clear then that $x_3(t)$ has its spectrum concentrated over the band $[\omega_c - \delta, \omega_c + \sigma] \cup [-\omega_c - \sigma, -\omega_c + \delta]$

Problem: Calculate the output of the coherent demodulator in terms of $m(t), \omega_c, \sigma, \delta$.

6.3 Noise in FM receivers when demodulation is done using the slope circuit detector

problem: Let

$$x(t) = A_c.cos(\omega_c t + K \int_0^t m(s)ds) + n_c(t)cos(\omega_c t) - n_s(t)sin(\omega_c t)$$

Express it in the form

$$x(t) = R(t)cos(\omega_c t + \psi(t))$$

where

$$R(t) = ((A_c cos(\phi(t)) + n_c(t))^2 + (A_c.sin(\phi(t) + n_s(t))^2)^{1/2},$$

$$\psi(t) = tan^{-1}(A_c sin(\phi(t)) + n_s(t))/(A_c cos(\phi(t)) + n_s(t)))$$

$$= A_c^2 + n_c^2 + n_s^2 + 2A_c(n_c.cos(\phi) + n_s sin(\phi))$$

where

$$\phi(t) = K \int_0^t m(s)ds$$

The output of the FM demodulator based on slope circuit is the envelope of the signal

$$dx(t)/dt = R'(t).cos(\omega_c t + \psi(t)) - R(t)(\omega_c + \psi'(t))sin(\omega_c t + \psi(t))$$

Show that its envelope is given approximately by

$$R(t)(\omega_c + \psi'(t))$$

Express this in terms of $A_c, n_c, n_s, m(t)$ and calculate using perturbation theory, its statistics in terms of the statistics of the processes $n_c(t), n_s(t)$. Show that without any approximation, the exact envelope of $dx(t)/dt$ is given by

$$(R(t)^2(\omega_c + \psi'(t))^2 + R'(t)^2)^{1/2}$$

Calculate its statistics again using perturbation theory.

6.4 Noise in phase locked loop demodulation of FM signals

FM signals are demodulated using a phased lock loop which involves multiplication by the output of a voltage controlled oscillator with input being the demodulated output followed by a lowpass filter. This scheme implies that the phase of the input FM signal is tracked by the phase of the VCO output and hence the frequency of the input message is tracked by the frequency of the VCO output or equivalently by the output of the lowpass filter. However if the FM signal gets corrupted by additive noise, this noise will enter into the demodulated output in a nonlinear way and one requires to develop some sort of perturbation theory to solve the resulting stochastic differential equation for the demodulated output as a functional of the input message signal and the noise.

Let $v(t)$ be the VCO input, ie, the output of the lowpass loop filter $H(\omega)$. The VCO output is $y(t) = A_c.sin(\omega_c t + \phi_1(t))$ where $\phi_1(t) = K_v \int_0^t v(s)ds$. The input to the PLL is the FM signal containing bandpass noise:

$$x(t) = s(t)+n(t), n(t) = n_c(t)cos(\omega_c t)-n_s(t)sin(\omega_c t), s(t)$$
$$= A_c cos(\omega_c t+\phi(t)), \phi(t) = K\int_0^t m(s)ds$$

where $n_c(t)$ and $n_s(t)$ are respectively the in phase and quadrature components of the noise within the FM band assumed to have approximately a finite band. Actually owing to the Bessel function expansion of single and multitone FM waves, and FM signal is not bandlimited. We can write

$$n(t) = r(t).cos(\omega_c t + \psi(t)), r(t) = \sqrt{n_c(t)^2 + n_s(t)^2}, \psi(t) = tan^{-1}(n_s(t)/n_c(t))$$

Thus, we can express

$$x(t) = A_c.Re(exp(j\omega_c t+\phi(t))(1+(r(t)/A_c).exp(j(\psi(t)-\phi(t))))$$
$$= A_c f(t)cos(\omega_c t+\phi(t)+\chi(t)),$$

where

$$f(t) = |1+(r/A_c).exp(j(\psi-\phi))|$$
$$= (1+(r/A_c)^2+2rcos(\psi-\phi)/A_c)^{1/2} \approx 1+r(t)cos(\psi(t)-\phi(t))/A_c$$

and

$$\chi(t) = tan^{-1}((r/A_c).sin(\psi-\phi)/(1+(r/A_c).cos(\psi-\phi))) \approx (r(t)/A_c)sin(\psi(t)-\phi(t))$$

where these expansions for $f(t)$ and $\chi(t)$ have been carried upto linear orders in the noise amplitude $r(t)$. The output of the loop filter is $h(t)$ convolved with the product $K_0 x(t)y(t)$ which since $h(t)$ is a lowpass filter that rejects frequencies around $2\omega_c$, can equivalently be expressed as the convolution of $h(t)$ with $(K_0 A_c f(t)/2).sin(\phi(t) - \phi_1(t) + \chi(t))$ or upto linear orders in the phase errors and the noise amplitude, simply as the convolution of $h(t)$ with $(K_0 A_c/2)(\phi(t) - \phi_1(t) + \chi(t))$. Thus we obtain the following stochastic differential equation for the phase $\phi_1(t)$ of the VCO input and output:

$$v(t) = Ah(t)*(\phi(t)-\phi_1(t)+\chi(t)), A = K_0 A_c/2, d\phi_1(t)/dt = K_v v(t)$$
$$= B.h(t)*(\phi(t)-\phi_1(t)+\chi(t))), B = AK_v = K_0 K_v A_c/2$$

In the absence of noise, this equation is

$$d\phi_1(t)/dt = Bh(t) * (\phi(t) - \phi_1(t))$$

Since $m(t)$ and $v(t)$ are lowpass signals, $\phi(t) = K\int_0^t m(s)ds$ and $\phi_1(t) = K_v \int_0^t v(s)ds$ which have Fourier transforms $KM(\omega)/j\omega$ and $K_v V(\omega)/j\omega$ are also lowpass signals and hence if the loop filter $h(t)$ is assumed to be ideal lowpass, the above differential equation simplifies to

$$d\phi_1(t)/dt = B(\phi(t) - \phi_1(t))$$

which shows that ϕ_1 will track ϕ because if $\phi_1(t) < \phi(t)$, then $d\phi_1(t)/dt = \phi(t) - \phi_1(t) > 0$, ie, $\phi_1(t)$ will increase while if $\phi_1(t) > \phi(t)$, then $d\phi_1(t)/dt = \phi(t) - \phi_1(t) < 0$, causing $\phi_1(t)$ to decrease. The general case taking noise into account gives under this approximation,

$$d\phi_1(t)/dt = B(\phi(t) - \phi_1(t) + \chi(t))$$

and this can be solved using integrating factors as

$$\phi_1(t) = B.\int_0^t exp(-B(t-s))(\phi(s) - \chi(s))ds$$

As $B \to \infty$, $B.exp(-Bt) \to \delta(t)$ which means that for large B, we obtain the approximation

$$\phi_1(t) - \phi(t) \approx \chi(t)$$

The mean square value of $\chi(t)$ can be evaluated using the above linearized approximations as

$$\mathbb{E}(\chi(t)^2) = \mathbb{E}(r(t)^2.sin^2(\psi(t) - \phi(t)))/A_c^2$$

Now

$$r(t)sin(\psi(t) - \phi(t)) = n_s(t)cos(\phi(t)) - n_s(t)sin(\phi(t))$$

so that

$\mathbb{E}(r(t)^2.sin^2(\psi(t)-\phi(t)))$

$$= R_{n_s}(0)cos^2(\phi(t))+R_{n_c}(0)sin^2(\phi(t))-R_{n_cn_s}(0)sin(2\phi(t))$$

Now

$$n_c(t) + jn_s(t) = (n(t) + j\hat{n}(t))exp(-j\omega_c t)$$

so

$$n_c(t) = n(t).cos(\omega_c t) + \hat{n}(t).sin(\omega_c t), n_s(t) = -n(t)sin(\omega_c t) + \hat{n}(t).cos(\omega_c t))$$

from which we deduce thata

$$R_{n_s}(\tau) = R_n(\tau)cos(\omega_c\tau) + \hat{R}_n(\tau)sin(\omega_c\tau) = R_{n_c}(\tau),$$

$$R_{n_cn_s}(\tau) = R_n(\tau)sin(\omega_c\tau) - \hat{R}_n(\tau)cos(\omega_c\tau)$$

and in particular,

$$R_{n_c(0)=R_{n_s}(0)=R_n(0),R_{n_cn_s}(\tau)=0}$$

since the Fourier transform of $\hat{R}_n(\tau)$ is $-j.sgn(\omega).S_n(\omega)$ and the latter being an odd function, its integral over all frequencies is zero. Thus,

$$\mathbb{E}(r(t)^2.sin^2(\psi(t) - \phi(t))) = R_n(0) = \mathbb{E}(n(t)^2)$$

This gives

$$\mathbb{E}(\chi(t)^2) = R_n(0)/A_c^2$$

in our approximate analysis. A more exact analysis would involve solving the non-linear stochastic integro-differential equation

$d\phi_1(t)/dt = B.h(t)*[f(t).sin(\phi(t)-\phi_1(t)+\chi(t))]$

$$= B\int h(\tau)f(t-\tau).sin(\phi(t-\tau)-\phi_1(t-\tau)+\chi(t-\tau))d\tau$$

In the special case when $h(t)$ is an RC lowpass filter with transfer function $H(s) = \tau/(1+s\tau)$ and impulse response $exp(-t/\tau)u(t)$, the above equation can be expressed as

$$d\phi_1(t)/dt = \phi_2(t), d\phi_2(t)/dt = -\phi_2(t)/\tau + B.f(t).sin(\phi(t) - \phi_1(t) + \chi(t))$$

Now observe that

$$f(t)exp(j\chi(t)) = (1 + r(t)exp(j(\psi(t) - \phi(t)))/A_c$$

and hence

$$f(t)cos(\chi(t)) = 1 + r(t)cos(\psi - \phi)/A_c, f(t)sin(\chi(t)) = r.sin(\psi - \phi)/A_c$$

This gives us

$$f(t)sin(\phi-\phi_1+\chi) = sin(\phi-\phi_1)(1+r.cos(\psi-\phi)/A_c)+cos(\phi-\phi_1)r.sin(\psi-\phi)/A_c$$

$$= sin(\phi - \phi_1) + (r/A_c).sin(\psi - \phi_1)$$

so that the above equations can be expressed as

$$\phi_1'(t) = \phi_2(t), \phi_2'(t) = -\phi_2(t)/\tau+B.sin(\phi(t)-\phi_1(t))+(Br(t)/A_c).sin(\psi(t)-\phi_1(t))$$

$$= -\phi_2(t)/\tau + B.sin(\phi(t) - \phi_1(t)) + (B/A_c)(n_s(t)cos(\phi_1(t)) - n_c(t)sin(\phi_1(t))$$

6.5 The Lichnerovicz formula for the square of the Dirac-Yang-Mills operator in curved space-time and application of Getzler scaling to this operator to obtain the Atiyah-Singer index theorem for the Dirac-Yang-Mills operator in curved space-time

The index of the Dirac operator can be computed using the difference of the traces of two heat kernels or equivalently by the supertrace of the heat kernel associated with the square of the Dirac-Yang- Mills operator. This supertrace can be evaluated by using the Getzler scaling method wherein the Fermion creation and annihilation operators that define the Dirac matrices of the Clifford algebra are replaced by scaled versions with the scaling factor coverging to zero. The result of this operation is that the supertrace becomes in the limit equal to the integral of an exterior differential form on the differential manifold on which the Dirac operator acts. This exterior form is defined in terms of functions of super-symmetric matrices, ie, matrices whose elements are differential forms. These matrices are respectively the curvature form and the Chern-Simons form obtained from the Dirac operator. The heat kernel of the square of the Dirac operator is evaluated using Mehler's formula for the evolution kernel of a harmonic oscillator but the matrices that define the harmonic oscillator have entries that are differential forms.

The Dirac-Yang-Mills operator in curved space-time is given by

$$D = \gamma^\mu(x)\nabla_\mu, \nabla_\mu = \partial_\mu + A_\mu + \Gamma_\mu, \gamma^\mu(x) = \gamma^a e_a^\mu(x)$$

where

$$A_\mu = A_\mu^a i(x)\tau_a$$

is the Yang-Mills gauge field,

$$\Gamma_\mu(x) = \Gamma_\mu^{ab}(x)[\gamma^a, \gamma^b]/8 = \Gamma_\mu^{ab}\gamma^a\gamma^b$$

is the spinor connection of the metric $g_{\mu\nu}(x)$, note that

$$\Gamma_\mu^{ab} = -\Gamma_\mu^{ba}$$

and hence the total curvature of the Gravity and Yang-Mills connection is given by

$$[\nabla_\mu, \nabla_\nu] = (\Gamma_{\nu,\mu} - \Gamma_{\mu,\nu} + [\Gamma_\mu, \Gamma_\nu]) + (A_{\nu,\mu} + A_{\mu,\nu} + [A_\mu, A_\nu])$$
$$= R_{\mu\nu} + K_{\mu\nu} = F_{\mu\nu}$$

where

$$R_{\mu\nu} = (\Gamma_{\nu,\mu} - \Gamma_{\mu,\nu} + [\Gamma_\mu, \Gamma_\nu]) = R_{\mu\nu ab}\gamma^a\gamma^b - - - (a)$$

and

$$K_{\mu\nu} = (A_{\nu,\mu} + A_{\mu,\nu} + [A_\mu, A_\nu]) = K^a_{\mu\nu} i\tau_a$$

Note that the spin generators $[\gamma^a, \gamma^b]/4$ commute with the Yang-Mills gauge Lie algebra generators τ_a since $[\gamma^a, \tau^b] = 0$. The total gauge field Lie algebra here is the direct sum of the Lie algebras of the spinor representation of the Lorentz group and the Yang-Mills gauge Lie-algebra. In obtaining (a), we have used the Lorentz Lie algebra representation commutation relations for the spinor representation generators $[\gamma^a, \gamma^b]/4$ of the Lorentz Lie algebra. Thus

$$(1/2)[\gamma^\mu(x), \gamma^\nu(x)]F_{\mu\nu} = \gamma^\mu\gamma^\nu F_{\mu\nu} = e^\mu_c e^\nu_d R_{\mu\nu ab}\gamma^c\gamma^d\gamma^a\gamma^d + e^\mu_c e^\nu_d K_{\mu\nu}\gamma^c\gamma^d$$
$$= R_{cdab}\gamma^c\gamma^d\gamma^a\gamma^d + K_{cd}\gamma^c\gamma^d$$

where

$$R_{abcd} = R_{\mu\nu ab}e^\mu_c e^\nu_d$$

is the gravitational Riemann curvature tensor in the spinor representation while

$$K_{cd} = K_{\mu\nu}e^\mu_c e^\nu_d$$

is the Yang-Mills curvature tensor in the spinor representation. Note that the total covariant derivative of a four vector field B_μ in the spinor representation is given by

$$[\nabla_\mu, B_\nu e^\nu_a \gamma^a] = [\nabla_\mu, B_\nu]e^\nu_a\gamma^a$$

because the spinor connection Γ_μ is chosen so that the covariant derivative of the tetrad e^ν_a is zero in the spinor representation:

$$0 = [\nabla_\mu, \gamma^\nu] = [\nabla_\mu, e^\nu_a\gamma^a] = [\partial_\mu + A_\mu + \Gamma_\mu, e^{\nu a}_a] = [\partial_\mu + \Gamma_\mu, e^\nu_a\gamma^a]$$
$$= e^\nu_{a,\mu}\gamma^a + [(1/4)\Gamma^{bc}_\mu\gamma^b\gamma^c, \gamma^a]e^\nu_a$$

Now from the Clifford anticommutation rules for the Dirac matrices,

$$\gamma^b\gamma^c\gamma^a = \gamma^b(2\eta^{ca} - \gamma^a\gamma^c) = 2\eta^{ca}\gamma^b - (2\eta^{ba} - \gamma^a\gamma^b)\gamma^c$$

or equivalently,

$$[\gamma^b\gamma^c, \gamma^a] = 2\eta^{ac}\gamma^b - 2\eta^{ab}\gamma^c$$

This gives the following condtion for the spinor connection:

$$e^\nu_{a,\mu}\gamma^a + (1/2)\Gamma^{bc}_\mu e^\nu_a(\eta^{ac}\gamma^b - \eta^{ab}\gamma^c) = 0$$

which in view of the linear independence of the γ^a, reduces to

$$e^\nu_{a,\mu} + (1/2)\Gamma^{ac}_\mu e^\nu_b\eta^{bc} - (1/2)\Gamma^{ba}_\mu e^\nu_c\eta^{cb} = 0$$

Now we introduce Lichnerowicz formula for the square of the Dirac operator in curved space time with Yang-Mills connection included:

$$D^2 = (\gamma^\mu(x)\nabla_\mu)^2 = \gamma^\nu\nabla_\nu\gamma^\mu\Gamma_\mu = \gamma^\nu[\nabla_\nu, \gamma^\mu]\nabla_\mu + \gamma^\nu\gamma^\mu\nabla_\nu\nabla_\mu$$

$$= \gamma^{\nu}\gamma^{\mu}\nabla_{\nu}\nabla_{\mu}$$
$$= (1/2)[\gamma^{\nu}, \gamma^{\mu}]_{+}\nabla_{\nu}\nabla_{\mu} + (1/2)\gamma^{\nu}\gamma^{\mu}[\nabla_{\nu}, \nabla_{\mu}]$$
$$= (1/2)g^{\mu\nu}(x)\nabla_{\nu}\nabla_{\mu} + (1/2)\gamma^{\nu}\gamma^{\mu}F_{\nu\mu}$$
$$= (1/2)\Delta + (1/2)R + (1/2)K$$

where

$$R = R_{abcd}\gamma^{a}\gamma^{b}\gamma^{c}\gamma^{d} = R_{\mu\nu\rho\sigma}\gamma^{\mu}(x)\gamma^{\nu}(x)\gamma^{\rho}(x)\gamma^{\sigma}(x), \gamma^{\mu}(x) = \gamma^{a}e_{a}^{\mu}(x)$$

so that

$$R_{abcd} = R_{\mu\nu\rho\sigma}e_{a}^{\mu}e_{b}^{\nu}e_{c}^{\rho}e_{d}^{\sigma}$$

and likewise

$$K = K_{ab}\gamma^{a}\gamma^{b} = K_{\mu\nu}\gamma^{\mu}(x)\gamma^{\nu}(x)$$
$$K_{ab} = K_{\mu\nu}(x)e_{a}^{\mu}(x)e_{b}^{\nu}(x)$$

Now we discuss the Getzler scaling idea: We have defined the Dirac operator

$$D = \gamma^{\mu}(x)\nabla_{\mu} = \gamma^{\mu}(\partial_{\mu} + A_{\mu} + \Gamma_{\mu})$$
$$\gamma^{0}D = \alpha^{\mu}\nabla_{\mu}$$

is a skew-Hermitian operator since the α^{a} matrices are Hermitian. Thus $D^{*}\gamma^{0} = -\gamma^{0}D$. Also, we can using the explicit form of the Dirac α matrices write

$$\gamma^{0}D = \begin{pmatrix} 0 & D_{+} \\ -D_{+}^{*} & 0 \end{pmatrix}$$

Now noting that

$$\gamma^{0} = \begin{pmatrix} I & 0 \\ 0 & -I \end{pmatrix}$$

we get

$$D = \gamma^{0}.\begin{pmatrix} 0 & D_{+} \\ -D_{+}^{*} & 0 \end{pmatrix} = \begin{pmatrix} 0 & D_{+} \\ D_{+}* & 0 \end{pmatrix}$$

Thus, D is a Hermitian operator and

$$D^{2} = \begin{pmatrix} D_{+}D_{+}^{*} & 0 \\ 0 & D_{+}^{*}D_{+} \end{pmatrix}$$

yielding thereby

$$str(exp(tD^{2})) = Tr(exp(tD_{+}D^{+*}) - Tr(exp(tD_{+}^{*}D_{+}))$$
$$= n(D_{+}D_{+}^{*}) - n(D_{+}^{*}D_{+}) = n(D_{+}^{*}) - n(D_{+}) = -Ind(D_{+})$$

where str denotes supertrace. To evaluate the supertrace, we use Getzler's scaling idea. We assume generally that we are working in n-dimensional space-time and we then have n Dirac γ-matrices $\gamma^{a}, a = 0, 1, ..., 9$ that satisfy the Clifford anticommutation relations

$$\gamma^{a}, \gamma^{b}]_{+} = 2\eta^{ab}$$

Then We take a finite dimensional inner product space $(V, <.,.>)$ with $dimV = r$ and consider the Fermion Fock space

$$\Lambda V = \bigoplus_k \Lambda^k V$$

where $\Lambda^k V$ is the kth skew-symmetric power of V. For any $v \in V$ define the creation operator $e(v) : \Lambda V \to \Lambda V$ by

$$e(v)(v_1 \wedge ... \wedge v_m) = v \wedge v_1 \wedge ... \wedge v_m$$

so its adjoint becomes

$$e(v)^*(v_1 \wedge ... \wedge v_m) - = \sum_k (-1)^{k-1} < v, v_k > v_1 \wedge ... \wedge \hat{v}_k \wedge ... \wedge v_m$$

Then, it is easy to verify that

$$[e(v), e(w)]_+ = 0, [e(v)^*, e(w)^*]_+ = 0, [e(v), e(w)^*]_+ =< v, w >$$

Thus $e(v)$ creates a Fermion of "polarization" v while $e(v)^*$ annihilates a Fermion of polarization v. Define on ΛV the operators

$$\gamma(v) = e(v) + e(v)^*$$

Then, in view of the above anticommutation relations, we have

$$[\gamma(v), \gamma(w)]_+ = 2 < v, w >$$

and hence if $e_a, a = 1, 2, ..., r$ is an onb for V, then $\gamma^a = \gamma(e_a), a = 1, 2, ..., r$ is a set of Dirac Gamma matrices. For $\epsilon > 0$, define the linear operator

$$S_\epsilon : \Lambda V \to \Lambda V$$

by

$$S_\epsilon \omega = \epsilon^{-\partial \omega} \omega$$

Suppose now that T is a linear operator on ΛV that is built out of composiions of a operators of the form $e(v_k), k = 1, 2, ..., a$ and b operators of the form $e(w_k)^*, k = 1, 2, ..., b$. Then it is clear that T maps $\Lambda^k V$ into $\Lambda^{k+a-b} V$ and hence

$$S_\epsilon T S_\epsilon^{-1} = \epsilon^{b-a} T$$

We write $degT = a - b$ and hence for such operators

$$S_\epsilon T S_\epsilon^{-1} = \epsilon^{-degT} T$$

In particular,

$$S_\epsilon \gamma(v) S_\epsilon^{-1} = \epsilon^{-1} e(v) + \epsilon.e(v)^*$$

or equivalently,

$$\epsilon S_\epsilon \gamma(v) S_\epsilon^{-1} = e(v) + \epsilon^2.e(v)^* \to e(v), \epsilon \to 0$$

Likewise, we can define for a Bosonic function $f(x)$ on the Riemannian manifold with values in $\mathfrak{g}$,

$$S_\epsilon f(x) = f(\epsilon x)$$

Then if L is a linear operator acting on such functions with kernel $L(x,y)$, so that

$$Lf(x) = \int L(x,y) f(y) dy$$

we get

$$S_\epsilon L S_\epsilon^{-1} f(x) = \int L(\epsilon x, y) f(\epsilon^{-1} y) dy = \epsilon^n \int L(\epsilon x, \epsilon y) f(y) dy$$

where now n is the dimension of the Riemannian manifold. This also follows from the following verification:

$$Tr(L) = \int L(x,x)dx) = \epsilon^n \int L(\epsilon x, \epsilon x) dx = Tr(S_\epsilon L S_\epsilon^{-1})$$

Now let L be a linear differential operator acting on functions on the Riemnannian manifold M with values in V of the form

$$L = \sum_k A_k(x) D^k$$

where $A_k(x)$ are matrices and $D^k = \Pi_{k_1,...,k_j} \partial_{x_1}^{k_1} ... \partial_{x_n}^{k_n}$ for $k = (k_1, ..., k_n)$. Then

$$S_\epsilon L f(x) = \sum_k A_k(\epsilon x)(D^k f)(\epsilon x) = \sum_k A_k(\epsilon x) \epsilon^{-|k|} D^k f(\epsilon x)$$

and hence

$$S_\epsilon L S_\epsilon^{-1} = \sum_k A_k(\epsilon x) \epsilon^{-|k|} D^k$$

6.6 On modeling EEG signals using the motion of electric dipoles in the cortical region of the brain

The dynamics of the electric dipoles moving in the brain cortex under mutual electromagnetic interactions is analyzed by constructing the Lagrangian. The distribution of time varying electric dipoles induces a volume polarization current density as a volume polarization charge density which generate electromagnetic fields and cause the dipoles to interact with each other. The movement of these dipoles causes variations in the electric and magnetic field on the surface of the head. By measuring this EEG electromagnetic fields on the surface of the head, it is possible using the methods of signal estimation theory to estimate the structure of the dipoles as well as their trajectories and orientation as a function of time.

Let the brain consist of infinitesimal electric dipoles $p_1(t), ..., p_N(t)$ located at positions $r_1(t), ..., r_N(t)$. Without taking into account relativistic effects, the polarization vector field at time t is given by

$$P(t,r) = \sum_{k=1}^{N} p_k(t)\delta(r - r_k(t))$$

The polarization charge density field is given by the well known formula

$$\rho_p(t,r) = -divP(t,r)$$

The polarization current density $J_p(t,r)$ then satisfies the equation of continuity

$$divJ_p = -\partial_t \rho_p$$

and hence we can take

$$J_p(t,r) = \partial_t P(t,r)$$

Another way to justify this formula for the polarization current density is to consider a charge q moving with a velocity v. If there are N such charges per unit volume, then the current density generated by the motion of these charges is $J_p = Nqv$ and since $P = Nqr$ represents the dipole moment of the se charges per unit volume and $v = dr/dt$, we get the relation $J_p = \partial_t P$. Now the magnetic vector potential and the electric scalar potential generated by these dipole charges are respectively given by the usual retarded potential formula

$$A(t,r) = (\mu/4\pi)\int J_p(t - |r - r'|/c, r')d^3r'/|r - r_{'}|,$$

$$V(t,r) = (1/4\pi\epsilon)\int \rho_p(t - |r - r'|/c, r')d^3r'/|r - r'|$$

If we ignore relativistic effects on the ground that the motion of the dipoles is not too rapid, then we can use the Coulomb and Biot-Savart formulae

$$A(t,r) = (\mu_0/4\pi)\int J_p(t,r')d^3r'/|r - r'|,$$

$$V(t,r) = (1/4\pi\epsilon_0)\int \rho_p(t,r')d^3r'/|r - r'|$$

to obtain

$$V(t,r) = \sum_k p_k(t).(r - r_k(t))/4\pi\epsilon_0|r - r_k(t)|^3$$

and

$$A(t,r) = (\mu_0/4\pi)\sum_k p'_k(t)/r - r_k(t)|$$

The electric field is then given by

$$E(t,r) = -\nabla V(t,r) - \partial_t A(t,r)$$

$$= -\sum_k \nabla_r(p_k(t).(r - r_k(t)/|r - r_k(t)|)$$

$$-(\mu_0/4\pi)\sum_k \partial_t(p'_k(t)/|r - r_k(t)|)$$

and the magnetic field by

$$B(t,r) = \nabla_r \times A(t,r) = (\mu_0/4\pi)\sum_k(\nabla_r(1/|r - r_k(t)|) \times p_k(t)$$

We can combine these equations in the form of a six component equation for the electromagnetic field vector $\psi(t,r) = [E(t,r), B(t,r)]$ expressible in the form

$$\psi(t,r) = F(p_k(t), r_k(t), v_k(t), k = 1,2,...,N)$$

Estimating the structure of the brain from electromagnetic field measurements then amount to estimating the dipole strengths $\{p_k(t)\}$, their positions $\{r_k(t)\}$ and their velocities $\{v_k(t)\}$ from the field measurements on a dynamic, ie, real time basis. For doing this, we must first derive the stochastic differential equations satisfied by $p_k(t), r_k(t), v_k(t)$. Assume that the motion of these dipoles takes place in an external electromagnetic field $E_e(t,r), B_e(t,r)$. To derive the Lagrangian for the interaction between the dipoles and the external fields, assume that the external fields are derivable from the electromagnetic potentials $V_e(t,r)$ and $A_e(t,r)$. A given dipole $p_k(t)$ located at $r_k(t)$ can be regarded as consisting of a pair of charges $-q_k$ and q_k located at $r_k(t)$ and $r_k(t) + d_k(t)$ where $p_k(t) = q_k d_k(t)$ with the separation vector $d_k(t)$ being very small. The contribution to the Lagrangian due to the interaction is then given by

$$U = q_k(V(t,r_k) - V(t,r_k + d_k) - q_k((v_k, A(t,r_k)) - (v_k + d'_k, A(t,r_k + d_k))$$

$$= -q_k(d_k, \nabla V(t,r_k)) - q_k(v_k,(d_k,\nabla)A(t,r_k)) - q_k(d'_k, A(t,r_k))$$

$$= -(p_k, \nabla V(t,r_k)) - (v_k,(p_k,\nabla)A(t,r_k)) - (p'_k, A(t,r_k))$$

upto linear orders in d_k, d'_k. The kinetic energy K of the dipoles is given by the sum of their translational kinetic energy and their rotational kinetic energy. Letting J_k denote the moment of inertia tensor of the kth dipole and $R_k(t)$ the 3-D rotation matrix applied to this dipole after time t around its centre of mass, we can express

$$K = \sum_k(1/2)m_k|v_k|^2 + (1/2)Tr(R'_k(t))J_k R'_k(t)^T)$$

Here, $d_k(t) = R_k(t)d_k(0)$ so that $d'_k(t) = R'_k(t)d_k(0)$ and hence, $p_k(t) = R_k(t)p_k(0)$ and $p'_k(t) = R'_k(t)p_k(0)$. Thus parametrizing the phase coordinates of the kth dipole by (r_k, v_k, p_k, p'_k) or equivalently by (r_k, v_k, R_k, R'_k) where by R_k, we mean the three Euler angles of R_k and by R'_k, we mean the rates of change of these three Euler angles, we can set up the Lagrangian of the system of dipoles as

$$L(r_k, r'_k, \phi_k, \theta_k, \psi_k, \phi'_k, \theta'_k, \psi'_k, k = 1, 2, ..., N)$$
$$= K - U$$

6.7 Quantum neural networks constructed by controlling the gravitational field interacting with the Dirac electron with the control forces for the gravitational field derived from the energy-momentum tensor of the matter and radiation field

The probability density associated with the Dirac wave function in curved space time can be made to track any given family of probability densities by controlling the metric tensor of the background gravity. This metric can be controlled using the energy momentum tensor of the other source fields in the Einstein field equations.

Exercise: The Dirac equation for an electron in curved space-time interacting with an electromagnetic field is given by

$$[\gamma^a e^\mu_a(x)(i\partial_\mu + eA_\mu(x) + i\omega^{ab}_\mu(x)\gamma_{ab}) - m]\psi(x) = 0$$

where $\gamma_{ab} = [\gamma_a, \gamma_b]/4$ define the standard generators of the spinor representation of the Lie algebra of the Lorentz group, $e^\mu_a(x)$ is a tetrad of the gravitational field, $\omega^{ab}_\mu(x)$ define the spinor connection of the gravitational field. They are constructed so that

$$D_\mu e^a_\nu = 0$$

where D_μ is the covariant derivative acting on a four vector with a spinor index. Specifically,

$$\partial_\mu e^a_\nu - \Gamma^\rho_{\mu\nu} e^a_\rho + \omega^{ab}_\mu e_{b\nu} = 0$$

Solving this equation gives us

$$\omega^{ab}_\mu = -e^{b\nu} e^a_{\nu:\mu}$$

It can be shown easily that under a local Lorentz transformation and also under general space-time diffeomorphisms, the Dirac equation is invariant by using the fact that under a local Lorentz transformation, $\Lambda(x)$, the tetrad transforms as

$$e^a_\mu(x) \rightarrow \Lambda^a_b(x) e^b_{|mu}(x)$$

while under a diffeomorphism, $x^\mu \to \bar{x}^\mu$, it transforms as

$$e^a_\mu(x) \to e^a_\nu(x)\frac{\partial x^\nu}{\partial \bar{x}^\mu}$$

We must of course also use the fact that under a local Lorentz transformation $\Lambda(x)$, the Dirac wave function $\psi(x)$ transforms in the spinor representation as

$$\psi(x) \to D(\Lambda(x))\psi(x)$$

where

$$D(\Lambda(x)) = exp(\gamma_{ab}\epsilon^{ab}(x))$$

where

$$\Lambda(x) = exp(\epsilon(x))$$

with $\epsilon(x)$ a Lie algebra element of the local Lorentz group, ie, $\epsilon(x)$ is a skewsymmetric real matrix of size 4×4. Now the

The Dirac equation can be derived from the action principle

$$\delta_{\chi,\psi}S = 0$$

where $\gamma_{ab} = [\gamma_a, \gamma_b]/4$ and

$$S[\psi, \chi] = Re\int \chi(x)^*\gamma^0(\gamma^a e^\mu_a(x)(i\partial_\mu + eA_\mu(x) + i\omega^{ab}_\mu(x)\gamma_{ab}) - m)\psi(x)\sqrt{-g(x)}d^4x = 0$$

From this action, we can easily construct the energy-momentum tensor of the Dirac field. More precisely, we can add to this action, the action $(-1/4)\int F_{\mu\nu}F^{\mu\nu}\sqrt{-g}d^4x$ of the electromagnetic field and construct the total energy-momentum tensor of the Dirac field and the Maxwell field by considering the variation of this total action w.r.t metric perturbations. This energy momentum tensor appears as a source on the rhs of the Einstein field equations and we can add to this source control source energy momentum tensor terms coming from the interaction between an external current source J^μ and the electromagnetic field A_μ described by an additional interaction Lagrangian of the form $-\int J_\mu A^\mu\sqrt{-g}d^4x$ and an interaction between an external electromagnetic field source B_μ and the Dirac current described by an interaction Lagrangian of the form $\int B_\mu(x)\psi(x)^*\gamma^0\gamma^\mu\psi(x)\sqrt{-g(x)}d^4x$. We then control these two sources J_μ and B_μ so that the metric tensor $g_{\mu\nu}(x)$ tracks a desired metric and simultaneously the Dirac probability density $Re(\chi(x)^*\psi(x))$ also tracks a given trajectory. Alternately, we can eliminate the metric tensor from this system of equations by perturbatively solving for it from the Einstein field equations in terms of the other fields and adapting the control sources so that the Dirac probability density alone tracks a given pdf.

Exercise: Show that if Λ is a Lorentz transformation, then

$$D(\Lambda)\gamma^a D(\Lambda)^{-1} = \Lambda^a_b\gamma^b - - - (a)$$

To prove this, you can consider $\Lambda = I + \epsilon$ as in infinitesimal Lorentz transformation, so that

$$D(\Lambda) = I + dD(\epsilon) = I + \epsilon^{ab}\gamma_{ab}$$

and then prove the infinitesimal form of (a), namely

$$\epsilon^{ab}[\gamma_{ab}, \gamma^c] = \epsilon^c_b\gamma^b = \epsilon^{cb}\gamma_b = \epsilon^{ab}\delta^c_a\gamma_b$$

for any antisymmetric matrix $((\epsilon^{ab}))$. This is equivalently to proving the commutation relations

$$[\gamma_{ab}, \gamma^c] = \delta^c_a\gamma_b - \delta^c_b\gamma_a$$

6.8 Synchronization of clocks

A and B are two neighbouring points in space with spatial coordinates x^r and $x^r + dx^r$ respectively. A light signal from A starting at time x^0 arrives at B at time dx^0. dx^0 is the positive solution to the quadratic equation

$$0 = g_{\mu\nu}dx^\mu dx^\nu = g_{00}(dx^0)^2 + 2g_{0r}dx^0dx^r + g_{rs}dx^rdx^s$$

Note that $g_{\mu\nu}(x)$ is assumed to be independent of x^0. Let $dx^0(1)$ and $dx^0(2)$ denote the two solutions to the above quadratic equation. These solutions are respectively given by

$$dx^0(1) = [-g_{0r}dx^r + (g_{0r}g_{0s} - g_{00}g_{rs})dx^rdx^s)^{1/2}]/g_{00},$$

$$dx^0(2) = [-g_{0r}dx^r - (g_{0r}g_{0s} - g_{00}g_{rs})dx^rdx^s)^{1/2}]/g_{00},$$

Since $g_{rs}dx^rdx^s < 0$, it is clear that $dx^0(1) > 0$ and $dx^0(2) < 0$. It is clear then that $dx^0(1)$ is the time taken for light to travel from A to B and $-dx^0(2)$ is the time taken for light to travel from B to A. The latter can be seen by changing dx^r to $-dx^r$ in the quadratic equation and noting that $-dx^0(2)$ is the positive solution to that equation. The time at B that is synchronized with the time at A at which the light signal left is then defined to be $x^0 + (dx^0(1) - dx^0(2))/2$. Specifically, the clock at A records the time taken for the light signal to reach B to be one half of the total time taken for the light signal to travel from A to B and back. This is the wave in which A must measure the spatial distance of B from A. This means that the spatial metric must be taken as

$$(dx^0(1) - dx^0(2))^2 = \gamma_{rs}(x)dx^rdx^s, \gamma_{rs} = (g_{0r}g_{0s} - g_{rs})/g_{00}^2$$

6.9 Design of a quantum neural network using Brehmstrallung radiation by charged particles

A charged particle moving along a trajectory generates an electromagnetic field given by the Lienard-Wiechert potentials obtained from the retarded potential solution to the Maxwell equations. By controlling the charged particle trajectory, we can therefore control the generated electromagnetic field in such a way that its interaction with an electron bound to a nucleus in an atom causes the electronic wave function to vary in such a way that the associated probability density of the electron's position tracks a given probability density. In this way, the information contained in an entire family of probability densities parametrized by time can be compressed into the trajectory of a charged particle. To ensure better approximations, we can increase the number of charged particles and control all their trajectories appropriately.

Let Q be a charge moving along a trajectory $R(t)$. At time t, its velocity is $v(t) = R'(t)$ and its acceleration is $w(t) = v'(t) = R''(t)$. The magnetic vector potential produced by this charge is

$$A(t,r) = \int J(t - |r - r'|)/c, r')d^3r'/|r - r'|$$

and the electric scalar potential produced by the same charge is

$$V(t,r) = \int \rho(t - |r - r'|/c, r')d^3r'/|r - r'|$$

where

$$\rho(t,r) = Qv(t)\delta^3(r - R(t), J(t,r) = Qv(t)\delta^3(r - R(t))$$

An approximate calculation gives

$$A(t,r) = \int J(t,r')d^3r'/|r - r'| - c^{-1}\partial_t \int J(t,r')d^3r'$$

$$= Qv(t)/|r - R(t)| - Qv'(t)/c$$

The next approximation term is

$$(1/2c^2)\partial_t^2 \int |r - r'|J(t,r')d^3r'$$

$$= (2c^2)^{-1}\partial_t^2(|r - R(t)|Qv(t))$$

More generally, for a system of charges $Q_1, ..., Q_M$ moving along trajectories $R_1(t), ..., R_M(t)$, the magnetic vector potential upto $O(1/c^2)$ terms is given by

$$A(t,r) = \sum_k Q_k v_k(t)/|r - R_k(t)| - \sum_k Q_k v_k'(t)/c + (1/2c^2)\sum_k Q_k \partial_t^2(|r - R(t)|v(t))$$

Likewise the electric scalar potential upto $O(1/c^2)$ terms is given by

$$V(t,r) = \sum_k Q_k/|r - R_k(t)| + (1/2c^2)\sum_k Q_k\partial_t^2(|r - R(t)|)$$

In general, we have the Taylor series

$$A(t,r) = \sum_{n\geq 0}(n!)^{-1}(\partial_t^n(Q_k v_k(t)|r - R_k(t)|^{n-1})$$

and

$$V(t,r) = \sum_k \sum_{n\geq 0}(n!)^{-1}(\partial_t^n)(Q_k|r - R_k(t)|^{n-1})$$

6.10 Design of a quantum neural network based on Feynman diagrammatic approach to non-Abelian gauge field theory

The three point function corresponding to a cubic term in the interaction Hamiltonian between the Dirac current of a non-Abelian matter field and the corresponding gauge field is evaluated using a triangle Feynman diagram. The vertices of this triangle are connected to gauge fields while the sides of the triangle define the matter field propagators. Two vertices of this triangle are connected to the ends of gauge field propagators and the third vertex to a control classical gauge field. this results in corrections to the gauge field propagator caused by interaction with a control classical gauge field. This corrected propagator yields an alteration in the scattering amplitude of two matter field particles that is controlled by the classical gauge field attached to the third triangle vertex. This control gauge field can be made adaptive so that the resulting scattering probability density in momentum space tracks any given family of probability densities.

Let $A_\mu = A_\mu^a i\tau_a$ be the $\mathfrak{g}$ valued gauge quantum field where τ_a are Hermitian matrices so that $\mathfrak{g}$ is a Lie subalgera of $\mathfrak{su}(N)$, the space of $N \times N$ skew-Hermitian matrices. The three point anomalous Feynman diagram involving a triangular graph with interior Fermions having currents $J_a^\mu(x) = \bar{\psi}(x)\gamma_{Ta}^\mu\psi(x), J_b^\nu(y), J_c^\rho(z)$ interacting with the three exterior interaction vertices at which non-Abelian c-number gauge fields $A_\mu^a(x), A_\nu^b(y)$ and $A_\rho^c(z)$ located at space-time points x, y, z respectively are connected, is given by

$$\Gamma_{abc}^{\mu\nu\rho}(x,y,z) =$$

$$Tr(\gamma^\mu T_a S(x-y)\gamma^\nu T_b S(y-z)\gamma^\rho T_c S(z-x))$$

where

$$S(x-y) = < T(\psi(x)\bar{\psi}(y)) >$$

is the Feynman propagator of the non-Abelian Fermion matter field $\psi(x)$. In the four-momentum domain, this vertex is represented by its Fourier transform:

$$\Gamma^{\mu\nu\rho}_{abc}(p,q,k) = \int \Gamma^{\mu\nu\rho}_{abc}(x,y,z)exp(-i(p.x+q.y+k.z))d^4xd^4yd^4z$$

The This vertex is shown in standard text-books to be not conserved, ie, anomalous. Note that

$$\int f(x-y,y-z)exp(-i(p.x+q.y+k.z))d^4xd^4yd^4z$$

$$= \int f(u,v)exp(-i(p.(z+u+v)+q.(z+v)+k.z))d^4ud^4vd^4z$$

$$= (\int f(u,v)exp(-i(p.u+(p+q).v))d^4ud^4v)\delta^4(k+p+q)$$

$$= \delta^4(k+p+q)\hat{f}(p,p+q)$$

Using this formula combining with the Non-Abelian matter field Fermion propagator in the absence of radiative corrections,

$$S(p) = \int S(x)exp(-ip.x))d^4x = (\gamma.p-m+i0)^{-1}I$$

$$= (\gamma.p+m)/(p^2-m^2)$$

we can easily calculate this anomalus current vertex. Now, the contribution of such a vertex to a physical process of say scattering of two Fermions by exchange of a gauge boson in the presence of a classical gauge boson field can be computed. TWo of the three vertices will then be attached to the two endpoints of a gauge boson propagator being exchanged by the Fermions and the third vertex will be attached to the classical external gauge boson field. The resulting correction to the scattering amplitude will have the form

$$\bar{u}(p_1')\gamma^\mu T_a u(p_1)\bar{u}(p_1')\gamma^\nu T_b u(p_1)(D(p_1-p_1')\Gamma^\rho_c)^{ab}_{\mu\nu}(p_1-p_1',q)D(q,p_2'-p_2))A^c_\rho(q)$$

6.11 Neumann problem within a sphere of radius R with applications to EEG signal modeling on the spherical brain surface

The electrostatic potential generated by electric dipoles within the brain cortex satisfies Poisson's equation with the polarization charge density. On the surface of the brain, the normal component of the electric field must vanish and hence solving the resulting boundary value problem amounts to calculating the

Green's function for the Neumann problem with the Laplacian operator. This is calculated most easily using the separation of the radial and angular variables by representing the angular part in terms of spherical harmonics and deriving a radial equation for the radial part of the Green's function and then applying the Neumann boundary condition as well as the delta function discontinuity condition on the radial part of the Green's function to solve for it.

Let $G(r, \hat{r}|r', \hat{r}')$ be the Green's function for $0 \leq r, r' \leq R$. Then G satisfies

$$\nabla^2 G(r, \hat{r}|r', \hat{r}') = \delta^3(r - r')$$

with Neumann boundary conditions

$$\partial G(r, \hat{r}|r', \hat{r}')/\partial r|_{r=R} = 0$$

For $r < r'$, we have $\nabla^2 G = 0$ and likewise for $r > r'$. Thus, we can write the equations in these two regions as

$$(\partial^2/\partial r^2 + (2/r)\partial/\partial r - L^2/r^2)G = 0$$

We write

$$G(r, \hat{r}|r', \hat{r}') = \sum_{lm} G_{lm}(r|r', \hat{r})Y_{lm}(\hat{r})$$

and derive

$$(\partial_r^2 + (2/r)\partial_r)G_{lm} - l(l+1)G_{lm} = 0$$

in these two regions. The solution to this equation that is non-singular at the origin is

$$G_{lm}(r|r', \hat{r}') = C(lm)(r, \hat{r})r^l$$

This solution is valid in the region $r < r'$ while in the region $R > r > r'$, the general solution is

$$G_{lm}(r|r', \hat{r}') = D_1(lm|r'\hat{r}')r^l + D_2(lm|r', \hat{r}')/r^{l+1}$$

The Neumann boundary condition at $r = R$ gives

$$\partial G_{lm}(r|r', \hat{r}')/\partial r|_{r=R} = 0$$

which implies

$$-lD_1(lm|r', \hat{r}')R^{l-1} - (l+1)D_2(lm|r'\hat{r}')/R^{l+2} = 0 - - - (1)$$

The three dimensional delta function in spherical polar coordinates is

$$\delta(r - r')\delta(cos(\theta) - cos(\theta'))\delta(\phi - \phi')/r^2$$

$$= r^{-2}\delta(r - r')\sum_{lm} Y_{lm}(\hat{r}).\bar{Y}_{lm}(\hat{r}')$$

and hence we can write the Green's function differential equation as

$$(\partial_r^2 + (2/r)\partial_r - l(l+1))G_{lm}(r|r'\hat{r}') = r^{-2}\delta(r - r')\bar{Y}_{lm}(\hat{r}')$$

and hence we obtain using the following discontinuity equation derived from this differential equation

$$\partial_r G_{lm}(r'+0) - \partial_r G_{lm}(r'-0) + (2/r')(G_{lm}(r'+0) - G_{lm}(r'-0)) = r'^{-2}\bar{Y}_{lm}(\hat{r}') - - - (2)$$

and

$$G_{lm}(r'+0) - G_{lm}(r'-0) = 0 - - - (3)$$

Equations (1),(2),(3) completely determine the functions $C(lm, r', \hat{r}')$, $D_k(lm, r', \hat{r}')$, k=1,2$andhencethesolutiontotheNeumannproblemfortheLaplaceoperator.$

In the context of EEG signals collected on the surface of the brain generated by dipoles within the brain cortex region, the normal component of the electric field must vanish on the head surface and hence the surface potential on the head is given by solving

$$\nabla^2 U(r, \hat{r}) = \sum_k p_k . \nabla \delta^3(r - r_k)$$

with the Neumann boundary condition on U at $r = R$. This solution is most conveniently expressed in terms of the Neumann Green's function calculated using the above method as

$$U(r, \hat{r}) = \sum_k p_k . \int G(r, \hat{r}|r', \hat{r}')\nabla\delta^3(r' - r_k)d^3r'$$

$$= -\sum_k p_k . nabla' G(r, \hat{r}|r_k, \hat{r}_k)$$

It is clear from this equation that by measuring U at $r = R$, ie, on the head surface, we can estimate $\{p_k, r_k, \hat{r}_k\}$.

6.12 A real time version of filtering for quantum image fields

Estimating the random window function obeying classical stochastic dynamics on a real time basis based on measurements of the transition probability of an atom that interacts with a quantum field after the field passes through the window.

Let the quantum Klein-Gordon field interact with a fan window $W(t, x, y)$ whose dynamics is defined by an sde

$$dW(t, x, y) = \sum_{n,m} a(n, m)\partial_x^n \partial_y^m W(t, x, y)dt + \sigma dB(t, x, y)$$

where B is a Brownian motion field, ie, a zero mean Gaussian field with

$$\mathbb{E}(dB(t,x,y)dB(t',x',y')) = Q(x,y|x',y')dt.\delta_{t,t'}$$

or equivalently,

$$\mathbb{E}(B(t,x,y)B(t',x',y')) = min(t,t')Q(x,y|x',y')$$

Let $\phi_n(x,y), n = 1,2,...$ and $\lambda_n, n = 1,2,...$ be the complete set of orthonormal eigenfunctions and eigenvalues of Q. Then, we define

$$\int B(t,x,y)\phi_n(x,y)dxdy = B_n(t)$$

Then

$$\mathbb{E}(B_n(t)B_m(t')) = \lambda_n\delta_{n,m}min(t,t')$$

implying thereby that $B_n(.)/\sqrt{\lambda_n}, n = 1,2,...$ are independent standard Brownian motion processes. We can write

$$B(t,x,y) = \sum_n B_n(t)\phi_n(x,y)$$

When the quantum Klein-Gordon field $\phi(t,x,y)$ passes through this random classical window, the output quantum field is

$$\psi(t,x,y) = \phi(t,x,y)W(t,x,y)$$

and when this field interacts with an atom, with H_0 as the unperturbed Hamiltonian of the atom, then the resulting perturbed atomic Hamiltonian becomes $H_0 + \int \psi(t,x,y)X(x,y)dxdy = H_0 + V(t)$ where $X(x,y)$ are atomic observables. The probability of the atom making a transition from a state $|n>$ to a state $|m>$ in the time duration $[0,t]$ under this perturbation with the field being in the coherent state $|\phi(u)>$ is then given upto first order perturbation theory by

$$P_t(m|n) = |\int_0^t < m\otimes\phi(u)|V(t)|n\otimes\phi(u) > exp(i(E_n - E_m)t)dt|^2$$

Note that

$$< m\otimes\phi(u)|V(t)|n\otimes\phi(u) >=$$

$$\int < m|X(x,y)|n >< \phi(u)|\psi(t,x,y)|\phi(u) > dxdy$$

6.13 Problems in communication and information theory

Rate distortion theory, equalizers for removing intersymbol interference, adaptive equalizer using decision device. In rate distortion problems of information theory, we encode the iid or more generally, ergodic source at a rate lesser than the source entropy with a distortion during the decoding process lesser than a prescribed threshold. The cost of encoding at a rate lesser than the entropy is that distortion is introduced and subject to the condition that the distortion should not exceed a given level, we minimize the code rate. Rate distortion theory is therefore a generalization of the Shannon noiseless compression theorem in that when the distortion level is set to zero, the minimum code rate coincides with the source entropy.

Give a physical interpretation of the algorithm in rate distortion theory. Given a channel with input output relation

$$y(t) = \sum_{k=1}^{d} a(k)x(t - t_k) + w(t)$$

design an equalizer of the form

$$\hat{x}(t) = \sum_{k=1}^{p} b(k)y(t - \tau_k)$$

ie, choose the parameters $\{b(k)\}$ so that

$$\mathbb{E}(x(t) - \hat{x}(t))^2$$

is a minimum. In case noise $w(t)$ is absent, under what conditions on $a(k), b(m), t_k, \tau_m$ can we get

$$\hat{x}(t) = x(t)$$

In an adaptive equalizer that uses a decision device, the transmitter first transmits a pilot sequence $x[n]$ through the channel where each $x[n]$ takes values in an M-ary alphabet A. The pilot sequence $x[n]$ is known to the receiver. Specfically, the input analog signal $x(t)$ is sampled at the Nyquist rate $2W = 1/T$ and each sample is quantized into N levels, then block of such samples, each block having length n-samples are encoded into $M = N^n$ signals which is transmitted over the channel by some sort of M-ary keying, for example MPSK in which the kth signal is $s_k(t) = A.cos(\omega t + 2\pi(k-1)/M), k = 1, 2, ..., M$. Here $\omega = 2\pi/\tau$ where $\tau = nT$ is the duration of each $s_k(t)$. The channel distorts the transmitted signal and couples it to noise, so that if $y[n]$ is the received signal, we can write $y[n] = f(x[n], x[n-1], ..., x[n-q], w[n])$, where

q is the channel memory. The adaptive equalizer at the receiver end generates an estimate $x_1[n] = g(y[n], y[n-1], ..., y[n-p], \theta[n])$ of $x[n]$ where θ is a vector adaptive parameter of the equalizer. Finally, the decision device uses some sort of a thresholding to generate an estimate $\hat{x}[n]$ of $x[n]$ from $x_1[n]$. Such a decision device is required because $x[n]$ is a finite level (M levels) quantized signal while the output $x_1[n]$ of the adaptive equalizer is not quantized. This forms the training process for the weights θ of the adaptive equalizer. The adaptive gradient scheme used for updating the parameters θ has the form

$$\theta[n+1] = \theta[n] - \mu.\nabla_{\theta[n]}(x[n] - x_1[n])^2$$

$$= \theta[n] + 2\mu(x[n] - g(y[n], ..., y[n-q], \theta[n]))\nabla_{\theta[n]}g(y[n], ..., y[n-q]))$$

The weights $\theta[n]$ of the adaptive equalizer are then readjusted based on minimizing the error $x[n] - \hat{x}[n]$ between the pilot sequence and the output of the decision device. This is some sort of a fine tuning of the equalizer parameters. After the training is complete, the transmitter and receiver switch over to the normal operation in which the transmitter transmits the desired message while the reciever's trained equalizer is cutoff from the decision device.

6.14 Performance analysis of the equalizer based on the LMS and RLS algorithm

Let $x(t), y(t)$ be two stochastic processes in discrete time with $y(t)$ an $\mathbb{R}^p$ valued vector process and $x(t)$ an $\mathbb{R}^q$ valued vector process. Assume that x, y are jointly statoinary, jointly Gaussian processes with zero mean. Consider the problem of estimating $x(t)$ based on $y(t)$, ie, calcualting a matrix A of size $q \times p$ so that

$$\sum_{t=0}^{n} \lambda^{n-t} \parallel x(t) - Ay(t) \parallel^2$$

is a minimum. Denote the resulting estimate by $A(n)$, ie,

$$A(n) = argmin_A \sum_{t=0}^{n} \lambda^{n-t} \parallel x(t) - Ay(t) \parallel^2 = R_{xy}(n).R_y(n)^{-1}$$

where

$$R_y(n) = \sum_{t=0}^{n} \lambda^{n-t} y(t)y(t)^T, R_{xy}(n) = \sum_{t=0}^{n} \lambda^{n-t} x(t)y(t)^T$$

Calculate using perturbation theory, the statistics of $A(n)$.

hint: Write

$$R_y(n) = R_{y0}(n) + \delta R_y(n), R_{xy}(n) = R_{xy0}(n) + \delta R_{xy}(n)$$

where

$$R_{y0}(n) = \sum_{t=0}^{n} \lambda^{n-t} \mathbb{E}(y(t)y(t)^T)/(1 - \lambda^{n+1}) = \mathbb{E}(y(t)y(t)^T)$$

and

$$R_{xy0}(n) = \sum_{t=0}^{n} \lambda^{n-t}\mathbb{E}(x(t)y(t)^T)/(1-\lambda^{n+1}) = \mathbb{E}(x(t)y(t)^T)$$

Now apply the perturbation expansion

$$(R_{y0}(n) + \delta R_y(n))^{-1} = [\sum_{k=0}^{\infty}(R_{y0}(n)^{-1}.\delta R_y(n))^k]R_{y0}(n)^{-1}$$

and use the formula for the joint higher order moments of a zero mean Gaussian random vector.

Chapter 7

Quantum Effective Action and Symmetry Breaking in Field Theories

Electromagnetic field of dipoles, stochastic quantum Boltzmann equation from the Belavkin quantum filter, supersymmetric field theory, Coupling between Boltzmann equation and the Einstein-Maxwell field equations in classical and quantum domains, plasma as a density operator field, Belavkin quantum filter in the Wigner distribution domain as a classical KK filter with quantum corrections, oscillations of a quantum plasma, Belavkin filter for Bosonic and Fermionic fields in a cavity interacting with a noisy bath, Belavkin filter for simultaneous state and parameter estimation, mode determination in nonlinear electromagnetics

7.1 Symmetry breaking in quantum field theory

The quantum effective action, relationship between the propagator and the quantum effective action, spontaneous and approximate symmetry breaking leading to massless Goldstone Bosons and pseudo-Goldstone Bosons.

Let $S[\phi]$ be the action functional of a field ϕ or a set of fields defined on the four dimensional space-time manifoldet J be an external classical current field and consider the Feynman path integral

$$W(J) = \int exp(iS(\phi) - iJ.\phi)D\phi$$

where

$$J.\phi = \int J(x)\phi(x)d^4x$$

The average value of the quantum field ϕ when the current J is applied is given by

$$<\phi>(x) = \int exp(iS(\phi) - iJ.\phi)\phi(x)$$

D$\phi/W(J) = i\delta logW(J)/\delta J(x) Let \phi_0(x)$ be a classical field and choose $J = J_{\phi_0}$ so that

$$<\phi(x)>_{J_{\phi_0}} = \phi_0(x), x \in \mathbb{R}^4$$

Then define the quantum effective action evaluated at ϕ_0 as

$$\Gamma(\phi_0) = -ilog(W(J_{\phi_0})) + J_{\phi_0}.\phi_0$$

It is easily seen (just as in the large deviation principle in classical probability theory) that

$$\Gamma(\phi_0) = ext_J[-ilog(W(J)) + J.\phi_0]$$

DOI: 10.1201/9781003728627-7

where ext denotes extremization over the classical current field J. We note the following quantum equations of motion:

$$\delta\Gamma(\phi_0)/\delta\phi_0(x) = -\phi_0(x).\delta J_{\phi_0}/\delta\phi_0(x) + \phi_0(x).\delta J_{\phi_0}/\delta\phi_0(x) + J_{\phi_0}(x) = J_{\phi_0}(x)$$

In particular, if $\phi_0(x)$ is the expected value of $\phi(x)$ in the absence of any current, ie

$$\phi_0(x) = i[\delta W(J)/\delta J(x)]|_{J=0}$$

then we get for the quantum equations of motion

$$\delta\Gamma(\phi_0)/\delta\phi_0(x) = 0$$

ie the quantum expected value of the quantum field ϕ in the absence of any external current source is an extremum of the quantum effective action, it can be shown that it is in fact a local minimum of the quantum effective action. We note that the propagator of the quantum field ϕ in the presence of an external classical current field J is given by

$$\Delta_J(x,y) = -[\delta^2 W(J)/\delta J(x)\delta J(y)].W(J)^{-1} = \frac{\int exp(iS(\phi) - iJ.\phi)\phi(x)\phi(y)D\phi}{\int exp(iS(\phi))D\phi}$$

Also

$$i\delta log(W(J))/\delta J(x) = <\phi(x)>_J,$$

$$-\delta^2 log(W(J))/\delta J(x)\delta J(y) = \Delta_J(x,y) - <\phi(x)>_J<\phi(y)>_J$$

In particular, if $S(\phi)$ is symmetric in ϕ, ie, $S(-\phi) = S(\phi)$, then $<\phi(x)>_0 = 0$ ($<.>_0$ means quantum expectation with zero classical current), in which case, we get

$$\Delta_0(x,y) = [-\delta^2 log(W(J))/\delta J(x)\delta J(y)]|_{J=0}$$

In analogy with the case of the vector Klein-Gordon field where the Lagrangian density if

$$L(\phi, \partial_\mu\phi) = (1/2)(\partial_\mu\phi(x))^T K(\partial^\mu\phi(x)) - (1/2)\phi(x)^T M^2.\phi(x)$$

with K as the kinetic and strain matrix and M as the mass matrix, where we find the matrix propagator at zero external current to be given by

$$\Delta_0(x,y) = <T(\phi(x)\phi(y)^T)> = -i(K\Box + M^2)^{-1}(x,y), \Box = \partial_\mu\partial^\mu$$

or equivalently in the four-momentum domain,

$$\Delta_0(q) = -(Kq^2 - M^2)^{-1}$$

it follows that the inverse of the propagator at zero four momentum equals the square of the mass matrix:

$$\Delta_0(0)^{-1} = M^2$$

To get an improved estimate of the mass matrix, we must use the quantum effective action in place of the action since in quantum field theory, the equations of motion must be written down for the quantum averaged field $\phi_0(x)$ and not for the classical field. The classical field equations for the action $S(\phi) - J.\phi$ are

$$\delta S(\phi)/\delta\phi(x) = J(x)$$

while as we saw above, the quantum field equations are

$$\delta\Gamma(\phi_0)/\delta\phi_0(x) = J_{\phi_0}(x)$$

or equivalently, if J is the classical current and $\phi_J]$ is the averaged quantum field in the presence of this current, then

$$\delta\Gamma_J(\phi_J)/\delta\phi_J(x) = J(x)$$

Thus the propagator should be taken as given by

$$\Delta^{-1}(x,y) = -\delta^2\Gamma(\phi_0)/\delta\phi_0(x)\delta\phi_0(y) = -\delta J_{\phi_0}(x)/\delta\phi_0(y) = -\delta J_{\phi_0}(y)/\delta\phi_0(x)$$

Note that we have that the average of $\phi(x)$ when the current is J is given by

$$\phi_J(x) = iW(J)^{-1}\delta W(J)/\delta J(x)$$

so that

$$\delta\phi_J(x)/\delta J(y)$$

$$= iW(J)^{-1}\delta^2 W(J)/\delta J(x)\delta J(y) - iW(J)^{-2}(\delta W(J)/\delta J(x)).(\delta W(J)/\delta J(y))$$

$$= -i\Delta_J(x,y) - iW(J)^{-2}(\delta W(J)/\delta\phi(x)).(\delta W(J)/\delta J(y))$$

where $\Delta_J(x,y)$ is the propagator at current J. Setting $J = J_{\phi_0}$ in this equation gives

$$(\delta\phi_J(x)/\delta J(y))|_{J=J_{\phi_0}} = -\Delta(x,y) + i\phi_0(x)\phi_0(y)$$

We choose $\phi_0 = 0$ corresponding to the vacuum and then if $J_0 = J_{\phi_0=0}$, we can write the above equation as

$$\delta\phi_0(x)/\delta J_0(y) = -\Delta_0(x,y) - - - (2)$$

Now the quantum equation of motion

$$\delta\Gamma(\phi_0)/\delta\phi_0(x) = J_{\phi_0}(x)$$

gives on another differentiation

$$\delta^2\Gamma(\phi_0)/\delta\phi_0(x)\delta\phi_0(y) = \delta J_{\phi_0}(x)/\delta\phi_0(y) - - - (3)$$

Comparing (2) with (3) at $\phi_0 = 0$ gives us

$$-\Delta_0^{-1}(x,y) = (\delta^2\Gamma(\phi_0)/\delta\phi_0(x)\delta\phi_0(y))|_{\phi_0=0}$$

as mentioned above. In other words, the Hessian of the quantum effective action in the ground state (zero field) equals the negative of the inverse propagator in the ground state which is the square of the mass matrix.

Symmetry breaking in quantum field theory:

Let $V_0(\phi)$ be the quantum effective action. Suppose that this potential is invariant under the infinitesimal gauge transformation

$$\phi \to \phi + \epsilon t.\phi$$

or more precisely, in terms of components

$$\phi_n \to \phi_n + \epsilon.\sum_m t_{nm}\phi_m$$

where $t = ((t_{nm}))$ is a skew symmetric real matrix so it becomes a purely imaginary Hermitian matrix. Then, we have

$$\sum_{n,m} \frac{\partial V_0(\phi)}{\partial \phi_n} t_{nm}\phi_m = 0 - - - -(1)$$

This is true for all fields $\phi = ((\phi_n)))$ with t running over the Lie algebra of the gauge group G which is assumed to be a Lie subgroup of $U(N)$. Since we are assuming the generators t of the gauge group to be real skew-symmetric matrices, it follows, in fact, that the gauge group here is, in fact, a subgroup of $O(N)$. Note that $U(N)$ can be embedded into $O(2N)$ as follows: Let $A \in U(N)$. Then, we can write

$$A = A_1 + iA_2$$

where A_1, A_2 are real. Now for any $\psi = \psi_1 + i\psi_2 \in \mathbb{C}^N$ with $\psi_1, \psi_2 \in \mathbb{R}^N$, we have

$$A\psi = (A_1\psi_1 - A_2\psi_2 + i(A_1\psi_2 + A_2\psi_1)$$

so we can equivalently regard the transformation $\psi \to A\psi$ in $\mathbb{C}^N$ as the transformation

$$[\psi_1^T, \psi_2]^T \to [(A_1\psi_1 - A_2\psi_2)^T, (A_1\psi_2 + A_2\psi_1)^T]$$

in $\mathbb{R}^{2N}$. The matrix of this transformation is then

$$\tau(A) = \begin{pmatrix} A_1 & -A_2 \\ A_2 & A_1 \end{pmatrix}$$

Since

$$\| \psi \|^2 = \psi_1^T\psi_1 + \psi_2^T\psi_2 = \| \psi_1 \|^2 + \| \psi_2 \|^2$$

it follows then that $A \in U(N)$ implies $\tau(A) \in O(2N)$. Further, suppose that A is an infinitesimal element of $U(N)$, ie, $A = I + \epsilon.X$ where $X^* = -X$. Then writing $X = X_1 + iX_2$ with X_1, X_2 real, it follows that

$$\tau(A) = I + \epsilon.\tau(X)$$

with

$$X_1^T = -X_1, X_2^T = X_2$$

ie

$$\tau(X)^T = -\tau(X)$$

and this implies in particular that the diagonal entries of $\tau(X)$ are zero and hence

$$0 = Tr(\tau(X)) = 2.Tr(X_1)$$

This means the the Lie algebra of $O(2N)$ is same as the Lie algebra of $SO(2N)$ which is the space of all $2N \times 2N$ real skew-symmetric matrices. Note that $O(2N) = SO(2N) \cup PSO(2N)$ as a disjoint union where P is any element of $O(2N)$ having determinant -1. Now if ϕ_0 is a local minimum of ϕ, ie

$$\partial V(\phi_0)/\partial\phi_k = 0, k = 1, 2, ..., N, ((\partial^2 V(\phi_0))/\partial\phi_k\partial\phi_m)) \geq 0$$

then it follows from (1) on differentiating w.r.t ϕ_k and then setting $\phi = \phi_0$ that

$$\sum_{n,m}(\partial V(\phi_0)/\partial\phi_k\partial\phi_n)t_{nm}\phi_{0m} = 0$$

which means that $t\phi_0$ is an eigenvector of the mass matrix square

$$M^2 = ((\partial V(\phi_0)/\partial\phi_k\partial\phi_m))$$

with zero eigenvalue. This means spontaneous symmetry breaking caused by the system falling to the ground/vacuum state ϕ_0 is accompanied by the generation of massless Bosons called Goldstone Bosons. Recall the definition of the mass matrix square above as the negative of inverse of propagator at zero four momentum. The above equation for the =eigenvector corresponding to zero mass Goldstone Bosons can also be therefore expressed as

$$\Delta(0)^{-1}(t\phi_0) = 0$$

where $\Delta(0)$ is the propagator at zero momentum.

7.2 On the relativistic calcuation of the electric and magnetic fields produced by electric and magnetic dipole's in translatory and rotational motion

Consider an electric dipole at time $t = 0$ consisting of a charge $-Q$ located at $r = -d/2$ and a charge $+Q$ located at $r = d/2$ where d is small, Q large so that $p(0) = Qd$ is fixed. After time t, the dipole moment is $p(t) = R(t)p(0)$ where $R(t)$ is the 3-D rotation matrix applied to the dipole around its centre of mass. Further, the CM of the diple after time t moves to the position $R_0(t)$. Thus, after time t, the CM of the dipole is at $r_0(t)$ with the charges $Q, -Q$ being located at at $r_0(t) + r_1(t)$, $\mathrm{R}_0(t) + r_2(t)$ respectively. We have approximately $r_1(t) = R(t)d/2, r_2(t) = -R(t)d/2$. Ideally speaking, once the rigid diple starts undergoing translatory and rotational motion, it no longer remains rigid in the lab frame, it is rigid only in the frame of the rotating dipole. The approximate kinetic contribution to the Lagrangian of the dipole is given by

$$T = -m_0c^2\sqrt{1 - |r_0'(t) + R'(t)d/2|^2/c^2} - m_0c^2\sqrt{1 - |r_0'(t) - R'(t)d/2|^2/c^2}$$

and the potential energy of interaction of this dipole with an external electric field $E(t,r)$ is given by

$$U = -(R(t)p(0), E(t, r_0(t)))$$

The aim is to determine the vector $r_0(t)$ as well as the rotation matrix $R(t)$ as a function of time. This can be achieved by applying the Euler-Lagrange equations to

$$L(t, r_0(t), R(t), r_0'(t), R'(t)) = T - U$$

after representing $R(t)$ using three Euler angles.

7.3 A more accurate computation based on taking relativity into account in computing the potential energy of interaction between the dipole and the electric field

Let $L(t)$ denote the 4×4 matrix corresponding to the Lorentz transformation of the space-time coordinates from the lab frame to the moving frame of the dipole at lab time t without taking rotation into account. Then, if (r, t_0) is the space-time coordinate vector relative to the lab frame (ie, moving frame of the dipole at time $t = 0$) and (r', t') is the space-time coordinate vector relative to the moving frame at time t, then

$$(r', t') = L(t)(r, t_0)$$

so the Lorentz transformation between the moving frame at time t and the moving frame at time $t + dt$ is given by

$$L(t + dt)L(t)^{-1} = I + dL(t).L(t)^{-1} = I + L'(t)L(t)^{-1}dt$$

Now if frame S' is obtained from frame S by translating with a velocity v over a time interval t as measured in the the first frame, then the spatial coordinates as measured in the two systems are related by

$$t' = \gamma(v)(t - v.r/c^2), r' = (r - (r.v)v/v^2) + \gamma(v)((r.v)v/v^2 - vt)$$
$$= r - \gamma(v)vt + (\gamma(v) - 1)(r.v)v/v^2$$

These equations define the general Lorentz transformation associated only with boost, ie translatory motion without rotation:

$$L = L(v), L(v)(r, t) = (r', t')$$

Now it is clear that if r_1, r_2 are two spatial vectors as seen in a frame S at the same time t and r_1', r_2' are the same vectors as seen in S' at the corresponding times t_1', t_2' then the Lorentz transformation gives

$$(r_1', t_1') = L(v)(r_1, t), (r_2', t_2') = L(v)(r_2', t)$$

Expressing the spatial part of these Lorentz transformation equations as

$$r_1' = L_{11}r_1 + L_{12}t, r_2' = L_{11}r_2 + L_{12}t, L_{11} = L_{11}(v), L_{12} = L_{12}(v)$$

we get

$$r_2' - r_1' = L_{11}(r_2 - r_1) = r_2 - r_+1 = L_{11}^{-1}(r_2' - r_1')$$

Note that

$$L_{11} = I_3 + (\gamma(v) - 1)vv^T/v^2, L_{12} = -\gamma(v)v$$

As applied to our dipole problem, in the moving frame of the dipole CM, the charge Q is located at $R(t)d/2$ and the charge $-Q$ at $-R(t)d/2$. It follows that the dipole moment as seen in the lab frame is given by

$$p(t) = QL_{11}(v(t))^{-1}R(t)p(0), v(t) = r_0'(t), \gamma(v(t)) = (1 - |v(t)|^2/c^2)^{-1/2}$$

where $R(t)$ is the 3-D rotation applied to the dipole in its moving frame, ie, in a frame relative to which it is at rest and is hence a rigid body in that frame. The interaction potential energy between the dipole and the electric field in the lab frame is then

$$U = -(p(t), E(t, r_0(t))$$

More precisely, the contribution to the Lagrangian of the dipole in the lab frame should be derived from that of a single point charge moving in an electromagnetic field. The interaction component of this energy is $-q(V - v.A)$ where V is the electric potential and A is the magnetic vector potential in the lab frame. We have to sum this contribution over both the charges $\pm Q$ in the dipole. The position of Q at time t as seen in the lab frame is

$$r_1(t) = r_0(t) + L(v(t))^{-1}R(t)d/2$$

because its position in the moving frame of the CM of the dipole is $R(t)d/2$ in view of the fact that the dipole is a rigid body in its rest frame. Likewise, the position of $-Q$ at time t as seen in the lab frame at time t

$$r_2(t) = r_0(t) - L(v(t))^{-1}R(t)d/2$$

It follows then that the Lagrangian of the dipole in the Lab frame is

$$L(t, r_0(t), v(t), R(t), R'(t)) = -m_0c^2(1-|r_1'(t)|^2/c^2)^{1/2} - m_0c^2(1-|r_2'(t)|^2/c^2)^{1/2}$$
$$-Q(V(t, r_1(t)) - V(t, r_2(t)) - (r_1'(t), A(t, r_1(t))) + (r_2'(t), A(t, r_2(t)))$$

and the resulting equations of motion of the dipole can be derived from this.

7.4 Some topics in algebra and algebraic geometry: Prime ideals, maximal ideals, closed sets in the Zariski topology on the spectrum of a ring, sheaf of rings

1.Maximal ideals and prime ideals of a ring. Proof that a maximal ideal is also a prime ideal.

Let M be a maximal ideal of a ring R. Suppose that $a, b \in R$ are such that $ab \in M$ but $a \notin M$. Then since M is a maximal ideal, it follows that the ideal generated by M and a is the whole of R. Hence, we can find $c \in R$ and $m \in M$ such that

$$ca + m = 1$$

It follows that

$$b = cab + bm \in M$$

since $ab, m \in M$ and M is an ideal and therefore $cab, bm \in M$.

Let $\mathfrak{a}_\alpha, \alpha \in I$ be a family of ideals of a ring R. For any ideal $\mathfrak{a}$ of R, let $V(\mathfrak{a})$ denote set of all prime ideals of R that contain $\mathfrak{a}$. Then show that

$$V(\sum_\alpha a_\alpha = \bigcap_\alpha V(\mathfrak{a}_\alpha)$$

Let $\mathfrak{a}_k, k = 1, 2$ be two ideals of a ring R. Then, show that

$$V(\mathfrak{a}_1.\mathfrak{a}_2) = V(\mathfrak{a}_1) \cup V(\mathfrak{a}_2)$$

If $\mathfrak{p}$ is any ideal containing either $\mathfrak{a}_k, k = 1, 2$, then $\mathfrak{p}$ will also contain $\mathfrak{a}_1.\mathfrak{a}_2$ because $\mathfrak{a}_1.\mathfrak{a}_2 \subset \mathfrak{a}_1 \cap \mathfrak{a}_2$ because $\mathfrak{a}_k, k = 1, 2$ are both ideals. Conversely, if $\mathfrak{p}$ is a prime ideal containing $\mathfrak{a}_1.\mathfrak{a}_2$ and suppose $\mathfrak{p}$ does not contain $\mathfrak{a}_1$. Then, we can choose an $a_1 \in \mathfrak{a}_1$ such that $a_1 \notin \mathfrak{p}$. But then $a_1.\mathfrak{a}_2 \subset \mathfrak{p}$ by hypothesis and hence $a_1 a_2 \in \mathfrak{p}$ for all $a_2 \in \mathfrak{a}_2$. By the definition of prime ideal, since $a_1 \notin \mathfrak{p}$, it follows then that $a_2 \in \mathfrak{p}$ for all $a_2 \in \mathfrak{p}$, ie, $\mathfrak{a}_2 \subset \mathfrak{p}$.

2.Grothendieck's notion of sheafs of functions, sheaves of rings, sheafs of groups on a topological space. Grothendieck's idea of a scheme as a generalization of the notion of a variety with functions defined on the variety. Notion of localization of a ring relative to a prime ideal. First, we must however discuss how to introduce the notion of fractions on a ring relative to a multiplier.

7.5 Composition of two Lorentz boosts can give rotations

Let $L(t)$ be the Lorentz transformation matrix from the lab frame K to the non-rotating frame $K(t)$ at time t attached to a particle moving along the trajectory $r_0(t)$. Compute $L(t)$ and show that the lorentz transformation over the infinitesimal time interval $[t, t + dt]$ given by $L(t + dt)L(t)^{-1}$ is a Lorentz transformation containing in general, both boosts and rotations.

7.6 Problems in electrostatics

N concentric spherical shells of radii $a_1 < a_2 < .... < a_N$ are arranged such that the first r spherical surfaces of radii $a_1, ..., a_r$ are at potentials $V_1, ..., V_r$ respectively and charges $Q_{r+1}, ...Q_N$ are uniformly distributed over the surfaces of the spheres of radii $a_{r+1}, ..., a_N$. Find the potential everywhere in space assuming that infinity is at zero potential.

Hint: Radially symmetric Laplace's equation for the potential $V(r)$ implies that the potential in the region $a_i < r < a_{i+1}$ has the form

$$V(r) = C_i/r + D_i, a_i < r < a_{i+1}, i = 0, 1, ..., N$$

where $a_0 = 0, a_{N+1} = \infty$. Since $V(r) \to 0$ as $r \to \infty$, it follows that $D_N = 0$. Since by the integral form of Gauss' law, the electric field for $0 \le r < a_1$ is zero, it follows that $C_0 = 0$. Also, the boundary conditions imply

$$V(a_i) = C_i/a_i + D_i = V_i = C_{i-1}/a_i + D_{i-1} i = 1, 2, ..., r$$

Further, the boundary condition that the difference between the normal components of the electric field on either side of a surface equals the surface charge density divided by ϵ_0 gives

$$-C_{i+1}/a_{i+1}^2 + C_i/a_{i+1}^2 = Q_{i+1}/4\pi\epsilon_0 a_{i+1}^2, i = r, r+1, ..., N-1$$

These equations are sufficient to determine the potential everywhere.

7.7 The quantum Boltzmann equation for a system of identical particles within a cavity resonator interacting with the quantum electromagnetic field of the cavity. How to use the Belavkin quantum filter to obtain an estimate of the marginal density of each particle

Let $A(t, r), V(t, r)$ be the quantum magnetic vector potential and the quantum electric potential within the cavity. We can expand

$$A(t, r) = \sum_k (a_k \psi_k(t, r) + a_k^* \psi_k(t, r)^*), V(t, r) = \sum_k (a_k \chi_k(t, r) + a_k^* \chi_k(t, r))$$

where $\psi_k(t, r)$ and $\chi_k(t, r)$ are respectively three vector and scalar valued functions of space-time. The $a_k's$ satisfy the canonical commutation relations

$$[a_k, a_m^*] = \delta(k, m), k, m = 1, 2, ..., M$$

The joint density of N particles and the field interacting with each other in this cavity satisfies

$$i\rho'(t) = [H_S + H_F + H_I, \rho(t)]$$

where

$$H_F = \sum_k \omega(k) a_k^* a_k$$

is the cavity field Hamiltonian,

$$H_S + H_I = \sum_{k=1}^N (p_k + eA(t, r_k))^2/2m - eU(r_k) - eV(t, r_k))$$

so that

$$H_S = \sum_k (p_k^2/2m - eU(r_k))$$

is the Hamiltonian of the system of particles modeled as atoms, each having one electron and

$$H_I = (e/2m)\sum_k ((p_k, A(t, r_k)) + (A(t, r_k), p_k)) - eV(t, r_k))$$

is the interaction Hamiltonian between the system of atoms and the cavity field. Note that the dynamical variables that describe the atom are $(p_k, r_k), k = 1, 2, ..., N$ while the dynamical variables that describe the field are $a_k, a_k^*, k = 1, 2, ..., N$. In addition, the total system of particles and field within the cavity can interact with the quantum noisy bath surrounding the cavity and this totality of cavity with field and particles and bath evolves according to a unitary evolution operator $U(t)$ in accordance with the Hudson-Parthasarathy-noisy Schrodinger equation

$$dU(t) = -(H_S + H_F + H_I(t) + P)dt + L_k dA_k + M_k dA_k^* + S_k d\Lambda_k)U(t)$$

where A_k, A_k^*, Λ_k are annihilation, creation and conservation processes of the bath and L_k, M_k, S_k are cavity space operators, expressible as functions of $\{a_k, a_k^*, r_m, p_m\}_{k,m}$. The state of the cavity field with atoms at time t is denoted by $\rho_{SF}(t)$ while the total state of the cavity field with atoms and the surrounding bath is denoted by $\rho(t) = \rho_{SFB}(t)$. Assume that initially the cavity atoms are in the state $\rho_S(0)$ while the cavity field is in the coherent state $|\phi(u)>$ and the bath in the coherent state $|\psi(v)>$. The initial state of system, field and bath is taken as the tensor product of these three:

$$\rho(0) = \rho_s(0) \otimes |\phi(u)><\phi(u)| \otimes |\psi(v)><\psi(v)|$$

Note that

$$a_k|\phi(u)>= u(k)|\phi(u)>, A_k(t)|\psi(v)>= (\int_0^t v_k(s)ds)|\psi(v)>$$

The overall Hilbert space is the tensor product of $L^2(\mathbb{R}^3)^{\otimes N}$, the Hilbert space of the atoms, the Hilbert space $\Gamma_s(\mathbb{C}^M)$ which is isomorphic to $L^2(\mathbb{R})^{\otimes M}$, the Hilbert space on which the cavity field operators $a_k, a_k^*, k = 1, 2, ..., M$ are defined and finally the bath Boson Fock space $\Gamma_s(L^2(\mathbb{R}_+) \otimes \mathbb{C}^d)$ on which the bath quantum stochastic processes $A_k(t), A_k(t)^*, \Lambda_k(t)$ are defined. Let M denote the number of atoms. Assume that each of the Lindblad operators L_k, M_k, S_k has the separable form, for example, L_k has the form

$$L_k = \sum_m L_{km1}^{\otimes M} \otimes L_{km2} \otimes ... \otimes L_{kmN}$$

we can based on non-demolition measurements in the Hudson-Parthasarathy-Belavkin formalism, write down the Belavkin filter equation for the real time estimate of the joint state of the atomic particles and cavity field modes. This is a stochastic Schrodinger equation. Then by taking the trace of this equation over all the field modes followed by a trace over all but the first of the atomic particle Hilbert spaces, we obtain the quantum Boltzmann equation for the estimate of the single particle state. Specifically, if $\hat{\rho}_{SF}$ denotes an estimate of the joint state of the system of particles and the cavity field modes, then the Belavkin filter assumes the form

$$d\hat{\rho}_{SF}(t) = \theta(\hat{\rho}_{SF}(t))dt$$
$$+(M(t)\hat{\rho}_{SF}(t)+\hat{\rho}_{SF}(t)M(t)^* - Tr(\hat{\rho}_{SF}(t)(M(t)+M(t)^*)\hat{\rho}_{SF}(t))(dY(t)$$
$$-Tr(\hat{\rho}_{SF}(t)(M(t)+M(t)^*))dt)$$

where $M(t)$ is a system operator (ie, a function of the atomic and cavity field observables only). θ is the generator of the system and cavity field space quantum dynamical semigroup and can be expressed in terms of the Hamiltonian and the Lindblad generators. We can include in this Hamiltonian, an interaction Hamiltonian between each pair of atomic particles $V_2(r_k, r_m)$ with all these pairwise interactions being identical copies. Taking the partial trace over the cavity field space and then followed by a trace over all but the first particles with the chaos assumption that $\hat{\rho}_{SF}(t) \approx \hat{\rho}_{S1}(t)^{\otimes M} \otimes \hat{\rho}_F(t)$, we get an equation for $\hat{\rho}_{S1}(t)$ in terms of the field state estimate $\hat{\rho}_F(t)$. The equation for $\hat{\rho}_F(t)$ in terms of $\hat{\rho}_{S1}(t)$ is obtained by taking a partial trace of the Belavkin filter equation over all the particle Hilbert spaces. Note that $M(t)$ is also of the separable form.

7.8 Demonstration that the Belavkin quantum filter for the filtered state estimate of the HPS equation based on non-demolition measurements, when viewed in the Wigner distribution domain, is the same as the Kushner-Kallianpur classical non-linear stochastic filter except for some extra quantum correction terms that can be expanded in powers of Planck's constant

The HPS equation is

$$dU(t) = (-(iH + LL^*/2)dt + LdA(t) - L^*dA(t)^*)U(t)$$

where H, L are system operators and $A(t), A(t)^*$ are the annihilation and creation processes in the HP quantum stochastic calculus satisfying the quantum Ito formula

$$dA.dA^* = dt, dA^*dA = -0, (dA)^2 = 0, (dA^*)^2 = 0$$

$U(t)$ is a unitary evolution operator on the tensor product of the system Hilbert space $\mathfrak{h}$ and the bath Boson Fock space $\Gamma_s(L^2(\mathbb{R}_+))$. Define

$$j_t(X) = U(t)^* X U(t)$$

where X is a system observable. Then j_t is a star unital homomorphism from the space of system operators in to the space of operators in the system and bath space upto time t, ie, in $\mathfrak{h} \otimes \Gamma_s(\mathcal{H}_{t]})$, $\mathcal{H} = L^2(\mathbb{R}_+)$, $\mathcal{H}_{t]} = L^2([0,t])$ From quantum Ito's formula,

$$dj_t(CX) = j_t(\theta_0(X))dt + j_t(\theta_1(X))dA(t) + j_t(\theta_2(X))dA(t)^*$$

where

$$\theta_0(X) = i[H,X] - (1/2)(LL*X + XLL^* - 2LXL^*), \theta_1(X) = [X,L], \theta_2(X) = [L^*,X]$$

Non-demolition measurement made is

$$Y_o(t) = U(t)^* Y_i(t) U(t), Y_i(t) = cA(t) + \bar{c}A(t)^*$$

$Y_o(t), t \geq 0$ is an Abelian family of observables in $\mathfrak{h} \otimes \Gamma_s(\mathcal{H})$ and also $Y_i(t)$ commutes with $j_s(X)$ for $s \geq t$, because unitarity of $U(t)$ combined with the fact that $dY_i(t)$ commutes with system observables and also with observables in $\Gamma_s(L^2[0,t])$ implies that

$$d_T(U(T)^* Y_i(t) U(T)) = 0, T \geq T$$

and therefore,

$$U(T)^* Y_i(t) U(T) = U(t)^* Y_i(t) U(t) = Y_o(t), T \geq t$$

and hence from the fact that $Y_i(t), t \geq 0$ forms an Abelian family, we get

$$[Y_o(t), Y_o(s)] = [U(T)^* Y_i(t) U(T), U(T)^* Y_i(s) U(T)]$$

$$= U(T)^* [Y_i(t), Y_i(s)] U(T) = 0, T \geq t, s$$

and

$$[Y_o(t), j_s(X)] = [U(s)^* Y_i(t) U(s), U(s)^* X U(s)] = U(s)^* [Y_i(t), X] U(s) = 0, s \geq t$$

A simple application of quantum Ito's formula combined with the fact that $dA(t), dA(t)^*$ commute with $Y_i(t)$ gives

$$dY_o(t) = dY_i(t) - j_t(\bar{c}L + cL^*)dt$$

which corresponds to measuring the system observable $-(\bar{c}L + cL^*)$ at time t with measurement noise $dY_i(t) = c.dA(t) + \bar{c}.dA(t)^*$. Let $\eta_o(t) = \sigma(Y_i(s) : s \leq t)$. Then $\eta_o(t), j_t(X)$ forms an Abelian family and hence in any state on $\mathfrak{h} \otimes \Gamma_s(\mathcal{H})$ and in particular, in the state $|f> \otimes |\phi(u)>$ which means that the system is in the pure state $|f>\in \mathfrak{h}$ and the bath is in the coherent state $|\phi(u)>$, we can talk of the conditional expectation

$$\pi_t(X) = \mathbb{E}(j_t(X)|\eta_o(t)) \in \eta_o(t)$$

Write

$$d\pi_t(X) = F_t(X)dt + G_t(X)dY_o(t), F_t(X), G_t(X) \in \eta_o(t)$$

Then the orthgonality principle

$$j_t(X) - \pi_t(X) \perp \eta_o(t)$$

gives

$$\mathbb{E}[(j_t(X) - \pi_t(X))C(t)] = 0, C(t) \in \eta_o(t) - - - (1)$$

In particular, this holds when $C(t)$ satisfies the sde

$$dC(t) = g(t)C(t)dY_o(t), C(0) = 1$$

where g is an arbitary real valued function of time. Taking the differential of (1) and using the arbitrariness of g, we get

$$\mathbb{E}(dj_t(X)-d\pi_t(X)|\eta_o(t))$$

$$= 0, \mathbb{E}((dj_t(X)-d\pi_t(X))dY_o(t)|\eta_o(t))+\mathbb{E}((j_t(X)-\pi_t(X))dY_o(t)|\eta_o(t)) = 0$$

which result the equations

$$F_t(X) = \pi_t(\theta_0(X))+u(t)\pi_t(\theta_1(X))+\bar{u}(t)\pi_t(\theta_2(X))-G_t(X)(cu(t)+\bar{c}\bar{u}(t)-\pi_t(\bar{c}L+cL^*)$$

$$G_t(X) = |c|^{-2}(\bar{c}\pi_t(\theta_1(X)) - \pi_t(X(\bar{c}L + cL^*)) + \pi_t(X)\pi_t(\bar{c}L + cL^*)) =$$

$$|c|^{-2}(\pi_t(X)\pi_t(\bar{c}L+cL^*)-\pi_t(\bar{c}LX+cXL^*))$$

$$= |c|^{-2}(-\pi_t(M+M^*)\pi_t(X)+\pi_t(MX+XM^*)), M = -\bar{c}L$$

The Belavkin filter in the observable domain is therefore

$$d\pi_t(X) = (\pi_t(\theta_0(X)) + u(t)\pi_t(\theta_1(X)) + \bar{u}(t)\pi_t(\theta_2(X)))dt$$

$$+|c|^{-2}(-\pi_t(M+M^*)\pi_t(X)+\pi_t(MX+XM^*))(dY_o(t)-(cu(t)+\bar{c}\bar{u}(t)+\pi_t(M+M^*))dt)$$

For simplicity of notation assume $|c| = 1$ and then with $u = 0$, ie, the bath is in the vacuum coherent state, the filter simpflies to give

$$d\pi_t(X) = \pi_t(\theta_0(X))dt+(\pi_t(MX+XM^*)-\pi_t(M+M^*)\pi_t(X))(dY_o(t)-\pi_t(M+M^*)dt)$$

or equivalently, writing

$$\pi_t(X) = Tr(\rho(t)X)$$

where $\rho(t)$ is the Belavkin filtered state, ie, $\rho(t)$ is a function of $\eta_o(t)$ with values in the space of system space density operators, we get the Belavkin filter in state domain:

$$d\rho(t) = \theta_0^*(\rho(t))dt+(\rho(t)M+M^*\rho(t)-Tr(\rho(t)(M+M^*))\rho(t))(dY_o(t)$$

$$-Tr(\rho(t)(M+M^*))dt)---(1)$$

and

$$\theta_0^*(\rho) = (-i/h)[H,\rho] - (1/2)(LL^*\rho + \rho LL^* - 2L^*\rho L)$$

Note that the output measurement model can also now be expressed as

$$dY_o(t) = dY_i(t) + j_t(M + M^*)dt = cdA(t) + \bar{c}dA(t)^* + j_t(M + M^*)dt$$

which corresponds to measuring the system observable $M + M^*$ with measurement noise. Note that for a general coherent state $|\phi(u)>$, not necessarily the vacuum, the Belavkin filter in the state domain is

$$d\rho(t) = (\theta_0^* + u(t)\theta_1^* + \bar{u}(t)\theta_2^*)(\rho(t))dt+$$

$$(\rho(t)M+M^*\rho(t)-Tr(\rho(t)(M+M^*))\rho(t))(dY_o(t)-(cu(t)+\bar{c}\bar{u}(t)+ Tr(\rho(t)(M+M^*)))dt)$$

Dynamics in the Wigner density domain: If ρ is a density operator for a single particle described by a canonical position-momentum pair (Q, P), we write

$$\rho(Q, Q') =< Q|\rho|Q' >$$

If $\rho(t)$ is a function of time, then we write

$$\rho(t, Q, Q') =< Q|\rho(t)|Q' >$$

Likewise for an observable X,

$$X(Q, Q') =< Q|X|Q' >$$

Then,

$$Tr(\rho X) = \int \rho(Q, Q')X(Q', Q)dQdQ'$$

For a state ρ, and real numbers Q, P, define the complex valued function of (Q, P)

$$\rho_1(Q, P) = (2\pi h)^{-1} \int \rho(Q + q/2, Q - q/2)exp(-iPq/h)dq$$

and for an observable X, define the complex valued function of (Q, P)

$$X(Q, P) = \int X(Q + q/2, Q - q/2)exp(-iPq/h)dq$$

where h is Planck's constant divided by 2π. Then, it is easy to verify the inversion formulas

$$\rho(Q, Q') = \int \rho_1((Q + Q')/2, P)exp(iP(Q - Q')/h)dP,$$

$$X(Q, Q') = (2\pi h)^{-1} \int X_1((Q + Q')/2, P)exp(iP(Q - Q')/h)dP$$

and further,

$$Tr(\rho X) = \int \rho(Q,Q')X(Q',Q)dQdQ' = \int \rho_1(Q,P)X_1(Q,P)dQdP$$

Note that since $Q, P = -ih\partial/\partial Q$ as operators, do not commute, it is not possible to speak about a joint probability density of Q, P. However, the complex valued function $\rho_1(Q,P)$ does have some nice properties which resemble what a joint density of (Q,P) in the state ρ should have, namely first, it is some sort of an average of the fuzzy density over a small region of area h in phase space and secondly, its marginals are exactly, the probability densities of Q and P in the state ρ:

$$\int \rho_1(Q,P)dP = \rho(Q,Q),$$

$$\int \rho_1(Q,P)dQ = (2\pi h)^{-1}\int \rho(Q,Q')exp(-i(P(Q-Q')/h)dQdQ' = \hat{\rho}(P,P)$$

where $\hat{\rho}$ denotes the Fourier transform of ρ from Q-representations to P-representations.

We now conmpute the Wigner transform of a product $H\rho$ of an observable H and a density operator ρ regarding this product as a density:

$$I = (H\rho)_1(Q,P) = (2\pi h)^{-1}\int (H\rho)(Q+q/2, Q-q/2)exp(-iPq/h)dq =$$

$$(2\pi h)^{-1}\int H(Q+q/2, Q')\rho(Q', Q-q/2)exp(-iPq/h)dQ'dq$$

$$= (2\pi h)^{-2}\int H_1((Q+Q')/2+q/4, P')$$

$$\rho_1((Q+Q')/2-q/4, P'')exp(iP'(Q-Q'+q/2)/h).exp(-iP''(Q'-Q+q/2)/h)exp(-iPq/h)dQ'dP'dP''dq$$

To evaluate this integral approximately, make the change of variables

$$(Q+Q')/2+q/4 = Q+hx/2, (Q+Q')/2-q/4 = Q+hy/2$$

These imply

$$(Q-Q')/2-q/4 = -hx/2, (Q-Q')/2+q/4 = -hy/2$$

Note that x, y are of order $1/P$ since q appears in factors like $exp(-iPq/h)$ which means that rapid phase oscillations will occur when $Pq >> h$. Only when Pq is comparable to h will phase oscillations not cancel. Equivalently when q/h is of order $1/P$ will phase oscillations not be too rapid. Then,

$$I = (2\pi)^{-2}\int H_1(Q+hx/2, P')\rho_1(Q+hy/2, P'')$$

$$exp(-iP'y)exp(iP''x).exp(-iP(x-y))dP'dP''dxdy$$

$$= (2\pi)^{-2}\int [H_1(Q,P') + H_{1,1}(Q,P)hx/2 + H_{1,11}(hx)^2/8]$$

$$.[\rho_1(Q,P'')+\rho_{1,1}(Q,P'')hy/2+\rho_{1,11}(Q,P'')(hy)^2/8]exp(i(P-P')y)$$

$$exp(i(P''-P)x)dP'dP''dxdy+O(h^3)$$

$$= H_1(Q,P)\rho_1(Q,P) + (ih/2)(H_{1,1}(Q,P)\rho_{1,2}(Q,P) - H_{1,2}(Q,P)\rho_{1,1}(Q,P))+$$

$$(h^2/8)(2H_{1,12}(Q,P)\rho_{1,12}(Q,P) - H_{1,22}(Q,P)\rho_{1,11}(Q,P) - H_{1,11}(Q,P)\rho_{1,22}(Q,P)) + O(h^3)$$

$$= H_1(Q,P)\rho_1(Q,P) + (ih/2)\{H_1,\rho_1\}(Q,P)+$$

$$(h^2/8)(2H_{1,12}(Q,P)\rho_{1,12}(Q,P) - H_{1,22}(Q,P)\rho_{1,11}(Q,P) - H_{1,11}(Q,P)\rho_{1,22}(Q,P)) + O(h^3)$$

In particular,

$$[H,\rho]_1(Q,P) = ih\{H_1,\rho_1\}(Q,P) + O(h^3)$$

since the $O(h^2)$ term in $(H\rho)_1$ is symmetric under interchange of H and ρ. Now let us look at the Lindblad term: Assume

$$L = g(Q)$$

where g is a complex valued function on $\mathbb{R}$, so that $g(Q)$ becomes a self-adjoint operator in $L^2(\mathbb{R})$. Then

$$\theta_0^*(\rho) = (-1/2)(g(Q)\bar{g}(Q) + g(Q')\bar{g}(Q') - 2g(Q)\bar{g}(Q'))\rho(Q,Q')$$

$$= (-1/2)|g(Q) - g(Q')|^2\rho(Q,Q') + i.Im(g(Q)\bar{g}(Q'))\rho(Q,Q')$$

It follows that the Wigner transform of $\theta_0^*(\rho)$ as a density is given by (with $v(Q,Q') = Im(g(Q)\bar{g}(Q'))$)

$$\theta_0^*(\rho)_1(Q,P) =$$

$$(-1/4\pi h)\int |g(Q+q/2)-g(Q-q/2)|^2\rho(Q+q/2,Q-q/2)exp(-iPq/h)dq$$

$$+(i/2\pi h)\int v(Q+q/2,Q-q/2)\rho(Q+q/2,Q-q/2)exp(-iPq/h)dq$$

$$= (-1/4\pi)\int |g(Q+hx/2)-g(Q-hx/2)|^2\rho(Q+hx/2,Q-hx/2)exp(-iPx)dx$$

$$+(i/2\pi)\int v(Q+hx/2,Q-hx/2)\rho(Q+hx/2,Q-hx/2)exp(-iPx)dx$$

$$= (1/2)h^2|g'(Q)|^2\rho_{1,22}(Q,P) - (h/2\pi)(Im(g'(Q)\bar{g}(Q) - g(Q)\bar{g}'(Q))\rho_{1,2}(Q,P) + O(h^3)$$

$$= (1/2)h^2|g'(Q)|^2\rho_{1,22}(Q,P) - 2h(Im(g'(Q)\bar{g}(Q))\rho_{1,2}(Q,P) + O(h^3)$$

Defining the functions

$$a(Q) = |g'(Q)|^2, b(Q) = -2Im(g'(Q)\bar{g}(Q))$$

we thus obtain the Belavkin filter (1) in the form after taking the Wigner transform on both sides,

$$d\rho_1(t,Q,P) = [\{H_1,\rho_1\}(Q,P) + h^2a(Q)/2)\rho_{1,22}(t,Q,P) + hb(Q)\rho_{1,2}(Q,P)]dt$$

$$+[\rho_1(M_1+\bar{M}_1) + (ih/2)\{\rho_1, M_1-\bar{M}_1\}+$$

$$(h^2/8)(2(\rho_{1,.12}(M_1+\bar{M}_1)_{,12} - \rho_{1,11}(M_1+\bar{M}_1)_{,22} - \rho_{1,22}(M_1+\bar{M}_1)_{,11})$$

$$-(\int \rho_1(M_1+\bar{M}_1)dQdP)\rho_1(t,Q,P)][dY_o(t) - (\int \rho_1(M_1+\bar{M}_1)dQdP)dt]$$

Equivalently, this equation is expressible in the form as the sum of the Kushner-Kallianpur component and quantum corrections containing first and second powers of Planck's constant:

$$d\rho_1(t,Q,P) = [\{H_1,\rho_1\}(Q,P) + h^2a(Q)/2)\rho_{1,22}(t,Q,P) + hb(Q)\rho_{1,2}(Q,P)]dt$$

$$+[(\rho_1.2.Re(M_1) - (\int \rho_1.2Re(M_1)dQdP)\rho_1)[dY_o(t) - (\int \rho_1.2Re(M_1)dQdP)dt]]$$

$$+[-h\{\rho_1, Im(M_1)\}+(h^2/8)(2(\rho_{1,12}(2ReM_1)_{,12}-\rho_{1,11}(2Re(M_1)),$$

$$_{22}-\rho_{1,22}(2Re(M_1))_{,11})][dY_o(t)-(\int \rho_1(2ReM_1)dQdP)dt]$$

7.9 Design of quantum unitary gates using the quantum gravitational field perturbed by classical control fields like the scalar field, the Dirac field, non-Abelian gauge fields and the Abelian $U(1)$ gauge electromagnetic field

Let $L(\phi_k(x), \phi_{k,\mu}(x)), k = 1,2,...,n, \mu = 0,1,2,3, J(x))$ denote the Lagrangian density of all the fields, ϕ_0 is the Dirac field, ϕ_1 is the electromagnetic field, ϕ_2 is the gravitational field and ϕ_4 are the non-Abelian gauge fields. The other $\phi's$ be correspond to superpartners of the above fields, like the gravitino, the scalar field and the gaugino fields. $J(x)$ are the control fields, like external classical currents and classical em and non-Abelian gauge fields. The Dirac equation is thus

$$\partial_\mu \partial L/\partial\phi_{0,\mu} - \partial L/\partial\phi_0 = 0,$$

The Other field equations are

$$\partial_\mu \partial L/\partial\phi_{k,\mu} - \partial L/\partial\phi_k = 0, k = 1,2,...,n$$

We assume that the control fields $J(x)$ are assumed to couple only to fields other than the Dirac field, ie, they appear only in the field equations for $\phi_k, k = 1,2,...,n$. We can express these field equations after spatial discretization as

$$\partial\phi_0(t) = F_0(\phi_0(t), \phi_k(t), k = 1,2,...,n),$$

$$\partial_t^2\phi_k(t) = F_k(\phi_0(t), \phi_m(t), \phi'_m(t), m = 1,2,...,n, J(t)), k = 1,2,...,n$$

We wish $|\phi_0(t)|^2$ (This is a vector whose components consist of the modulus square of each component of ϕ_0) to track a given spatially discretized pdf vector $p_0(t)$. Thus, we can use the adaptation rule

$$J(t+h) = J(t) - \mu.\frac{\partial}{\partial J(t)} \parallel p_0(t+h) - |\phi_0(t+h)|^2 \parallel^2$$

where

$$\phi_0(t+h) = \phi_0(t) + h.F_0(\phi_0(t), \phi_k(t+h), \phi'_k(t+h), k = 1, 2, ..., n)$$

with the dependence of $\phi_k(t+h), \phi'_k(t+h)$ upon $J(t)$ coming via the equations

$$\phi_k(t+h) = \phi_k(t) + h.\phi'_k(t+h),$$

$$\phi'_k(t+h) = \phi'_k(t) + h.F_k(\phi_0(t), \phi_m(t), \phi'_m(t), m = 1, 2, ..., n, J(t))$$

for $k = 1, 2, ..., n$. Thus by exploiting the dependence of the Dirac field dynamics upon the other fields and the dependence of the other field dynamics upon the control fields thereby causing the Dirac field dynamics to depend indirectly upon the control fields, we can design a quantum neural network.

7.10 Some aspects of quantum gravity: Spinor connection of the gravitational field

The spinor connection of the gravitational field is $\omega_\mu^{mn}[\gamma_{m,n}]/4 = \Gamma_\mu$. Since $\gamma_{mn} = [\gamma_m, \gamma_n]/4, 0 \le m < n \le 3$ satisfy the Lorentz Lie algebra commutation relations, they define a basis for a Lie algebra that is a representation of the Lorentz Lie algebra $\int o(3,1)$. This representation is called the spinor representation of the Lorentz group Lie algebra. The Lorentz spinor covariant derivative is then given by

$$\nabla_\mu = \partial_\mu + \Gamma_\mu(x)$$

and the curvature of this connection is

$$R_{\mu\nu} == [\Gamma_\mu, \Gamma_\nu] = R_{\mu\nu mn}[\gamma_m, \gamma_n/4] = R_{\mu\nu mn}\gamma_{mn}$$

$R_{\mu\nu mn}$ is the spinor representation of the Riemann curvature tensor and the standard four vector representaion of the curvature tensor is then

$$R_{\mu\nu\rho\sigma} = R_{\mu\nu mn}e^m_\rho e^n_\sigma$$

where e^n_μ is a tetrad basis for the metric:

$$g_{\mu\nu} = \eta_{mn}e^m_\mu e^n_\nu$$

The Einstein-Hilbert action $\int R\sqrt{-g}d^4x$ can therefore be expressed as a quadratic form the in the first order partial derivaives of the spinor connection ω_μ^{mn}. Now there exists a starndard isomorphism between the proper orthochronous Lorentz group $SO(3,1)$ (ie, the connected component of $SO(3,1)$ containing the identity)and $SL(2,\mathbb{C})$. This isomorphism is obtained by representing a point (t, x, y, z) is the space-time manifold by the 2×2 Hermitian matrix

$$\Phi(t,x,y,z) = \begin{pmatrix} t+z & x+iy \\ x-iy & t-z \end{pmatrix}$$

then an element $L \in SO(3,1)$ is represented by the element $g \in SL(2,\mathbb{C})$ where

$$\Phi(t',x',y',z') = Phi(L(t,x,y,z)) = g\Phi(t,x,y,z)g^*$$

Taking determinants on both sides verifies that $g \in SL(2,\mathbb{C})$ implies that $g\Phi(t,x,y,z)g^*$ will be Hermitian and hence of the form $\Phi(t',x',y',z')$ and that

$$t'^2 - x'^2 - y'^2 - z'^2 = t^2 - x^2 - y^2 - z^2$$

implying that L arising from $g \in SL(2,\mathbb{C})$ is indeed an element of $SO(3,1)$. The fact that as g runs over the wbole of $SL(2,\mathbb{C})$ implies that L will run over the connected component of $SO(3,1)$ containing the idenity follows from the fact that $SL(2,\mathbb{C})$ is connected and contains the identity. Actually g and $-g$ give the same L and hence we can say that $SL(2,\mathbb{C})/\mathbb{Z}_2$ is isomorphic as a group to the connected component of $SO(3,1)$ containing the identity. This means that the spinor connection matrix $((\omega_\mu^{mn}))$ can equivalently be viewed as an element of the Lie algebra $\mathfrak{s}L(2,\mathbb{C})$ which is the complexification of the real Lie algebra $\mathfrak{s}u(2,\mathbb{C})$ of 2×2 skew-Hermitian matrice of trace zero. We can thus say that the connection of the gravitational field can be represented by a vector field $A_\mu(x)$ that takes values in the Lie algebra $\mathfrak{s}u(2,\mathbb{C})$, ie, as $A_\mu^a(x)i\tau_a$ where $\tau_a, a = 1,2,3$ is a set of 3 linearly independent Hermitian matrices of trace zero, for example, we can take $\tau_a, a = 1,2,3$ as the three Pauli matrices. Equivalently, we can represent the gravitational connection locally by elemens of the group $SU(2,\mathbb{C})$ obtained by parallely translating along the connection. This leads us to loop quantum gravity wherein we divide the whole of the space-time manifold into a graph having edges of small lengths $l_k, k = 1,2,....$ Parallel displacement along a given line l specified by the path $s \to x^\mu(s), 0 \le s \le 1$ yields an element o $g(1)$ of $SU(2,\mathbb{C})$ obtained by solving

$$dg(s)/ds = g(s)A_\mu(x(s))dx^\mu(s)/ds, 0 \le s \le 1, g(0) = I_2$$

Formally, this can be represented as a Dyson series

$$g(1) = T\{exp(\int_0^1 A_\mu(x(s))dx^\mu(s))\}$$

The resulting algebra of $SU(2,\mathbb{C})$ transformation matrices obtained by translating along edges by the $\mathfrak{s}u(2,\mathbb{C})$ connection field $A_\mu(x)$ is called the $SU(2,\mathbb{C})$-holonomy flux algebra. In this formalism, we require complex valued connections $A_\mu(x)$. In order to do the same using real valued connections, we must use the ADM formalism of separaing out the spatial and temporal coordinates (See for reference : Thomas Thiemann, "Modern canonical quantum general relativity", Cambridge University Press). In this formalism, we embed at each time t a spatial three dimensional surface Σ_t into four dimensional spac-time $\mathbb{R}^4$ and represent the metric in separable form as

$$g^{\mu\nu} = q^{\mu\nu} + N^2 n^\mu n^\nu, q^{\mu\nu} == q_{ab}X^\mu_{,a}X^\nu_{,b}, g_{\mu\nu}n^\mu n^\nu = 12$$

where a, b run over $1, 2, 3$, the spatial indices. The metric is thus parametrized by the ten independent functions $q_{ab}, 1 \leq a \leq b \leq 3, N^a, a = 1, 2, 3, N$ where

$$X^{\mu}_{,a}N^a + Nn^{\mu} = X^{\mu}_{,0}$$

with N^a, N chosen so that n^{μ} has length one and is orthogonal to $X^{\mu}_{,a}, a = 1, 2, 3$:

$$g_{\mu\nu}(X^{\mu}_{,0} - X^{\mu}_{,a}N^a)X^{\nu}_{,b} = 0$$

This equation is solved for N^a and then $Nn^{\mu} = X^{\mu}_{,0} - X^{\mu}_{,a}N^a$ is calculated and N then chosen as the length of $X^{\mu}_{,0} - X^{\mu}_{,a}N^a$. The ADM action involves expressing the Einstein-Hilbert action in terms of the 3-D metric q_{ab} and the functions N^a, N rather than the original metric $g_{\mu\nu}$. The spatial component of this action is expressed in terms of the spatial curvature of q_{ab} and the time component in terms of the time derivatives of q_{ab} and the functions N^a, N. The spatial metric q_{ab} then generates an $SO(3)$ Lie algebra connection just as the full spatio-temporal metric $g_{\mu\nu}$ generated an $SO(3, 1)$ Lie algebra connection. The spatial curvature of q_{ab} is expressible in terms of this $SO(3)$ connection just as the full curvature was expressble in terms of the $SO(3, 1)$ spinor connection. The generators $\tau_a, a = 1, 2, 3$ of the Lie algebra of $SO(3)$ are used to express this $SO(3)$ connection just as the generators of the Lie algebra of the spinor representation of the Lorentz group $SO(3, 1)$ were used to express the spinor connection of the gravitational field. Thus, if $\tau_a, a = 1, 2, 3$ are three real skew-symmetric 3×3 matrices that are linearly independent (as for example, the standard generators of the Lie algebra of $\mathfrak{so}(3)$), then we can express the connection of q_{ab} as linear combinations of τ_a with real coefficient functions. We can thus express this spatial connection as $A^k_a(x)\tau_k$ where A^k_a are real functions and then write down the holonomy flux matrix fields as $SO(3)$ valued quantities, each quantity defined for a given edge in a spatial graph discretization of three dimnensional space These, $SO(3)$ valued quantities for spatial edge l will have the form $exp(\int_0^1 A^k_a(x)\tau_k)$ and the wave function of the metric will be expressible as a function of these $SO(3)$ matrices or equivalently in terms of $SU(2, \mathbb{C})$ Lie algebra generators with real coefficients.

7.11 Supersymmetric field theories based on the supervector field representation of the Fermionic generators of the super-algebra

Develop the theory of supersymmetry using superfields, super-vector fields and Chiral superfields and construct Lagrangian densities as functions of the components of superfields and their bosonic partial derivatives so that the action functionals associated with these Lagrangian densities are invariant under supersymmetry transformations defined infinitesimally by super-vector fields. The complete supersymmetry algebra consists of the Bosonic generators in the form of four momenta and four angular momenta, namely generators of the Poincare group which consists of space-time translations along with spatial rotations and Lorentz boosts and along with these Poincare group genenrators, there are the Fermionic generators and generators of internal symmetries of the particles. The anticommutators between two Fermionic generators is a superposition of Bosonic generator plus generators corresponding to the internal degrees of freedom, the commutator between two Bosonic generators is either zero or a superposition of Bosonic generators. Finally, the commutator between a Bosonic genrators and a Fermionic generator is either zero or a superposition of Fermionic generators plus generators corresponding to internal degrees of freedom. Salam and Strathdeee proposed representing these generators of of the supersymmetry algebras by super-vector fields with their action defined on the space of superfields. This is a special kind of super-Lie algebra represenation. let $\theta^a, a = 0, 1, , 2, 3$ be Majorana Fermionic variables and $x^\mu, \mu = 0, 1, 2, 3$ the Bosonic space-time variables. The Fermionic parameters anticommute with each other and commute with the Bosonic variables:

$$\{\theta^a, \theta^b\} = 0, a, b = 0, 1, 2, 3$$

It follows that any superfield (namely a function of θ, x) can be represented as a polynomial of degree four in θ with coefficients being functions of x:

$$S(x, \theta) = C_0(x) + \sum_a C_{1a}(x)\theta^a + \sum_{0 \leq a < b \leq 3} C_{2ab}(x)\theta^a\theta^b$$

$$+ \sum_{0 \leq a < b < c \leq 3} C_{3abc}(x)\theta^a\theta^b\theta^c + C_4(x)\theta^0\theta^a\theta^2\theta^3$$

Consider super-vector fields defined by

$$F(\mu, b)\theta^b.\partial/\partial x^\mu + G(a)\partial/\partial\theta^a, \partial/\partial x^\mu, x^\mu\partial^\nu - x^\nu\partial^\mu$$

where $F(\mu, b), G(a)$ are arbitrary complex scalars. In other words, $\theta^b\partial/\partial x^\mu = \theta^b\partial_\mu, \partial/\partial\theta^a, \partial_\mu, x^\mu\partial^\nu - x^\nu\partial^\mu$ form a basis for super vector field defined on a supermanifold in which the coordinates of a point are $(x, \theta) = (x^\mu, \theta^a : \mu, a = 0, 1, 2, 3)$. We verify the Lie algebra commutation-anticommutation relations using the identites

$$\{\theta^a, \partial/\partial\theta^b\} = \delta^a_b$$

For example,

$$\{\theta^a\partial_\mu, \partial/\partial\theta^b =^a_b \partial_\mu,$$

$$\{\theta^a\partial_\mu, \theta^b\partial_\nu\} = 0,$$

$$[\theta^a\partial_\mu, x^\nu\partial^\rho - x^\rho\partial^\nu] =$$

$$\delta^\nu_\mu\theta^a\partial^\rho - \delta^\rho_\mu\theta^a\partial^\nu$$

showing explicitly that the anticommutator between two Fermionic generators is a Bosonic generator and the commutator between between a Fermionic and a Bosonic generator is a Fermionic generator. This is a special case of a super-Lie algebra with generators $X^a, a = 1, 2, ..., N$ some of which are of even parity and some of which are odd. If X has definie parity $p(X)$ and Y has definite parity

$p(Y)$ ($p((X)$ is zero if X is even and is one if it is odd), then their super Lie bracket

$$XY - (-1)^{p(X)p(Y)}YX$$

is in the super-Lie algebra, ie is expressible as a complex linear combination of $X^a, a = 1, 2, ..., N$. Note that this super-Lie bracket is an ordinary commutator Lie bracket if either one of X is or Y is even and is other wise an anticommutator. Thus' if we assume that $X^a, a = 1, 2, ..., M$ are even and $X^a, a = M + 1, ..., N$ are odd, then

$$[X^a, X^b] \in span(X^1, ..., X^M), a, b = 1, 2, ..., M,$$

$$\{X^a, X^b\} \in span(X^1, ..., X^M), a, b = M + 1, ..., N,$$

$$[X^a, X^b] \in span(X^{M+1}, ..., X^N), a = 1, 2, ..., M, b = M + 1, ..., N$$

In general any X, Y in the super-Lie algebra are expressible as

$$X = Z_1 + Z_2, Y = U_1 + U_2$$

where Z_1, U_1 are even and Z_2, U_2 are odd. Then the super-Lie bracket between X, Y is defined by

$$[X, Y]_s = [Z_1, U_1] + \{Z_2, U_2\}[Z_1, U_2] + [Z_2, U_1]$$

This is an again an element of the super Lie algebra.

7.12 Quantum neural networks based on the Schrodinger equation by adapting the control parameters in the Hamiltonian so that the Schrodinger pdf tracks a given pdf while simultaneously ensuring via controlled variation in the adaptation constant that the average of a set of observables w.r.t the Schrodinger wave function is strongly correlated with a corresponding set of classical signals

Consider the time-discretized approxmation to Schrodinger's equation in the form

$$\psi(n + 1, x) = F(n, x, \psi(n, x), \psi(n, x - 1), \psi(n, x + 1), \theta(n))$$

or more generally, when the spatial components of the wave function are arranged in the form of a state vector:

$$\psi(n) = ((\psi(n, x)))_x$$

we can express the dynamics as

$$\psi(n+1) = F(n, \psi(n), \theta(n))$$

The dynamics is unitary although not linear in general since the adaptive control parameters $\theta(n)$ are generally wave function dependent. $—\psi(0)|^2 = 1$, then $|\psi(n)|^2 = 1$ for all $n \geq 1$, where $|\psi(n)|^2 = \sum_x |\psi(n,x)|^2$ is the norm of the state vector at time n. For example, time-space discretization of the standard Schrodinger equation with potential $V(t,x)$:

$$\psi_{,t}(t,x) = (-1/2)\psi_{,xx}(t,x) + V(t,x)\psi(t,x)$$

gives

$$\psi(n+1) = \psi(n) + (i/2)\delta.D^2\psi(n) - i\delta.V(n).\psi(n)$$

where D^2 is a tridiagonal real Hermitian matrix representing the operator $\partial^2/\partial x^2$. To make this dynamics unitary, we replace it by the Cayley transform dynamics:

$$\psi(n+1) = [(I + iQ(n, \theta(n)))/(I - iQ(n, \theta(n)))]\psi(n) = G(n, \theta(n))\psi(n)$$

say, where

$$Q(n, \theta(n)) = \delta.D^2/4 - \delta.V(n, \theta(n))/2$$

is a real Hermitian matrix. We usually assume that the potential is a linear function of the control parameters:

$$V(n, \theta(n)) = \sum_k theta_k(n)V_k(n)$$

and therefore, we can write

$$Q(n, \theta(n)) = \sum_{k \geq 1} \theta_k(n)Q_k(n) + Q_0(n)$$

Let $p_0(n) = ((p_0(n,x)))$ be the desired probability density to be tracked by the wave function modulus square $|\psi(n,x)|^2$. The weighted error energy between these two pdf's can be expressed as

$$E(p_0(n), \psi(n), \bar{\psi}(n)) = \sum_x w(x)(p_0(n,x) - |\psi(n,x)|^2)^2$$

The adaptive stochastic gradient algorithm for updating the control parameters for reducing the error energy would thus read

$$\theta(n+1) = \theta(n) - \mu(n).\nabla_{\theta(n)}E(p_0(n+1), \psi(n+1), \bar{\psi}(n+1)) =$$

$$\theta(n) - \mu(n).\nabla_{\theta(n)}E(p_0(n+1), G(n, \theta(n))\psi(n), \bar{G}(n, \theta(n))\bar{\psi}(n))$$

The "adaptation constant" $\mu(n)$ is made to be an increasing function of the prediction error energy between the average value of a set of observavbles

$O_k, k = 1, 2, ..., p$ and a given set of signals $s_k(n), k = 1, 2, ..., p$ which are assumed to be strongly correlated with the signals whose pdf's are those of $O_k, k = 1, 2, ..., p$ in the state $\psi(n)$. This prediction error energy can be expressed as

$$f(n) = \| e(n) \|^2$$

where $e(n) = ((e_k(n)))_{k=1}^p$,

$$e_k(n) = s_k(n) - \sum_{m=0,...,D, j=1,2,...,p} a(k, j, m, n) < \psi(n-m)|O_j|\psi(n-m) >, k = 1, 2, ..., p$$

or even

$$e_k(n) =< \psi(n)|O_k|\psi(n) - \sum_{m=0,...,D, j=1,2,...,p} b(k, j, m, n)s_j(n-m), k = 1, 2, ..., p$$

where $a(k, j, m, n)$ are chosen to minimize $\sum_{r=1}^M f(n-r)$ with M a fixed positive integer. In other words, we are predicting the vector valued process $s(n) = ((s_k(n)))$ linearly based on the quantum averages of $((O_k))$ and its past D samples.

Problem: Carry out a statistical performance analysis of this algorithm, ie, if $p_0(n) \to p_0(n) + \delta p_0(n)$ where $\delta p_0(n)$ is a small random vector with known statistics, then what will the the corresponding statistics of the perturbed parameter vector process $\theta(n) + \delta\theta(n)$. This problem of statistical analysis, is a special case of the following problem: Given a dynamical system

$$\theta(n+1) = F(n, \theta(n), x(n)) + w(n), n \geq 0$$

if $x(n) \to x(n) + \delta x(n)$ where $\delta x(n)$ is a "small" random process, then the linearized perturbed dynamics is given by

$$\delta\theta(n+1) = F_{,1}(n, \theta_0(n), x(n))\delta\theta(n) + F_{,2}(n, \theta_0(n), x(n))\delta x(n)$$

where

$$\theta_0(n+1) = F(n, \theta_0(n), x(n)) + w(n)$$

is the unperturbed dynamics. Note that the exact perturbed dynamics is

$$\delta\theta(n+1) = \sum_{k,m \geq 0, k+m \geq 1} (k!m!)^{-1} F^{(k,m)}(n, \theta_0(n), x(n))(\delta\theta(n))^{\otimes k} \otimes (\delta x(n))^{\otimes m}$$

where

$$F^{(k,m)}(n, \theta, x) = (\partial^k / \partial\theta^k)(\partial^m / \partial x^m)F(n, \theta, x)$$

is the mujltivariate-higher order Jacobian matrix.

7.13 Quantum Boltzmann equation for a plasma within a cavity resonator: Determining the oscillation frequencies of the plasma

The quantum Boltzmann equation for the one particle density operator in momentum space for a plasma of N indistinguishable charged particles with external electromagnetic field interactions by partial tracing out of the N-particle quantum Liouville equation. The expression for the avearge current density and charge density in such a plasma is substituted into the Maxwell equations for the electromagnetic field. We thus obtain a coupled system of differential equations for the one particle plasma density operator in position space and the classical electromagnetic field. The quantum Boltzmann equation is a quantum Liouville equation with Hamiltonian containing interaction terms of the charged particle with the electromagnetic field as well as a nonlinear collision term coming from the partial trace of the N particle density with the molecular chaos hypothesis. These two terms cause perturbations in the characteristic oscillation frequencies of the quantum plasma. In the absence of any field interactions and nonlinear collision terms, if H is the single particle Hamiltonian with spectral representation $H = \sum_n E_n|n><n|$, then its one particle density operator will evolve as $\rho(t) = exp(-itH/h)\rho(0)exp(itH/h) = \sum_{n,m} <n|\rho(0)|m>$ $exp(-it(E_n - E_m)/h)$ so that the unperturbed characteristic plasma oscillation frequencies are $(E_n - E_m)/h$. Interactions with the electromagnetic field and the presence of the nonlinear collision term cause perturbation in these oscillation frequencies.

Let ρ be density operator for a system of N particles, each of charge $-e$, each having the same mass m moving in the electromangnetic field within a resonator. It satisfies the Schrodinger-Von-Neumann-Liouville equation

$$i\partial_t \rho(t) = [H, \rho]$$

where

$$H = \sum_{k=1}^{N}(p_k + eA(t, r_k))^2/2m - \sum_{k=1}^{N} e\Phi(t, r_k) + \sum_{k<j} V(r_k - r_j), p_k = -i\nabla_{r_k} = -i\nabla_k$$

where $V(r - r')$ is the interaction potential energy between any two of the plasma particles located at r, r'.

We can write

$$H = \sum_k H_k + \sum_{k<j} V_{kj}$$

where

$$\begin{aligned} H_k &= (p_k + eA(t, r_k))^2/2m - e\Phi(t, r_k) \\ &= p_k^2/2m - e\Phi(t, r_k) + e((p_k, A(t, r_k)) + (A(t, r_k), p_k)) + e^2 A(t, r_k)^2/2m \end{aligned}$$

Here, $A(t, r)$ and $\Phi(t, r)$ are respectively the magnetic vector potential and the electric scalar potential fields corresponding to the cavity em field. The quantum

Boltzmann for each particle is obtained by partial tracing out of the Schrodinger equation over the remaining $N-1$ particles. Thus we get

$$\rho_1(t) = Tr_{23...N}(\rho(t)), \rho_{12}(t) = Tr_{34...N}(\rho(t)), \rho_{123}(t) = Tr_{45...N}(\rho(t))$$

$$i\partial_t\rho_1(t) = [H_1, \rho_1(t)] + (N-1)Tr_2[V_{12}, \rho_{12}(t)]$$

$$i\partial_t\rho_{12}(t) = [H_1 + H_2 + V_{12}, \rho_{12}(t)] + (N-2)Tr_3[V_{13} + V_{23}, \rho_{123}(t)]$$

Write

$$\rho_{12} = \rho_1 \otimes \rho_1 + g_{12}$$

where g_{12} is small like the interaction potential. Then, we get from the above,

$$i\partial_t g_{12}(t) + [H_1, \rho_1]\otimes\rho_1 + \rho_1\otimes[H_1, \rho_1] + (N-1)[Tr_2[V_{12}, \rho_{12}(t)],$$

$$\rho_1] + (N-1)[\rho_1, Tr_2[V_{12}, \rho_{12}(t)]]$$

$$-(N-2)Tr_3[V_{13} + V_{23}, \rho_{123}]$$

$$= [H_1 + H_2 + V_{12}, g_{12}] + [V_{12}, \rho_1 \otimes \rho_1] + [H_1, \rho_1] \otimes \rho_1 + \rho_1 \otimes [H_1, \rho_1]$$

or equivalently,

$$i\partial_t g_{12}(t) + (N-1)[Tr_2[V_{12}, \rho_{12}(t)], \rho_1] + (N-1)[\rho_1, Tr_2[V_{12}, \rho_{12}(t)]]$$

$$-(N-2)Tr_3[V_{13} + V_{23}, \rho_{123}]$$

$$= [H_1 + H_2 + V_{12}, g_{12}] + [V_{12}, \rho_1 \otimes \rho_1]$$

So far everything is exact without any approximations. We now make the approximation of neglecting terms of second order of smallness like the product of $V's$ and g_{12}. We then get the approximate equation

$$i\partial_t g_{12}(t) + (N-1)[Tr_2[V_{12}, \rho_1 \otimes \rho_1], \rho_1] + (N-1)[\rho_1, Tr_2[V_{12}, \rho_1 \otimes \rho_1]]$$

$$-(N-2)Tr_3[V_{13} + V_{23}, \rho_1 \otimes \rho_1 \otimes \rho_1]$$

$$= [H_1 + H_2 + V_{12}, g_{12}] + [V_{12}, \rho_1 \otimes \rho_1]$$

Another frequently used approximation is to neglect product of three terms like three $\rho_1' s$ and an interaction potential V. This gives us

$$i\partial_t g_{12} = [H_1 + H_2 + V_{12}, g_{12}] + [V_{12}, \rho_1 \otimes \rho_1]$$

Then, substituting this into the equation for ρ_1 gives us

$$i\partial_t\rho_1 = [H_1, \rho_1] + (N-1)Tr_2[V_{12}, \rho_1 \otimes \rho_1] + (N-1)Tr_2[V_{12}, g_{12}]$$

with

$$g_{12}(t) = \int_0^t exp(-i(t-s)ad(H_1 + H_2 + V_{12}))ad(V_{12})(\rho_1(s) \otimes \rho_1(s))ds$$

7.14 Problem: How to describe the plasma by a field of wave operators, or equivalently by a field of density operators

If the plasma consists of Dirac Fermions interacting with each other and with an external electromagnetic field, then we can write down the Hamiltonian for this Fermionic liquid in the second quantized picture after taking into account Hartree-Fock interaction terms as

$$H = \int \psi(t,r)^*(\alpha, -i\nabla + eA(t,r)) + \beta m - eV(t,r))\psi(t,r)d^3r$$
$$+ \int V_1(t,r,r')\psi(t,r)^*\psi(t,r)\psi(t,r')^*\psi(t,r')d^3rd^3r'$$

with the canonical anticommutation relations

$$\{\psi_a(t,r), \psi_b(t,r')^*\} = \delta_{ab}\delta^3(r-r')$$

Let ρ denote the initial state of this second quantized field. The Heisenberg equations of motion are

$$\partial_t\psi(t,r) = i[H, \psi(t,r)]$$

and the commutator can be evaluated using the above anticommutation relations to yield the Schrodinger equation for the wave field operators. Now let $X(r,r')$ be any first quantized observable in the position representation. Its second quantized density at time t is then given by

$$\rho_X(t,r) = \int \psi(t,r)^* X(r,r')\psi(t,r')d^3r'$$

For example, the second quantized charge density is obtained by setting $X(r,r') = -e\delta^3(r-r')$ and the second quantized current density is given by setting $X(r,r')$ to be $-i(e/m)\nabla_r\delta^3(r-r')$. The average value of this density is the classical density of X given by $Tr(\rho.\rho_X(t,r))$. For example taking X to be the charge and current densities, we can substitute these into the classical Maxwell equations for $A(t,r), V(t,r)$ to obtain the desired coupling between the plasma of Fermions and the classical electromagnetic field.

7.15 Wave-guides and resonator cavities filled with an material having inhomogeneous and anisotropic permittivity and permeability

The general case when the permittivity and permeability tensors are functions of ω, x, y, z. The aim is to derive the Maxwell equations in such a medium within a cavity/guide by separating the longitudinal and transverse field components and to arrive at a form wherein perturbation theory can be used to obtain approximate solutions for the field and the characteristic modes/oscillation frequencies by assuming that the dependence of the medium on z is small. The method involves expanding the medium properties as well as the fields as a Fourier series in the z variable.

We assume that the length of the cavity along the z axis is d and expand the tensors as Fourier series in z:

$$\epsilon(\omega, x, y, z) - = \sum_{n \in \mathbb{Z}} \epsilon_n(\omega, x, y) exp(-jn\beta z), \mu(\omega, x, y, z)$$

$$= \sum_{n \in \mathbb{Z}} \mu_n(\omega, x, y) exp(-jn\beta z), \beta = 2\pi/d$$

where ϵ, μ are 3×3 matrix valued functions. Assume that the electric and magnetic fields are also Fourier series expandable as

$$E(\omega, x, y, z) = \sum_n E_n(\omega, x, y) exp(-(\gamma + jn\beta)z), , H(\omega, x, y, z)$$

$$= \sum_n H_n(\omega, x, y) exp(-(\gamma + jn\beta)z)$$

The transverse components of the Maxwell curl equations are

$$\nabla_\perp E_{nz} \times \hat{z} - (\gamma + jn\beta)\hat{z} \times E_{n\perp} = -j\omega. \sum_m \mu_{m\perp} H_{n-m\perp} - j\omega \sum_m \nu_m H_{n-mz}$$

$$\nabla_\perp H_{nz} \times \hat{z} - (\gamma + jn\beta)\hat{z} \times H_{n\perp} = j\omega. \sum_m \epsilon_{m\perp} E_{n-m\perp} + j\omega \sum_m \rho_m E_{n-mz}$$

where

$$\mathrm{E}_{n\perp} = [E_{nx}, E_{ny}]^T, H_{n\perp} = [H_{nx}, H_{ny}]^T, \mu_{n\perp} = \begin{pmatrix} \mu_{nxx}) & \mu_{nxy} \\ \mu_{nyx} & \mu_{nyy} \end{pmatrix}, \nu$$

$[\mu_{nxz}, \mu_{nyz}]^T, \epsilon_{n\perp} = \begin{pmatrix} \epsilon_{nxx}) & \epsilon_{nxy} \\ \epsilon_{nyx} & \epsilon_{nyy} \end{pmatrix}, \rho_n = [\epsilon_{nxz}, \epsilon_{nyz}]^T, THelongitudinalcomponents, ie, zcomponentsofthetw_n$
$= oMaxwellcurlequationsare \nabla_\perp \times$
$E_{n\perp} = -j\omega \sum_m \sigma_m^T E_{n-m\perp} - j\omega. \sum_m \mu_{mzz} H_{n-mz} \hat{z} \nabla_\perp \times H_{n\perp} = j\omega \sum_m \kappa_m^T E_{n-m\perp} +$
$j\omega. \sum_m \epsilon_{mzz} E_{n-mz} \hat{z} where \sigma_= [\mu_{nzx}, \mu_{nzy}]^T, \kappa = [\epsilon_{nzx}, \epsilon_{nzy}]^T Wenowmaketheassumptionthat \epsilon_n, n =$
0 and $\mu_m, m = 0$ are of first order of smallness compared to μ_0, ϵ_0 respectively. Note that here μ_0, ϵ_0 Let $((\mu 1_{n-m\perp}))$ denote the inverse of the matrix $((\mu_{n-m\perp}))$ and likewise $((\epsilon 1_{n-m\perp}))$ denote the inverse of $((\epsilon_{n-m\perp}))$. Note that these two infinite matrices are block structured with each block being a 2×2 matrix. From the transverse equations,

$$\nabla_\perp E_{nz} \times \hat{z} - (\gamma + jn\beta)\hat{z} \times E_{n\perp} = -j\omega. \sum_m \mu_{m\perp} H_{n-m\perp} - j\omega \sum_m \nu_m H_{n-mz}$$

$$\nabla_\perp H_{nz} \times \hat{z} - (\gamma + jn\beta)\hat{z} \times H_{n\perp} = j\omega. \sum_m \epsilon_{m\perp} E_{n-m\perp} + j\omega \sum_m \rho_m E_{n-mz}$$

we derive

$$\nabla_\perp H_{nz} + (\gamma + jn\beta) H_{n\perp} = j\omega. \sum_m \hat{z} \times \epsilon_{m\perp} E_{n-m\perp} + j\omega \sum_m \hat{z} \times \rho_m E_{n-mz}$$

and hence

$$\nabla_\perp E_{nz} \times \hat{z} - (\gamma + jn\beta)\hat{z} \times E_{n\perp} =$$

$$j\omega \sum_m \mu_{m\perp}(\gamma+j(n-m)\beta)^{-1}\nabla_\perp H_{n-mz} +$$

$$\omega^2 \sum_{m,k}(\gamma+j(n-m)\beta)^{-1}\mu_{m\perp}\hat{z}\times\epsilon_{k\perp}E_{n-m-k\perp}$$

$$-j\omega\sum_m \nu_m H_{n-mz} + \omega^2\sum_{m,k}(\gamma + j(n-m)\beta)^{-1}\mu_{m\perp}\rho_k E_{n-m-k\perp}$$

$$= [j\omega(\gamma+jn\beta)^{-1}\mu_{0\perp}\nabla_\perp H_{nz} + \omega^2(\gamma+jn\beta)^{-1}\mu_{0\perp}\hat{z}\times\epsilon_{0\perp}E_{n\perp} + \omega^2(\gamma+jn\beta)^{-1}\mu_{0\perp}\hat{z}\times\rho_0 E_{n\perp}]$$

$$+\delta.F_n(H_z, \nabla_\perp H_z, \nabla_\perp E_\perp)$$

where $\delta.F(.)$ are terms of the first order of smallness. The perturbation parameter δ has been introduced to keep track of the first order of smallness terms. Specifically,

$$\delta.F_n(H_z, \nabla_\perp H_z, \nabla_\perp E_\perp) =$$

$$j\omega\sum_{m\neq 0}(\gamma+j(n-m)\beta)^{-1}\mu_{m\perp}\nabla_\perp H_{n-mz} + \omega^2\sum_{m,k\neq 0}(\gamma+j(n-m)\beta)^{-1}\mu_{m\perp}\hat{z}\times\epsilon_{k\perp}E_{n-m-k\perp}$$

$$-j\omega\sum_{m\neq 0}\nu_m H_{n-mz} + \omega^2\sum_{(m,k)\neq(0,0)}(\gamma + j(n-m)\beta)^{-1}\mu_{m\perp}\hat{z}\times\epsilon_{k\perp}E_{n-m-k\perp} +$$

$$\omega^2\sum_{(m,k)\neq(0,0)}(\gamma + j(n-m)\beta)^{-1}\mu_{m\perp}\hat{z}\times\rho_k E_{n-m-k\perp}$$

Taking the cross product on both sides of this equation with $\hat{z}$ gives us

$$\nabla_\perp E_{nz} + (\gamma + jn\beta)\times E_{n\perp} =$$

$$= [j\omega(\gamma+jn\beta)^{-1}\hat{z}\times\mu_{0\perp}\nabla_\perp H_{nz} + \omega^2(\gamma+jn\beta)^{-1}\hat{z}\times\mu_{0\perp}\hat{z}\times\epsilon_{0\perp}E_{n\perp} - j\omega\hat{z}\times\nu_0 H_{nz}]$$

$$+\omega^2(\gamma + jn\beta)^{-1}\hat{z}\times\mu_{0\perp}\hat{z}\times\rho_{0\perp}E_{n\perp}]$$

$$+\delta.G_n(H_z, \nabla_\perp H_z, \nabla_\perp E_\perp)$$

where

$$G_n = \hat{z}\times F_n = j\omega\sum_{m\neq 0}(\gamma + j(n-m)\beta)^{-1}\hat{z}\times\mu_{m\perp}\nabla_\perp H_{n-mz} +$$

$$\omega^2\sum_{(m,k)\neq(0,0)}(\gamma + j(n-m)\beta)^{-1}\hat{z}\times\mu_{m\perp}\hat{z}\times\epsilon_{k\perp}E_{n-m-k\perp}$$

$$-j\omega\sum_{m\neq 0}\hat{z}\times\nu_m H_{n-mz} + \omega^2\sum_{(m,k)\neq(0,0)}(\gamma + j(n-m)\beta)^{-1}\hat{z}\times\mu_{m\perp}\hat{z}\times\rho_k E_{n-m-k\perp}$$

Note that ρ_k is of the first order of smallness since it contains nondiagonal entries of the permittivity matrix.

7.16 A remark concerning the determination of the characteristic frequencies of oscillation within a cavity resonator upto (δ^2) having anisotropic permittivity and permeability tensors that are functions of (ω, x, y), not of z

The permittivity and permeability tensors have the following forms:

$$\epsilon(\omega, x, y) = \epsilon_0(I + \delta.\chi_e(\omega, x, y)), \mu(\omega, x, y) = \mu_0(I + \delta.\chi_m(\omega, x, y))$$

where χ_e and χ_m are 3×3 matrix valued functions. The final generalized Helmholtz equations for the z components of the electric and magnetic field assume upto $O(\delta^2)$ the following form:

$$((\nabla_\perp^2 + h_0^2 + \delta.h_1^2 + \delta^2.h_2^2)(\psi^0(x, y) + \delta\psi^1(x, y) + \delta^2\psi^2(x, y))+$$

$$\delta(G_1(\omega, \gamma, x, y) + F_{1x}(\omega, \gamma, x, y)\partial_x + F_{1y}(\omega, \gamma, x, y)\partial_y)(\psi^0(x, y) + \delta\psi^1(x, y))$$

$$+\delta^2(G_2(\omega, \gamma, x, y) + F_{2x}(\omega, \gamma, x, y)\partial_x + F_{2y}(\omega, \gamma, x, y)\partial_y)\psi^0(x, y) = 0 - - - (1)$$

In this equation, we are to substitute $\gamma = j\pi p/d$ with p an integer and

$$\omega = (\mu_0\epsilon_0)^{-1/2}[(h_0^2 + \delta.h_1^2 + \delta^2.h_2^2) + (\pi p/d)^2]^{1/2}$$

Thus, determining the modal eigenvalues $h^2 = h_0^2 + \delta.h_1^2 + \delta^2.h_2^2$ upto $O(\delta^2)$ would enable us to determine the characteristic frequencies ω of oscillation upto $O(\delta^2)$. (1) can therefore also be expressed upto $O(\delta^2)$ as

$$((\nabla_\perp^2 + h_0^2 + \delta.h_1^2 + \delta^2.h_2^2)(\psi^0(x, y) + \delta\psi^1(x, y) + \delta^2\psi^2(x, y))+$$

$$\delta(G_1(h_0^2+\delta h_1^2, p, , x, y)+F_{1x}(h_0^2+\delta.h_1^2, p, x, y)\partial_x+F_{1y}(h_0^2+\delta.h_1^2, p, x, y)\partial_y)(\psi^0(x, y)+\delta\psi^1(x, y))$$

$$+\delta^2(G_2(h_0^2, p, x, y) + F_{2x}(h_0^2, p, x, y)\partial_x + F_{2y}(h_0^2, p, x, y)\partial_y)\psi^0(x, y) = 0 - - - (2)$$

We get on equating the coefficients of $\delta^m, m = 0, 1, 2$ in (1) the following set of three equations:

$$(\nabla_\perp^2 + h_0^2)\psi^0 = 0 - - - (3a)$$

$$(\nabla_\perp^2+h_0^2)\psi^1+h_1^2\psi^0+(G_1(h_0^2, p, x, y)+F_{1x}(h_0^2, p, x, y)\partial_+x+F_{1y}(h_0^2, p, x, y)\partial_y)\psi^0 - - - (3b),$$

$$(\nabla_\perp^2+h_0^2)\psi^2+h_1^2\psi^1+h_2^2\psi^2+(G_1(h_0^2, p, x, y)+F_{1x}(h_0^2, p, x, y)\partial_x+F_{1y}(h_0^2, p, x, y)\partial_y)\psi^1$$

$$+h_1^2(G_{11}(h_0^2, p, x, y) + F_{11x}(h_0^2, p, x, y)\partial_x + F_{11y}(h_0^2, p, x, y)\partial_y)\psi^0$$

$$+(G_2(h_0^2, p, x, y) + F_{2x}(h_0^2, p, x, y)\partial_x + F_{2y}(h_0^2, p, x, y)\partial_y)\psi^0 = 0 - - - (3c)$$

(3a) has eigenvalues $h_{0n}^2, n = 1, 2, ...$ with respective degeneracies $d_n, n = 1, 2, ...$ and we can choose an orthonormal basis $\{u_{nk}(x, y), k = 1, 2, ..., d_n\}$ for $N(\nabla_{\perp\}^2+h_{0n}^2)}$. Thus, ψ^0 can be expanded as $\sum_{n,k} c(n, k)u_{nk}$. In particular, the general unperturbed n^{th} mode has the form

$$\psi^0 = \sum_{k=1}^{d_n} c(n, k)u_{nk}$$

7.17 The classical Boltzmann equation in classical curved space-time

Let $f(t, r, v)$ be the Boltzmann particle distribution function in a plasma where each particle has charge q. Here, $r = (x^1, x^2, x^3)$ is the position of the particle and $v = (v^1, v^2, v^3)$ is the three velocity of the particle and $t = x^0$ is coordinate time. If $g_{\mu\nu}(x)$ is the metric of space-time and u^μ the four velocity, then we have

$$v^r = dx2/dt, u^r = dx^r/d\tau = (dx^r/dt)(dt/d\tau) = v^r u^0, u^\mu = dx^\mu/d\tau$$

Thus the equation

$$g_{\mu\nu} u^\mu u^\nu = 1$$

can be expressed as

$$g_{00}(u^0)^2 + 2g_{0r}(u^0)^2 v^r + g_{rs}(u^0)^2 v^r v^s = 1$$

or equivalently.

$$u^0 = (g_{00}+2g_{0r}v^r+g_{rs}v^r v^s)^{-1/2}, u^r = u^0 v^r = v^r(g_{00}+2g_{0r}v^r+g_{rs}v^r v^s)^{-1/2} --- (a)$$

Note that u^r is a function of both $v = (v^r)$ and of the space-time coordinates $x = (t, r)$ of the particle. The energy-momentum tensor of the plasma is

$$T^{\mu\nu} = m\int f(t, r, v)u^\mu u^\nu d^3v = m\int f(x, v)u^\mu u^\nu d^3v$$

The Boltzmann equation taking into account the collision term using the relaxation time approximation can be expressed as

$$\partial_t f(t, r, v)+(v, \nabla_r)f(t, r, v)+(F(t, r, v), \nabla_v)f(t, r, v) = (f_0(r, v)-f(t, r, v))/\tau(v)$$

where f_0 is the equilibrium density, ie, the Maxwell-Boltzmann-Gibbs density. $F(t, r, v)$ is the force per mass on a plasma particle due to both the electromagnetic field and the gravitational field. This force can be computed by noting that the equation of motion of a plasma particle can be derived from the action principle

$$\delta[-m\int d\tau - \int A_\mu(x)u^\mu d\tau] = 0$$

where

$$d\tau = \left((g_{\mu\nu}(x)dx^\mu dx^\nu\right)^{1/2}$$

and this gives the geodesic equation in an electromangnetic field

$$du^\mu/d\tau + \Gamma^\mu_{\alpha\beta}u^\alpha u^\beta = qF^{\mu\nu}u_\nu --- (1)$$

which is the general relativistic tensor generalization of the Lorentz equation for a particle in an electromagnetic field and a gravitational field described by a potential $U(x)$:

$$mdv/dt = q(E(t,r) + v \times B(t,r)) - m\nabla U(t,r)$$

Note that (1) can be expressed as

$$u^0 d(u^0 v^r)/dt + \Gamma^r_{00}(u^0)^2 + 2\Gamma^r_{0s}(u^0)^2 v^s + \Gamma^r_{km}(u^0)^2 v^k v^m = q(F^r_0 u^0 + F^r_k u^0 v^k)$$

or equivalently as

$$d(u^0 v^r)/dt + \Gamma^r_{00}u^0 + 2\Gamma^r_{0s}u^0 v^s + \Gamma^r_{km}u^0 v^k v^m = q(F^r_0 + F^r_k v^k) --- (2)$$

where u^r is given by (a). This equation gives us an equation for dv^r/dt as a function of x, v on noting that the electric field F^r_0 and magnetic field F^r_k are functions of x. We can thus write (2) as

$$dv^r_d t = F^r(x,v) = F^r(t,r,v), r = 1,2,3$$

and hence write down the Boltzmann equation in the presence of gravity described by the metric tensor $g_{\mu\nu}(x)$ and the electromagnetic field $F^\mu_\nu(x)$. The Einstein field equations are

$$R^{\mu\nu} - (1/2)Rg^{\mu\nu} = -K(T^{\mu\nu} + S^{\mu\nu})$$

where $T^{\mu\nu}$ as described above is the energy-momentum tensor of the plasma fluid and $S^{\mu\nu}$ is the energy-momentum tensor of the electromagnetic field. Note that if $< u^\mu >$ denotes the average four velocity field of the plasma:

$$< u^\mu > (x) = \int u^\mu f(x,v)d^3v / \int f(x,v)d^3v$$

with

$$\rho(x) = m\int f(x,v)d^3v$$

denoting the plasma matter density field in position space, then we can write

$$T^{\mu\nu}(x) = m\int (u^\mu - < u^\mu(x) + < u^\mu > (x))(u^\nu - < u^\nu > (x) + < u^\mu > (x))f(x,v)d^3v =$$

$$= \rho(x) < u^\mu > (x) < u^\nu > (x) + p^{\mu\nu}(x)$$

where $p^{\mu\nu}(x)$ is the pressure tensor defined by

$$p^{\mu\nu}(x) = m\int (u^\mu - < u^\mu > (x))(u^\nu - < u^\nu > (x))f(x,v)d^3v$$

The Maxwell equations in the plasma are given by

$$F^{\mu\nu}_{:\nu}(x) = \mu_0 J^\mu(x)$$

where

$$J^\mu(x) = q\int u^\mu f(x,v)d^3v$$

is the four current field. Thus we get a coupled set of equations for the metric field $g_{\mu\nu}(x)$, the Boltzmann distribution function $f(x,v)$ and the electromagnetic four potential $A_\mu(x)$.

7.18 Emerging trends in signal analysis: A summary

Study signals defined on an n-dimensional differentiable manifold described by partial differential equations with noisy field terms. Incorporate unknown parameters into the model and control these parameters so that the multidimensional signal field tracks a desired signal field.

Functions defined on each open subset of a differentiable manifold are replaced with sheaves of Abelian groups, vector spaces or rings on a manifold with the restriction map. Varieties described as the common set of zeroes of a set of polynomials in n complex variables are replaced by schemes in which the topological space becomes the spectrum of a ring, ie, the set of all prime ideals of the ring and for each open subset U of this space in the Zariski topology wherein a set is closed iff it consists of all the prime ideals containing a given ideal, a function on this set evaluated at a prime ideal $\surd$ assumes values in the local ring at $\surd$.

Quantum Boltzmann equation and its application to modeling the refractive index of a material. When the driving electromagnetic field in the quantum Boltzmann equation is quantized, then we assume that the resuling wave field is a field of Heisenberg operators for which the temperature Green's function is computed in an initial state of the quantum field, like a Gibbs state corresponding to the unperturbed Hamiltonian or with respect to a coherent state of the electromagnetic field and this temperature Green's function is used to calculate things like the average current density, average charge density or more generally, averages of any observable like the momentum of a particle. These ideas have been described in the book "Quantum many particle systems" by Fetter and Walecka, Dover.

Evaluating the effect of quantum noise on the state of a quantum system when the Hamiltonian consists of those of N charged particles interacting with a superposition of first and second quantized electromagnetic fields, for example, the first quantized electromagnetic field can come from the nuclei of atoms interacting with N electrons and the second quantized electromagnetic field can come from the cavity field. The quantum noise fields come from the bath surrounding the cavity.

Formulating the classical and quantum Boltzmann equations in classical and quantum general relativity.

For Fermionic operator valued fields satisyfing the Schrodinger/Dirac equation in curved space-time taking into account nonlinearties in the Hamiltonian like Hartree-Fock corrections, we derive the two point temperature Green's function using the initial state of the field and the dynamics of the wave field operators. This temperature Green's function can be used to evaluate averages of one particle observables yielding things like the medium density, the medium current/particle density and more generally, the density of a one particle observable. The dynamics of the Green's function used to derive these densities is governed by the field Hamiltonian in curved space-time which can also contain external electromagnetic field interaction terms.

Let $\psi(t,r)$ be the wave field operators of the quantum Fermionic plasma. Let ρ be the state of the field at time $t = 0$. The plasma two point function is given by

$$G(t,r|t',r') = Tr(\rho.\psi(t,r)\psi(t',r')^*)$$

Suppose $O(r,r')$ is the position space representation of an observable O in first quantized Hilbert space which we take as $L^2(\mathbb{R}^3)$. For example, we can take $O = i\nabla_r$ to be the particle momentum in which case its average value over all the plasma particles at time t is given by

$$< O >_\rho = Tr(\rho. \int \psi(t,r)^* O(r,r')\psi(t,r')d^3rd^3r') = \int G(t,r'|t,r)O(r,r')d^3rd^3r'$$

Let m denote the mass of a plasma particle. Then the average density of particles in position space at time t and at the position r is given by

$$\sigma(t,r) = m.G(t,r|t,r)$$

Likewise the average flux of particles at (t,r) is given by

$$J(t,r) = Tr(\rho.\psi(t,r)^*(-i\nabla)\psi(t,r)) = -i\nabla_r G(t,r|t,r')|_{r'=r}$$

It follows that the average velocity field at (t,r) of the plasma is given by

$$v(t,r) = J(t,r)/\sigma(t,r) = -i(\nabla_r G(t,r|t',r')|_{r'=r}/mG(t,r|t,r)$$

Remark: Consider the density operator in the position representation $\rho(t,r,r')$ to be a second quantized field of operartors satisfying the Boltzmann equation

$$i\partial_t \rho_1(t) = [H_1(t), \rho_1(t)] + (N-1)Tr_2[V_{12}, \rho_1(t) \otimes \rho_1(t)]$$

where

$$V_{12} = V(r_1, r_2)$$

is a first quantized operator while

$$H_1(t) = -(1/2m)\nabla^2 - eU(r) + (ie/m)(2(A(t,r), \nabla_r) + divA(t,r))$$

contains both first and second quantized observavbles. The first quantized observables are $r, U(r), -i\nabla_r$ while $A(t,r)$ is a second quantized field. In fact it can be expressed as

$$A(t,r) = \sum_k (a(k)f_k(t,r) + a(k)^*\bar{f}_k(t,r))$$

This a function of second quantized operators $a_k, a_k^*, k = 1,2,...,p$ and first quantized operators $f_k(t,r), \bar{f}_k(t,r), k = 1,2,...,p$. It follows then that $\rho_1(t,r,r')$ will also be a function of t, r, r' and the second quantized operators $a_k, a_k^*, k = 1,2,...,p$. Now given any first quantized observable O like particle position, momentum, angular momentum or energy in say the position representation, its average density at (t,r) would be given by

$$Tr(\rho_0\psi(t,r)^*O\psi(t,r)) = Tr(\rho_0\psi(t,r)^*\int O(r,r')\psi(t,r')d^3r')$$

$$= \int O(r,r')Tr(\rho_1(t,r',r))d^3r'$$

For example, the average particle flux field is given by

$$J(t,r) = Tr(\rho_0\psi(t,r)^*(-i\nabla_r)\psi(t,r)) = -i\nabla_r Tr(\rho_1(t,r,r'))|_{r'=r}$$

In this way, we can evaluate all the averaged density field for any given one particle operator once we know $\rho_1(t,r,r')$. Note that $Tr(\rho_1(t,r,r'))$ is the density operator for a single plasma particle in the position representation and is obtained by tracing out ρ_1 over the second quantized field degrees of freedom. In other words, ρ_1 acts in the Hilbert space $\mathfrak{h} \otimes \mathcal{H}$ where $\mathfrak{h} = L^2(\mathbb{R}^3)$ is the Hilbert space of a single plasma particle while $\mathcal{H}$ is the field Hilbert space, ie, the Hilbert space in which the operators $a(k), a(k)^*, k = 1,2,...,p$ act.

7.19 Signal analysis problems in nonlinear electromagnetics

By setting up the equation of motion of the electron in the nuclear potential field, taking into account the effective electron mass and the effective damping coefficients, both of which are non-uniform as well as the external electric field, we determine the electron dipole moment as a nonlinear functional of the electric field and hence the polarization as a functional of the electric field. Likewise by setting up the torque equation for the precession of a magnetic dipole moment in a magnetic field taking into account the linear inhomogeneous relationship between the angular momentum of the dipole and its magnetic moment, we obtain the magnetic moment as a nonlinear functional of the magnetic field and hence also the magnetization of the medium as a functional of the magnetic field. These formulas result in expressions for the permittivity and permeability of the medium as nonlinear anisotropic functionals of the electromagnetic field. These expressions are used in the Maxwell equations and a numerical method based on the method of moments is used to obtain approximate solutions to the generalized wave equations that govern the electromagnetic field in a nonlinear (ie field dependent) inhomogeneous and anisotropic medium. This sort of perturbation theory based on moment expansions is applied to these nonlienar field equations to determine the approximate shift in the oscillation frequencies when the fields are enclosed within a finite boundary with vanishing boundary conditions on the field due to the nonlinearity, inhomogeneity and anisotropy of the medium. The final equations that govern the oscillation frequencies of the field form a nonlinear eigenvalue problem.

Consider a medium with field dependent permittivity, ie, a nonlinear relationship between the polarization field and the electric field. This can be explained from the microscopic structure of matter as follows. First the classical model in which an electron moves in a material under the influence of an anisotropic mass matrix $M(r)$, an anisotropic damping coefficient matrix $\Gamma(r)$ and a nuclear binding potential $U(r)$. The equation of motion taking into account both the external electric and magnetic field is given by

$$M(r(t))r''(t) + \Gamma(r(t))r'(t) + \nabla U(r(t)) = -e(E(t, r(t)) + r'(t) \times B(t, r(t)))$$

Likewise the precession of the magnetic moment $-eL/2m$ or more generally taking into account the anisotropicity of the mass matrix, this magnetic moment can be expressed as $F(r)L$ where $F(r)$ is a known 3×3 matrix valued function of the position and L is the angular momentum of the electron. This expression can be justified by noting that in a constant external magnetic field B, the magnetic vector potential is $B \times r/2$ and its contribution to the Hamiltonian $(1/2)(P+eB\times r/2)^T M(r)^{-1}(P+eB\times r/2)$ has the form $(e/2)[M(r)^{-1}]_{ij}P_i(B\times r)_j = (e/2)(B \times r)_j[M(r)^{-1}P]_j = (e/2)B.r \times M(r)^{-1}P$, so that the magnetic moment of the electron is $(e/2)r \times M(r)^{-1}P$ It i^{th} component can be expressed as

$$(e/2)\epsilon(ijk)x_i[M(r)^{-1}]_{jk}P_k = (e/2)\epsilon(ijk)[M(r)^{-1}]_{jk}x_iP_k$$

which in certain special cases can be expressed as $F_{ik}(r)L_k$ where $L_i = \epsilon(ijk)x_jP_k$. Thus the precession equation of the magnetic moment in the electromagnetic field can be expressed as

$$dL/dt = F(r)L \times B$$

ie the torque equation. Of course there will also be a contribution to the torque produced by the electric field but that will be small. These equations can be solved using perturbation theory yielding Volterra expansions of the form. If we assume that the electric and magnetic fields are functions of only time, then these expansions assume the form

$$r(t) = \sum_{n\geq 1, 0\leq r\leq n} \int K_1(n, r, t-t_1, ..., t-t_n)E(t_1)\otimes...\otimes E(t_r)\otimes B(t_{r+1})\otimes...\otimes B(t_n)dt_1...dt_n$$

Likewise for the angular momentum

$$L(t) = \sum_{n\geq 1} \int M_1(n, t-t_1, ..., t-t_n)B(t_1) \otimes ... \otimes B(t_n)dt_1...dt_n$$

The dipole moment per unit volume, ie, polarization, is

$$P(t,r) = -N(r)er(t)$$

and the magnetic moment per unit volume, ie, magnetization is

$$M(t,r) = N(r)F(r)L(t)$$

where $N(r)$ is the number of electrons per unit volume. Thus we obtain relations for the polarization and magnetization in the frequency domain as

$$P(\omega,r) = \sum_{n,r} \delta(\omega-\omega_1-...\omega_n)K(n,r,\omega_1,...,\omega_n)E(\omega_1)\otimes...\otimes E(\omega_r)\otimes B(\omega_{r+1})\otimes...\otimes B(\omega_n)d\omega_1...d\omega_n$$

$$M(\omega,r) = \sum_n \int \delta(\omega-\omega_1-...-\omega_n)B(\omega_1)\otimes...\otimes B(\omega_n)d\omega_1...d\omega_n$$

Now consider a cavity resonator filled with such a nonlinear medium. The basic equations are

$$curlE(\omega,r) = -j\omega B(\omega,r), curlH(\omega,r) = j\omega D(\omega,r)$$

where

$$B(\omega,r)/\mu_0 - M(\omega,r) = H(\omega,r), D(\omega,r) = \epsilon_0(E(\omega,r+P(\omega,r))$$

Using the above nonlinear relations, we obtain

$$B(\omega,r) = \mu(\omega,H(.,r)H(\omega,r), D(\omega,r) = \epsilon(\omega,E(.,r),B(.,r))E(\omega,r)$$

and hence we can derive using perturbation theory generalizations of the Helmholtz equations for E and H in the form,

$$(\nabla^2+\omega^2/c^2)E(\omega,r) + \delta.F_1(\omega,r,E(.,r),H(.,r))E(\omega,r) = 0,$$

$$(\nabla^2+\omega^2/c^2)H(\omega,r) + \delta.F_2(\omega,r,E(.,r),H(.,r))H(\omega,r) = 0,$$

where δ is a small perturbation parameter. The perturbation in the oscillation frequencies due to the nonlinear terms can be computed using perturbation theory. Specifically, suppose that the unperturbed basis functions corresponding to the eigenfrequencies $\omega(0,n), n = 1,2,...$ are given by $\psi_n(r) = [E_n(r),H_n(r)], n = 1,2,....$ More generally, assume that the nth mode has a degeneracy of d_n, so that corresponding to the eigen-frequency $\omega(0,n)$, we have an orthonormal basis $\psi_{nk}(r), j = 1,2,...,d_n$. Then, in the presence of the nonlinear perturbations, assume that $\omega(0,n)^2$ gets perturbed by $\delta.\omega_1^2$ and correspondingly, the unperturbed eigenmode $\psi_n(r) = \sum_{k=1}^{d_n} c(n,k)\psi_{nk}(r)$ gets perturbed by $\delta.\psi^1$. Then we get using first order perturbation theory,

$$[\nabla^2+\omega(0,n)^2/c^2+\delta.\omega_1^2/c^2][\psi_n(r)+\delta.\psi^1(r)]+\delta.F(\omega,r,\psi(.,r))[\psi_n(r)+\delta.\psi^1(r)] = 0$$

(Note that F assumes values in the space of 6×6 matrices. This gives on equation coefficients of δ,

$$[\nabla^2 + \omega(0,n)^2/c^2]\psi^1(r) + \omega_1^2\psi_n(r)/c^2 + F(\omega(0,n), r, \psi_n(., r))\psi_n(r) = 0$$

Taking the inner product on both sides with $\psi_{ms}(r)$ gives

$$(\omega(0,n)^2 - \omega(0,m)^2) < \psi_{ms}, \psi^1 > /c^2 + \omega_1^2 c(n,k)\delta_{mn}\delta_{sk}$$
$$+ < \psi_{ms}(r), F(\omega(0,n), r, \sum_k c(n,k)\psi_{nk}(r)) \sum_k c(n,k)\psi_{nk} >= 0$$

For $m \neq n$, this gives equations for $< \psi_{ms}, \psi^1 >$ in terms of $c(n,k)$ and ω_1. In order to determine the former, we put $m = n$ to get

$$\omega_1^2 c(n,s)\delta_{sk} + \sum_r c(n,r) < \psi_{ns}, F(\omega(0,n), r,$$
$$\sum_k c(n,k)\psi_{nk}(r))\psi_{nr} >= 0, s = 1, 2, ..., d_n$$

This equation is a nonlinear eigenvalue equation for the "eigenvector" $((c(n,k)))_{k=1}^{d_n}$ and the "eigenvalue ω_1^2 which can be solved perturbatively.

7.20 A summary of some signal analysis problems in quantum mechanics

Wigner distribution formulation of the Belavkin filter equation.

Belavkin filter when the Hamiltonian and Lindblad operators in the Hudson-Parthasarathy noisy Schrodinger equation contain classical stochastic parameters evolving according to a classical Markovian law. The problem is to find the filtered state on a real time basis in terms of non-demolition measurements so that this filtered state also contains information about the conditional probability density of the parameters at any given time given the non-demolition measurements upto that time.

Belavkin filter for the quantum cavity field resonator problems with a noisy bath outside the cavity. When the cavity contains elementary particles, the Belavkin filtered state of the cavity field an particles is to be partially traced out yielding theregby the one particle state dynamics in the presence of measurement noise in the form of a nonlinear quantum Boltzmann equation with measurement noise.

N identical particles interact with each other and with an external electromagnetic field. The quantum nonlinear Boltzmann equation is to be formulated for the single particle evolution by partial tracing and then translated to the Wigner distribution domain and then a comparison of this equation with the classical Boltzmann equation with collision terms is to be made. In the absence of the nonlinear term, in the Wigner distribution domain, the one particle evolution equation approximates to a classical Fokker-Planck equation.

Using the Wigner distribution, express the Belavkin quantum filter equation for the mixed state estimate of a noisy quantum system based on non-demolition measurements in the form of the Kushner-Kallianpur nonlinear stochastic filtering equations with extra quantum correction terms expressed in powers of Planck's constant due to non-commutativity of the system observables and the density operator.

When the Hudson-Parthasarathy-noisy Schrodinger equation contains classical parameters in the system operators like the Hamiltonian and the Lindblad operators and these parameters satisfy classical stochastic differential equations drived by Brownian motion and Poisson processes, or more generally, when these classical parameters form a classical Markov process with known infinitesimal generator and the non-demolition measurements include functions of these classical parameters plus white Gaussian noise, then generalize the Belavkin filter equation to estimate on a real time basis the state of the system. The Belavkin filtered state of the system should be expressed in the form $\rho(t, \theta)$ where θ is the classical parameter vector so that $Tr(\rho(t, \theta)) = p(t, \theta|Y_t)$ equals the conditional probability density of the parameter at time t given measurements Y_t upto time t and $\int \rho(t, \theta)d\theta = \rho(t)$ equals an estimate of the state of the quantum part of the system at time t given non-demolition measurements Y_t upto time t.

When the electromagnetic field within a cavity resonator is quantized with the medium being inhomogeneous, anisotropic and nonlinear, then to calculate the resulting field Hamiltonian as a nonlinear function of the free photon creation and annihilation operators in momentum space. When the cavity also contains electrons, positrons and non-Abelian gauge particles and the medium outside the cavity is a noisy quantum bath described in the Hudson-Parthasarathy formalism in terms of creation, annihilation and conservation processes, then to write down the Hudson-Parthasarathy-Schrodinger equation for the unitary evolution operator on the tensor product of the system and bath Hilbert spaces. Based on non-demolition measurements, derive the Belavkin filtered state of the cavity fields. Extend this filter to include the presence of N indistinguishable particles, say electrons within the cavity. Derive the Belavkin filtered state of a single charged particle by partially tracing out the state of the cavity fields and charged particles over the field variables. This would be a nonlinear quantum Boltzmann equation for the single particle state and would contain apart from the interaction potential energy between each pair of charged particles, the interaction energy between each charged particle and the quantum cavity field. The initial state of the cavity field is assumed to be a coherent state for the photons, electrons, positrons and non-Abelian matter and gauge fields.

Consider the quantum Boltzmann equation for the marginal state of a single charged particle in a plasma of N identical particles interacting with an external electromagnetic field. This equation is a nonliear state evolution equation which approximates to a quadratic order differential equation. By representing this state using the Wigner-distribution when the Hamiltonian is a function of the position and momentum variables, show that the resulting differential equation for the Wigner distribution function is like a classical Boltzmann equation with a quadratic collision terms and in addition, containing quantum correction terms expandable in powers of Planck's constant. Evaluate these quantum correction terms. In the absence of nonlinear collision terms, this reduces to a classical Fokker-Planck equation for the Wigner distribution of position and momentum containing in addition, quantum correction terms.

7.21 Deriving the quantum Boltzmann equation for the Belavkin filtered state estimate when the state evolution satisfies the Hudson-Parthasarathy noisy Schrodinger (HPS) equation and the measurement noise is non-demolition

The Hamiltonian in the HPS differential equation is assumed to be the sum of identical Hamiltonians for N indistinguishable particles plus identical interaction potentials between the different pairs of particles and the Lindblad noise coupling operators are also assumed to have a separable form. By partial tracing out of the resulting Belavkin filtered state equations over the remaining particles, we obtain after making the assumption of molecular chaos a nonlinear stochastic quantum Boltzmann equation for the one particle filtered state. This equation is generalization of the standard quantum Boltzmann equation in that it contains the Lindblad terms coming from the theory of open quantum systems and also contains measurement noise terms. The hard part of this problem is to cast this generalized stochastic quantum Boltzmann equation in the language of the Wigner distribution function when the one particle Hamiltonians are functions of the position and momentum of each particle and the interaction potentials are likewise functions of the position-momentum pairs of the individual particle pairs.

The Belavkin filter equation for the filtered state estimate of the Hudson-Parthasarathy-noisy Schrodinger state based on non-demolition measurements $Y_o(t)$ is given by

$$d\rho(t) = \theta_0(\rho(t))dt+[(\rho(t)M+M^*\rho(t))-Tr(\rho(t)(M+M^*))\rho(t)][dY_o(t) \\ -Tr(\rho(t)(M+M^*))dt]$$

where

$$\theta_0(\rho) = -i[H,\rho] - (1/2)(LL^*\rho + \rho LL^* - 2L^*\rho L)$$

We are herein regarding this Belavkin filtered state as a stochastic Schrodinger equation wherein the $Y_o(.)$ process is regarded as a classical stochastic process. Now assume that the system consists of N identical indistinguishable particles with Hamiltonian

$$H = \sum_k H_k + \sum_{k<j} V_{kj}$$

where $H_k's$ are identical copies acting on different tensor components of the tensor product Hilbert space and that V_{kj} a are also identical copies acting in the tensor product of two components. Assume that the Lindblad operator L has the separable form

$$L = \sum_k a(k) L_k^{\otimes N}$$

corresponding to the fact that the noise from the bath interacts with each particle in the system in the same way, for indeed, otherwise, we could not regard the particles as being indistinguishable. Likewise, we write

$$M = \sum_k b(k) M_k^{\otimes N}$$

We have

$$Tr_{23...N}\rho(t) = \rho_1(t), Tr_{34..N}\rho(t) = \rho_1(t) \otimes \rho_1(t) + g_{12}(t) = \rho_{12}(t)$$

$$Tr_{23...N}(\rho(t)M) = \sum_k b(k) Tr_{23...N}(\rho(t)M_k^{\otimes N}) \approx \sum_k b(k)(Tr(\rho_1(t)M_k))^{N-1}(\rho_1(t)M_k)$$

where we have made the "molecular chaos" assumption of approximating $\rho(t)$ by $\rho_1(t)^{\otimes N}$ and neglecting cross coupling terms in the density operator since these cross terms will contribute very little to the above trace because the Lindblad operators M are themselves very small. Likewise,

$$Tr_{23....N}(M\rho(t)) \approx \sum_k b(k)(Tr(\rho_1(t)M_k))^{N-1}(M_k\rho_1(t))$$

To proceed further, we also have to evaluate

$$Tr_{23...N}(\rho(t)M.dY_o(t))$$

Now, we have

$$dY_o(t) = dY_i(t) + j_t(M + M^*)dt, Y_i(t) = cA(t) + \bar{c}A(t)^*$$

Note that we are taking the partial trace over the particles $2, 3, ..., N$, not over the bath space. Note that $\rho(t)$ herein is being regarded as a function of $\eta_o(t) = (Y_o(s), s \leq t)$, which is Abelian and hence the above partial trace approximates on making the molecular chaos assumption to

$$Tr_{23...N}(\rho(t)M.dY_o(t)) = Tr_{23...N}(\rho(t)M).dY_o(t)) \approx \sum_k b(k) Tr(\rho_1(t)M_k)^{N-1} M_k\rho_1(t).dY_o(t)$$

Also

$$Tr_{23...N}[H, \rho(t)] \approx [H_1, \rho_1(t)] + Tr_2[V_{12}, \rho_1(t) \otimes \rho_1(t)],$$

$$Tr_{23...N}[LL^*\rho(t)] = \sum_{k,m} a(k)\bar{a}(m) Tr_{23...N}((L_k L_m^*)^{\otimes N}\rho(t)) \approx \sum_{k,m} a(k)\bar{a}(m) Tr(L_k L_m^* \rho_1(t))^{N-1} L_k L_m^* \rho_1(t)$$

and likewise,

$$Tr_{23...N}[\rho(t)LL^*] = \sum_{k,m} a(k)\bar{a}(m) Tr_{23...N}((\rho(t)L_k L_m^*)^{\otimes N}) \approx \sum_{k,m} a(k)\bar{a}(m) Tr(\rho_1(t)L_k L_m^*)^{N-1}\rho_1(t)L_k L_m^*$$

$$Tr_{23...N}[L^*\rho(t)L] = \sum_{k,m} a(k)\bar{a}(m) Tr_{23...N}((L_m^*)^{\otimes N}\rho(t)L_k^{\otimes N}) \approx \sum_{k,m} a(k)\bar{a}(m) Tr(L_m^*\rho_1(t)L_k)^{N-1} L_m^*\rho_1(t)L_k$$

Note that in conformity with the indistinguishability of the interaction of the particles with the bath, we could more generally assume that the Lindblad operators have the general form

$$L = \sum_{k} a(k) \sum_{\sigma \in S_N} L_{k,\sigma 1} \otimes ... \otimes L_{k,\sigma n}$$

$$M = \sum_{k} b(k) \sum_{\sigma \in S_N} M_{k,\sigma 1} \otimes ... \otimes M_{k,\sigma n}$$

where $L_{k,j}, M_{k,j}$ are one particle Lindblad operators and σ runs over all permutations of $\{1, 2, ..., N\}$

Chapter 8

Algebraic Geometry, Robotics, Electromagnetics, Cosmology, and Filtering Theory Topicss

8.1 An example of a covariant functor from the theory of differentiable manifolds involving pullback of a differential form from one manifold to the other given a differentiable map from the first manifold to the second

Let $A \to A' \to A''$ be an exact sequence of Abelian groups. Let F be a covariant functor on the space of Abelian groups. Thus, if A, B are Abelian groups then F naturally induces a map from $Hom(A, B)$ into $Hom(FA, FB)$. Specifically, take $T \in Hom(A, B)$. Then since F is injective, the map induced by F denoted by F_1 is defined by $T_1 = F_1 T = FTF^{-1}$. Actually covariance of F is defined by the property $T_1 F = FT$. As an example of this, suppose M and N are two differentiable manifolds and $x \in M$. Let $f : M \to N$ be a differentiable map and let ω be a one-form on N. Then f induces on M a one form by $(f^*\omega)(x)(X) = \omega(f(x))(df(x)(X))$, ie, $f^*\omega(x) = \omega(f(x))odf(x)$. If f is a diffeomorphism, then the map f^* defines a one-one correspondence between one forms on N and one forms on N. We can in this case write the above covariant transformation law as

$$f^*\omega(x)df(x)^{-1} = \omega(f(x))$$

8.2 Some applications of master-slave robots operating in teleoperation to control the slave robot having a camera attached to its end effector for calculating the relative position of a construction site so that it can deposit construction material at that site by generating a feedback force that reduces the error between the position of the construction site and the its own location

The dynamics of the robot are governed by stochastic differential equations with noise comging in the form of Poisson jerks applied at the hand operator of the

DOI: 10.1201/9781003728627-8

master robot and continuous Brownian motion noise coming from the motion of the slave in its environment, ie, on a rugged road with friction and also air friction

Suppose we are given a set of n construction sites and we are given a slave robot S controlled by a master robot M. Assume that the job is to collect the waste products from these n sites and then deposit construction material at these sites. A camera attached to the end-effector of the slave robot takes a picture of a given site. Let $f(x, y)$ be the resultant image recorded. By use of radar-like technology, ie, by transmitting electromagnetic waves and recording the scattered field pattern, the robot also determines the range an bearing of the site relative to its own location. Let r_0 be the location of the slave robot and ξ the position vector of the site relative to the slave robot. This information is sent to the master robot which accordingly generates a force process proportional to the relative position ξ. This force is transmitted to the the slave from the master using teleoperation techniques. Thus the equation of motion of the slave robot becomes

$$d^2r_0(t)/dt^2 = F(r_0, dr_0/dt) + K(r_1 - r_0(t))$$

where r_1 is the position vector of the site. When teleoperation is used, this equation gets modified to

$$d^2r_0(t)/dt^2 = F(r_0(t), dr_0(t)/dt) + \sum_m K_m(r_1 - r_0(t - t_m))$$

Further, the amount of construction material that is loaded onto the slave robot would now be a function of the image field $f(X, Y)$ recorded by the slave camera. In the presence of noise, the above equation becomes the following delay stochastic differential equation:

$$dr_0(t) = v_0(t)dt, dv_0(t) = F(r_0(t), v_0(t))dt + \sum_m K_m(r_1 - r_0(t - t_m))dt + \sigma.dV(t)$$

where $V(t)$ is a Levy process, ie, a linear superposition of independent Brownian motion processes and Poisson processes. The problem is to analyze the statistical performance of this system by statistical methods such as large deviation principles. Assuming that the process $V(t)$ has rate function $\int_0^T I(xi'(t))dt$, it follows that the rate function of the robot position process becomes

$$\int_0^T [I(r_0''(t) - F(r_0(t), r_0'(t)) - \sum_m K_m(r_1 - r_0(t - t_m))]dt$$

and the problem is to control the feedback coefficients $\{K_m\}$, so as to ensure that the probability that the trajectory would deviate from a desired trajectory $r_d(t)$ by an amount more than a given threshold ϵ is minimized. This amount to choosing the $K_m's$ so that

$$inf_{max(|r_0(t)-r_d(t)|, 0\le t\le T)>\epsilon} \int_0^T [I(r_0''(t) - F(r_0(t), r_0'(t)) - \sum_m K_m(r_1 - r_0(t - t_m))]dt$$

is a maximum.

Further the construction material loading process is a dynamical function of the camera image field of the site and this image will also depend upon the robot position $r_0(t)$ and the orientation $\theta(t)$ of its links. Accordingly if the recorded image is $f(X, Y|r_0(t), \theta(t))$, we generate a load quantity and composition dependent on this field at each stage of its motion. This dynamical load is denoted by $M(r_0(t), \theta(t)) = M(f(X, Y|r_0(t), \theta_0(t)) : (X, Y) \in D)$, where D is the spatial region of the camera coordinates. We wish that this load should fall within a given range R with a large probability during the motion of the robot. This can be achieved using the large deviation principle by approximating the probability that at some stage, the load falls outside the region R by using the LDP and minimizing this probability w.r.t the control parameters K_m. This amounts to maximizing

$$min(I(r_0(.), \theta(.)) : (M(r_0(t), \theta(t)) \notin Rforsomet \in [0, T])$$

Publication plans: A thorough study of large deviation principles as applied to nonlinear systems driven by weak noise and then formulating the problem in terms of stochastic differential equations followed by summarizing the mathematical derivations and software simulation of the problem in the form of articles will be our goal.

Details of applications: During metro rail constructions, we have to construct the platforms, railway track as well as several support pillars. In each of this objects, we require a different quantity and composition of materials like cement, iron and others. Let $C_k, k = 1, 2, ..., d$ denote the composition for the d kinds of objects involved here. While loading the slave robot to deposit these materials there could be random fluctuations, so that the dynamics of these quantities will follow a system of coupled stochastic differential equations of the form

$$dC_k(t) = f_k(C_1(t), .., C_d(t)|\theta)dt + \sum_m \sigma(k, m)dV_m(t), k = 1, 2, ..., d$$

The drift coefficients f_k are determined by the constraints involved during the loading process like size of the slave robot, continual deposit at the sites along the robot trajectory etc. Here, θ are control parameters. The objective would be to choose these parameters so that $C_k(t)$ is as close as possible to a desired trajectory $d_k(t)$. This can be achieved by fine tuning the parameters θ using large deviation methods, ie, maximizing w.r.t θ, the minimum of

$$\int_0^T I(\sigma^{-1}(C'(t) - f(C(t)|\theta))dt$$

over all trajectories $C(.)$ for which $max(|C(t) - d(t)| : 0 \leq t \leq T) > \epsilon$.

8.3 Examples of left exact functors based on homomorphisms of Abelian groups

This scheme enables us to construct exact sequences on the space of homomorphisms of Abelian groups. Then, given a complex with differential maps, we can use a short exact sequences of homomorphisms of Abelian groups to construct a long exact sequence of cohomology groups which can in turn be applied to the problem of discretizing the Atiyah Singer index theorem for the Chiral anomaly in the zero modes of the Dirac equation and casting this theorem in a more abstract framework.

Let $0 \to B \to B' \to B'' \to 0$ be an exact sequence of three Abelian groups. Here, $T : B \to B'$ and $T' : B' \to B''$ are the corresponding morphisms. Thus, by the definition of exact sequence, $Ker(T) = 0, Ker(T') = Im(T)$ and $Im(T') = B''$. Let A be an object. We claim that the induced sequence

$$0 \to Hom(A, B) \to Hom(A, B') \to Hom(A, B'')$$

is exact, ie it forms a left exact sequence. Let $T_1 : Hom(A, B)(A, B')$ and $T_2 : Hom(A, B') \to Hom(A, B'')$ be the induced maps. Thus, for $f \in Hom(A, B)$ and $g \in Hom(A, B')$, we have

$$T_1(f) = Tof, T_1'(g) = Tog$$

To see this, we take any $f \in Hom(A, B)$ such that $T_1(f) = 0$. Then $Tof = 0$, ie, $T(f(a)) = 0 \forall a \in A$ and hence, by the injectivity of T, ie, $Ker(T) = 0$, we get $f(a) = 0 \forall a \in A$, ie, $f = 0$. This shows that $Ker(T_1) = 0$. Now take $g \in Hom(A, B')$ and suppose $T_1'(g) = 0$. Then $T'(g(a)) = 0 \forall a \in A$, ie, $g(a) \in Ker(T')$ forall $a \in A$. Since $Ker(T') = Im(T)$, it follows then that $g(a) = T(S(a)) \forall a \in A$ where for each $a \in A$, $S(a) \in B$. It is clear that since T is injective, $S(a) \in B$ is uniquely determined by $a \in A$. It is easy to see that $S \in Hom(A, B)$ and hence $g = ToS = T_1(S)$, ie we have proved that $Ker(T_1') \subset Im(T_1)$. Conversely suppose $g \in Im(T_1)$. Then, $g = T_1(S)$ for some $S \in Hom(A, B)$. Thus, $g(a) = T(S(a)) \forall a \in A$. It follows then that since $Im(T) = Ker(T')$, we have $T'(g(a)) = 0$ forall $a \in A$, ie, $T_1'(g) = 0$, ie, $g \in Ker(T_1')$. Thus, $Ker(T_1') = Im(T_1)$. This proves that $0 \to Hom(A, B) \to Hom(A, B') \to Hom(A, B'')$ is exact. Right exactness however is not true in general, ie, we do not have in general, $Im(T_1') = Hom(A, B'')$ even though $Im(T') = B''$. Indeed, let $g \in Hom(A, B'')$. For $Im(T_1') = Hom(A, B'')$, we require that there exist $f \in Hom(A, B')$ such that $T_1'(f) = g$, ie, we require that $g(a) = T'(f(a)) \forall a \in A$. However, suppose T' was not injective while g was chosen to be injective. Specifically, suppose that T' mapped the entire of B' to a single point in B''. Then, no matter how we choose $f \in Hom(A, B')$, we would not ba able to satisfy $g(a) = T'(f(a)) \forall a \in A$. Thus, the map $B \to Hom(A, B)$ is a left exact functor, not right exact.

Now we shall prove that exactness of $0 \to B \to B' \to B'' \to 0$ via the maps T, T'implies that

$$0 \to Hom(B'', A) \to Hom(B', A) \to Hom(B, A) - - - (1)$$

is exact, ie, the sequence is left exact. Define $T_1' : Hom(B'', A) \to Hom(B', A)$ by $T_1'(f) = foT', f \in Hom(B'', A)$ and define $T_1 : Hom(B', A) \to Hom(B, A)$ by $T_1(f) = foT, f \in Hom(B', A)$.

Let $f \in Hom(B'', A)$ be such that $T_1'(f) = 0$. Then, $f(T'(b')) = 0 \forall b' \in B'$. Since $T' : B' \to B''$ is surjective, it follows then that $f' = 0$ proving $Ker(T_1') = 0$. Now, let $f \in Hom(B', A)$ be such that $T_1(f) = 0$. Then, it follows that $f(T(b)) = 0 \forall b \in B$. Now observe that since T' is surjective and since $Ker(T') = Im(T), B''$ is isomorphic to $B'/T(B)$ as groups under the mapping $T' : B'/T(B) \to B''$ given by $T'(b' + T(B)) = T'(b')$. We can define $g : B'/T(B) \to A$ by $g(b' + T(B)) = f(b'), b' \in B'$. This map is well defined since we have already seen that $f(T(B)) = 0$. Then, g is a group Homomorphism (Note that $T(B) = Ker(T')$ is a normal subgroup of B'). Since then, $Hom(B'', A)$ is isomorphic to $Hom(B'/T(B), A)$, it follows that f is contained in the image of $Hom(B'', A)$, thereby proving that $Ker(T_1) = Im(T_1')$ and thereby exactness of (1). Note that right exactness does not hold in general, ie, $Im(T_1) \neq Hom(B, A)$. Indeed, for $f \in Hom(B', A)$, we have $T_1(f) = foT$ and if T is not surjective (Note that T is injective), then it follows that foT cannot cover the whole of $Hom(B, A)$ as f varies over the whole of $Hom(B', A)$.

8.4 Problems in analog and digital communication and electromagnetics

Delta modulation of a signal based on adaptive linear prediction of the continuous time signal based on a finite number of past samples of the modulated staircase signal. At each time iteration, the prediction error being of a small dynamic range can be quantized using a one bit quantizer and the modulated signal value at the next time iteration is obtained by adding this quantized prediction error to the present value of the modulated signal. It is an important problem to determine the statistics of the error between the original continuous time signal and the delta modulated signal in the case when the original signal is a stochastic process. For this, we derive a nonlinear stochastic difference equation for the error process driven by the given stochastic signal to be modulated. This difference equation can be solved approximately to any degree of accuracy using perturbation theory for nonlinear difference equations.

Cavity resonator filled with an inhomogeneous, anisotropic, nonlinear, ie, field dependent medium. Derivation of the general form of the oscillation frequency modal equations from a variational principle followed by a method of moments approximate numerical calculation. The modal frequencies appear as solutions of a nonlinear matrix eigenvalue problem.

Construction of the delta-modulator of a signal based on one bit quantization of a finite order linear prediction error sequence of the given signal.

Consider a signal $m(t)$ for which a delta-modulator is to be constructed based on p^{th} order linear prediction. At time nT, let $m_1[n]$ be the delta-modulated signal. To construct $m_1[n+1] = m_1((n+1)T)$, we predict $m((n+1)T)$ based on $m_1((n-k)T), k = 0, 1, ..., p$ linearly using the scheme

$$\hat{m}((n+1)T) = \sum_{k=0}^{p} a(k) m_1((n-k)T)$$

The LPC coefficients $a(k), k = 0, 1, ..., p$ are chosen so that

$$\sum_{j=0}^{p}(m((n+1-j)T) - \hat{m}((n+1-j)T))^2$$

is minimized. Let

$$e(n+1) = m((n+1)T) - \hat{m}((n+1)T)$$

denote the associated prediction error at time $n+1$. Let $Q(e(n+1)) = e_1(n+1)$ denote the one bit quantization of $e(n+1)$, ie, $Q(e) = +\delta$ if $e \geq 0$ and $Q(e) = -\delta$ if $e < 0$. Then, we define

$$m_1(n+1) = \hat{m}(n+1) + e_1(n+1) = \sum_{k=0}^{p} a(k)m_1((n-k)T) + e_1(n+1)$$

This constitutes the δ-modulated signal. it is a staircase signal that increases or decreases in steps of size δ.

Problem: Let $m(t)$ be a random signal whose joint probability distributions at all times $(t_1, t_2, ..., t_k).k = 1, 2, ...$ are known. Then, derive an approximate algorithm to obtain the joint probability distributions of the delta-modulated signal $m_1(n), n = 1, 2, ...$ The above minimization problem amounts to solving for the $a(k)'s$

$$\frac{\partial}{\partial a(m)}\sum_{j=0}^{p}(m(n+1-j) - \sum_{k=0}^{p} a(k)m_1(n-j-k))^2 = 0, m = 0, 1, ..., p$$

Equivalently, the $a(k)'s$ satisfy the linear equations

$$\sum_{k=0}^{p} a(k).[\sum_{j=0}^{p} m_1(n-j-k)m_1(n-j-k)m_1(n-j-m)]$$

$$= \sum_{j=0}^{p} m(n+1-j)m_1(n-j-m), m = 0, 1, ..., p$$

This gives us the $a(k)'s$ as function of $m_1(n-r), 0 \leq r \leq 2p$ and $m(n+1-r), 0 \leq r \leq p$. Formally, we can thus write

$$\hat{m}(n+1) = \sum_{k=}^{p} a(k)m_1(n-k) = F(n, p, m_1(n-k), k = 0, 1, ..., 2p)$$

and hence

$$m_1(n+1) = \hat{m}(n+1) + Q(m(n+1) - \hat{m}(n+1)) =$$

$$F(n, p, m_1(n-k), k = 0, 1, ..., 2p) + Q(m(n+1) - F(n, p, m_1(n-k), k = 0, 1, ..., 2p))$$

This equation is a $2p + 1^{th}$ order non-linear difference equation for $m_1(.)$ from which its joint statistics can be derived.

Cavity resonator filled with an inhomogeneous, anisotropic, nonlinear, ie, field dependent medium The equations for the modes can be derived from a variational principle of the form

$$\delta \int_B (\nabla \otimes V(\omega, r), F(\omega, r, V(\omega', r), \omega' \in \mathbb{R}) \nabla \otimes V(\omega, r)) d^3r$$

$$+\delta\omega^2 \int_B (V(\omega, r), G(\omega, r, V(\omega', r), \omega' \in \mathbb{R}) V(\omega, r)) d^3r = 0$$

Here V is assumed to be a d-component complex vector field, so that $\nabla \otimes V$ becomes a $3n$-component complex vector field. The variational principle then gives Noting that we can also write

$$\int_B (\nabla \otimes V(\omega, r), F(\omega, r, V(\omega', r), \omega' \in \mathbb{R}) \nabla \otimes V(\omega, r)) d^3r =$$

$$\int \bar{V}_{i,j} F_{ijkm} V_{k,m} d^3r$$

and

$$\delta_{\bar{V}} \int \bar{V}_{i,j} F_{ijkm} V_{k,m} d^3r = -\int \delta \bar{V}_i (F_{ijkm} V_{k,m})_{,j} d^3r$$

our nonlinear eigen-equation for the eigenfrequencies ω becomes

$$[F_{ijkm} V_{k,m}]_{,j} + \omega^2.[G_{ij} V_j] = 0$$

which in tensor notation, reads

$$div(F(\nabla \otimes V)] + \omega^2 V = 0$$

This nonlinear eigenvalue equation can be solved approximately using perturbation theory: Write

$$F = F_0 + \delta.F_1(\omega, r, V)$$

where F_0 is a constant matrix and F_1 is a matrix valued function of $\omega, r, V(r', \omega'), r' \in B, \omega' \in \mathbb{R}$ and likewise write

$$G = G_0 + \delta.G_1$$

where G_0 is a constant matrix and G_1 is a matrix valued function of $\omega, r, V(r', \omega'), r' \in B, \omega' \in \mathbb{R}$. Then, writing

$$\omega^2 = \omega_0^2 + \delta.\omega_1^2, V = V_0 + \delta V_1,$$

we get from first order perturbation theory,

$$div[F_0 \nabla \otimes V_0] + \omega^2 G_0 V_0 = 0 - - - (1)$$

$$div[F_0 \nabla \otimes V_1] + div[F_1(\omega_0, r, V_0] + \omega_0^2 G_0 V_1 + \omega_1^2 G_1(\omega_0, r, V_0) = 0 - - - (2)$$

(1) is a linear eigenvalue PDE for V. It is to be solved subject to the boundary conditions on V_0. It can be expressed in expanded form as

$$F_0(kmij) V_{0i,jm} + \omega^2 G_0(k, m) V_{0m} = 0, k = 1, 2, ..., d - - - (2)$$

Expand $V_0(r)$ in terms of an orthonormal basis $U_{in}, n = 1, 2, ...$ for the cavity that is compatible with the boundary conditions:

$$V_{0i}(r) = \sum_n C(n) U_{in}(r)$$

That gives

$$\sum_n C(n)[F_0(kmij) < U_{sr}, U_{in,jm} > + \omega^2 G_0(k, m) < U_{sr}, U_{mn} >] = 0$$

Define the matrices $Q_1.Q_2$ with entries

$$Q_1(sr, kn) = F_0(kmij) < U_{sr}, U_{in,jm} >, Q_2(sr, kn) = G_0(k, m) < U_{sr}, U_{mn} >$$

Note that the Einstein summation convention is being used, ie, summation over the repeated indices. Then, the possible eigenfrequencies of oscillation in the absence of inhomogeneities, anisotropicity and nonlinearity in the medium are given by the solutions to the characteristic equation

$$det(Q_1 + \omega^2 Q_2) = 0$$

Denote the solutions by $\omega_{0n}^2, n = 1, 2, ...$

8.5 Maximally symmetric spaces in general relativistic cosmology

The metric of $N + 1$ space-time is taken to be a quadratic form with constant coefficients. This metric is invariant under a group of rotations. By restricting this metric to the corresponding ellipsoid/hyperboloid obtained by setting the quadratic form equal to a constant, we obtain a metric on an N-dimnensional manifold that is invariant under rotations and quasitranslations of the N-dimensional manifold. This group has dimension $N(N+1)/2$ and so the restricted metric is maximally symmetric. We learn how to set up the Maxwell equations in such a metric.

Let $\mathbb{R}^{N+1}$ be endowed with the Riemannian metric

$$d\tau^2 = C_{\mu\nu} dx^\mu dx^\nu + K^{-1}.dz^2$$

where $((C_{\mu\nu}))$ is a constant $N \times N$ symmetric matrix and K is a positive real number. The coordinates in $\mathbb{R}^{N+1}$ are $\{x^\mu\}_{\mu=1}^N, z$. We now define an N-dimensional submanifold S of $\mathbb{R}^{N+1}$ by the constraint equations $- \mathrm{C}_{\mu\nu} x^\mu x^\nu + K^{-1} z^2 = K^{-1} Morequivalently, bysettingz = (M - KC_{\mu\nu} x^\mu x^\nu))^{1/2} - - - (0) Nowconsiderthefollowinglineartransformationon \mathrm{R}^{N+1}$: $(x^\mu, z) \rightarrow (y^\mu, v)$ where

$$y^\mu = R_\nu^\mu x^\nu + R_z^\mu z, v = R_\nu^v x^\nu + R_z^v z$$

The coefficients $R^\mu_\nu, R^\mu_z, R^v_\nu, R^v_z$ are chosen so as to satisfy

$$C_{\mu\nu}y^\mu y^\nu + K^{-1}v^2 = C_{\mu\nu}x^\mu x^\nu + K^{-1}z^2 - - - (1)$$

for all x^μ, z. This is equivalent to requiring that

$$C_{\mu\nu}R^\mu_\rho R^\nu_\sigma + K^{-1}R^v_\rho R^v_\sigma = C_{\rho\sigma},$$

$$C_{\mu\nu}R^\mu_\rho R^\nu_z + K^{-1}R^v_\rho R^v_z = 0,$$

$$C_{\mu\nu}R^\mu_z R^\nu_z + K^{-1}(R^v_z)^2 = K^{-1}$$

Note that these are precisely $N(N+1)/2+N+1$ conditions on the N^2+2N+1 coefficients $\mathrm{R}^\mu_\nu, R^\mu_z, R^z_\mu, R^v_z$ and this defines a Lie group of transformations on $\mathbb{R}^{N+1}$ of dimension $(N^2+2N+1)-(N(N+1)/2+N+1) = N(N+1)/2$. This group is the orthogonal group $G = SO(N+1, g)$ with respect to the metric matrix

$$g = \begin{pmatrix} ((C_{\mu\nu})) & ((R^\mu_z)) \\ ((R^v_\mu)) & R^v_z \end{pmatrix}$$

Consider now the N dimensional manifold S as defined above. We wish to show that this surface is left invariant by the group G. In fact, we see that if $(x^\mu, z) \in S$, then for $(y^\mu, z) \in S$, we require that

$$C_{\mu\nu}y^\mu y^\nu + K^{-1}v^2 = K^{-1}M$$

But this is obvious by (1). Now it follows that G leaves the metric on $\mathbb{R}^{N+1}$ invariant and also preserves S. Hence, the the group G when restricted to S must necessarily leave the metric induced on S by that on $\mathbb{R}^{N+1}$ also invariant. Note that the metric induced on S is given by

$$g_{\mu\nu}(x)dx^\mu dx^\nu$$

where

$$g_{\mu\nu} = C_{\mu\nu} + K^{-1}(\partial z/\partial x^\mu).(\partial z/\partial x^\nu)$$

where $z = z(x)$ is defined by (0). It is clear that S endowed with this metric is maximally symmetric since it is a group of dimension $N(N+1)/2$.

Problem: Formulate Maxwell's equations on S in terms of the above metric.

Remark: Consider a Riemannian metric $g_{\mu\nu}(x)$ on any differentiable manifold M. Let $x^\mu \to x^\mu + \xi^\mu(x) = \bar{x}^\mu$ be an infinitesimal diffeomorphism that leaves invariant the metric, ie, $\bar{g}_{\mu\nu}(x) = g_{\mu\nu}(x)$ for all x. Then, we can easily show that

$$\xi_{\mu:\nu} + \xi_{\nu:\mu} = 0$$

To see this first observe that

$$g_{\rho\sigma}(x) = \bar{g}_{\mu\nu}(\bar{x})(\partial\bar{x}^\mu/\partial x^\rho).(\partial x^\nu/\partial x^\sigma)$$

or equivalently, on neglecting terms of second order of smallness, in order that $\bar{g}_{\mu\nu}(x) = g_{\mu\nu}(x)$, we require that

$$g_{\rho\sigma}(x) = g_{\mu\nu}(x+\xi)(\delta^{\mu}_{\rho} + \xi^{\mu}_{,\rho}).(\delta^{\nu}_{\sigma} + \xi^{\nu}_{\sigma})$$

$$= (g_{\mu\nu}(x) + g_{\mu\nu,\alpha}\xi^{\alpha})(\delta^{\mu}_{\rho} + \xi^{\mu}_{,\rho})(\delta^{\nu}_{\sigma} + \xi^{\nu}_{,\sigma})$$

$$sothatwithneglectofsecondorderofsmallnessterms, thisequationbecomes$$

$g_{\rho\nu}\xi^{\nu}_{,\sigma} + g_{\mu\sigma]xi^{\mu}_{,\rho}+g_{\rho\sigma,\alpha}\xi^{\alpha}=0} whichcanberearrangedas \xi_{\mu:\nu} + \xi_{\nu:\mu} = 0$

Maxwell equations in a maximally symmetric N-dimensional Riemannian manifold: Consider the above metric

$$d\tau^2 = C_{\mu\nu}dx^{\mu}dx^{\nu} + K^{-1}dz^2, z = (1 - KC_{\mu\nu}x^{\mu}x^{\nu})^{1/2}$$

Thus,

$$dz = (KC_{\mu\nu}x^{\nu})/z$$

and this maximally symmetric metric can be expressed as

$$d\tau^2 = (C_{\mu\nu} + KC_{\mu\alpha}C_{\nu\beta}x^{\alpha}x^{\beta}/z^2)dx^{\mu}dx^{\nu} = g_{\mu\nu}(x)dx^{\mu}dx^{\nu}$$

where

$$g_{\mu\nu} = C_{\mu\nu} + KC_{\mu\alpha}C_{\nu\beta}x^{\alpha}x^{\beta}/(1 - KC_{\alpha\beta}x^{\alpha}x^{\beta})$$

$$= \frac{[C_{\mu\nu} + K(C_{\mu\alpha}C_{\nu\beta} - C_{\mu\nu}C_{\alpha\beta})x^{\alpha}x^{\beta}]}{1 - KC_{\alpha\beta}x^{\alpha}x^{\beta}}$$

In matrix notation,

$$g = (C + KCxx^TC)/(1 - Kx^TCx)$$

If K is infinitesimal, then upto first order of smallness,

$$g = C + K(x^TCx)C + KCxx^TC = C + K(Cxx^TC + (x^TCx)C)$$

$$det(g) = det(C) + K.Tr(C^{-1}(Cxx^TC + (x^TCx.C)) = det(C) + K(N+1)(x^TCx)$$

We live in a universe that has more than four space-time dimensions. Let N be the dimension of our universe. A bosonic string exists in a 26 dimensional universe while a Fermionic stri ng exists in a ten dimensional universe. When the two strings are put together, we get a superstring that propagates in a ten dimenionsional universe. After compactification of six out of these ten dimensions, we are left with the observable four dimensional universe of space-time in which the conventional fields, namely the scalar, electromagnetic, Dirac and non-Abelian gauge and matter fields propagate. How then does one formulate the actual laws of physics in an N dimensional maximally symmetric universe ? Let A_{μ} be the em N-potential and the em antisymmetric field tensor then becomes

$$F_{\mu\nu} = A_{\nu,\mu} - A_{\mu,\nu}$$

The field equations are

$$(F^{\mu\nu}\sqrt{g})_{,\nu} = 0$$

Now suppose that the curvature K is small, say of the first order of smallness. Then, upto $O(K)$, Maxwell's equations are

$$det(C)div(F(x)) + K(N+1)div(x^T Cx.F(x)) = 0 - - - (2)$$

where $F(x)$ is the $N \times N$ antisymmetric matrix field $((F^{\mu\nu}(x)))$ and $divF(x)$ is the vector N-vector field $(())F^{\mu\nu}_{,\nu}))$. Let us try to solve (2) using perturbation theory for partial differential equations. However we should first express F in terms of $A = (A_\mu)$ in matrix notation, apply the Lorentz gauge condition $(A^\mu\sqrt{-g})_{,\mu} = 0$ and hence reduce (2) to a second order pde for A which is an approximate maximally symmetric generalization of the wave equation in N dimensions.

8.6 Time dependent Hartree-Fock equations for a system of N interacting particles with application to the two electron problem

The Hamiltonian of the N-electron atom consists of the sum of the individual kinetic energies, the electrostatic interaction energy of each electron with the nucleus and the pairwise mutual electrostatic interaction of the electrons. The wave function is assumed to be an antisymmetrized version of the product of one particle wave functions. This wave function is substituted into the Hamiltonian and the corresponding action functional is extremized w.r.t the one particle wave functions resulting thereby in a system of N one particle nonlinear Schrodinger wave equations. These equations can be derived from a Hamiltonian for each particle in the smeared potential field generated by the other particles and also in addition taking into account the exchange interaction of a pair of particles.

The Hamiltonian is

$$H = \sum_{a=1}^{N} H_a + \sum_{1 \le a < b \le N} V_{ab}$$

where H_a acts in $\mathcal{H}_a$ while V_{ab} acts in $\mathcal{H}_a \otimes \mathcal{H}_b$. The Hilbert space of the entire system is $\mathcal{H} = \bigotimes_{a=1}^{N} \mathcal{H}_a$. The Schrodinger equation can be derived from the variational principle

$$\delta_{\psi,\bar{\psi}} S = 0, S = \int L dt, L = i < \psi|\partial_t\psi > - < \psi|H|\psi >$$

In the Hartree-Fock approximation, we restrict to variations over only antisymmetric separable wave functions of the form

$$\psi = \sum_{\sigma \in S_n} sgn(\sigma) \bigotimes_{k=1}^{n} \psi_{\sigma k}$$

Note that we are assuming herein that all the Hilbert spaces $\mathcal{H}_a, a = 1, 2, ..., N$ are identical copies of each other. Now

$$< \psi|\psi >= \sum_{\sigma,\rho} sgn(\sigma\rho)\Pi_{k=1}^{N} < \psi_{\sigma k}|\psi_{\rho k} >= \sum_{\sigma,\rho} sgn(\sigma\rho)\Pi_k < \psi_k|\psi_{\rho\sigma k} >$$

$$= N!. \sum_{\sigma} sgn(\sigma)\Pi_k < \psi_k|\psi_{\sigma k} >= N!det((< \psi_k|\psi_m >))$$

Further,

$$\partial_t|\psi >= \sum_{k,\sigma} sgn(\sigma)\psi_{\sigma 1} \otimes ... \otimes \partial_t\psi_{\sigma k} \otimes ... \otimes \psi_{\sigma n}$$

$$< \psi|\partial_t|\psi >= \sum_{k,\sigma,\rho} sgn(\sigma.\rho) < \psi_{\rho 1}|\psi_{\sigma 1} > ... < \psi_{\rho k}|\partial_t|\psi_{\sigma k} > ... < \psi_{\rho n}|\psi_{\sigma n} >$$

$$= \sum_{k,m} \sum_{\sigma,\rho:\rho k=m} sgn(\sigma\rho)[\Pi_{j\neq k} < \psi_{\rho j}|\psi_{\sigma j} >] < \psi_m|\partial_t|\psi_{\sigma k} >$$

The coefficient of $\delta\psi_m^*$ in the variation of this quantity is thus given by

$$\sum_{k,m} \sum_{\sigma,\rho:\rho k=m} sgn(\sigma\rho)[\Pi_{j\neq k} < \psi_{\rho j}|\psi_{\sigma j} >]\partial_t|\psi_{\sigma k} >$$

$$+ \sum_{\sigma,\rho,k,l:k\neq l,\rho l=m} sgn(\sigma\rho)[\Pi_{j\neq k,l} < \psi_{\rho j}|\psi_{\sigma j} >] < \psi_{\rho k}|\partial_t|\psi_{\sigma k} > |\psi_{\sigma l} >$$

Another way to calculate the same is

$$< \psi|\partial_t|\psi >= \sum_{\rho,\sigma} sgn(\sigma\rho) < \psi_{\rho 1} \otimes ...\psi_{\rho n}|\partial_t|\psi_{\sigma 1} \otimes ...\psi_{\sigma n} >$$

4

$$= \sum_{\sigma,\rho} sgn(\rho) < \psi_{\rho\sigma 1} \otimes ...\psi_{\rho\sigma n}|\partial_t|\psi_{\sigma 1} \otimes ...\psi_{\sigma n} >$$

$$= \sum_{\sigma,\rho,k} sgn(\rho)[\Pi_{j\neq k} < \psi_{\rho\sigma j}|\psi_{\sigma j} >]. < \psi_{\rho\sigma k}|\partial_t|\psi_{\sigma k} >$$

Next,

$$< \psi|H_a|\psi >= \sum_{\sigma,\rho} sgn(\sigma\rho)[\Pi_{k\neq a} < \psi_{\rho k}|\psi_{\sigma k} >]. < \psi_{\rho a}|H_a|\psi_{\sigma a} >$$

The coefficient of $\delta\psi_m^*$ in the variation of this is

$$\sum_{\sigma,\rho:\rho a=m} sgn(\sigma\rho)[\Pi_{k\neq a} < \psi_{\rho k}|\psi_{\sigma k} >]H_a|\psi_{\sigma a} >$$

$$+ \sum_{\sigma,\rho,k:k\neq a,\rho k=m} sgn(\sigma\rho)[\Pi_{j\neq k,a} < \psi_{\rho k}|\psi_{\sigma k} >] < \psi_{\rho a}|H_a|\psi_{\sigma a} > |\psi_{\sigma k} >$$

and we also calculate for $a < b$,

$$< \psi|V_{ab}|\psi >= \sum_{\sigma,\rho} sgn(\sigma\rho) < \psi_{\rho 1} \otimes ...\psi_{\rho n}|V_{ab}|\psi_{\sigma 1} \otimes ...\psi_{\sigma n} >$$

$$= \sum_{\sigma,\rho} sgn(\sigma\rho)[\Pi_{k\neq a,b} < \psi_{\rho k}|\psi_{\sigma k} >] < \psi_{\rho a} \otimes \psi_{\rho b}|V_{ab}|\psi_{\sigma a} \otimes \psi_{\sigma b} >$$

The coefficient of $\delta\psi_m^*$ in the variation of this is given by

$$\sum_{\sigma,\rho:\rho a=m} sgn(\sigma\rho)[\Pi_{k\neq a,b} < \psi_{\rho k}|\psi_{\sigma k} >] < I \otimes \psi_{\rho b}|V_{ab}|\psi_{\sigma a} \otimes \psi_{\sigma b} >$$

$$+ \sum_{\sigma,\rho:\rho b=m} sgn(\sigma\rho)[\Pi_{k\neq a,b} < \psi_{\rho k}|\psi_{\sigma k} >] < \psi_a \otimes I|V_{ab}|\psi_{\sigma a} \otimes \psi_{\sigma b} >$$

$$+\sum_{\sigma,\rho,k:k\neq a,b,\rho k=m} sgn(\sigma\rho)[\Pi_{j=a\!\!/,b,k} < \psi_{\rho k}|\psi_{\sigma k} >] < \psi_{\rho a}\otimes\psi_{\rho b}|V_{ab}|\psi_{\sigma a}\otimes\psi_{\sigma b} > |\psi_{\sigma k} >$$

Exercise: [a] Work out the special case of the Hartree-Fock equations for a two electron atom by taking the Hamiltonian as

$$H = H_1 + H_2 + V_{12}, \psi = (\psi_1 \otimes \psi_2 - \psi_2 \otimes \psi_1)$$

by evaluating

$$< \psi|\partial_t|\psi >, < \psi|H|\psi >, < \psi|\psi >$$

assuming that H_1, H_2 are identical copies and taking into account wave function normalization, the Lagrangian density to be extermized w.r.t ψ_1, ψ_2 is then given by

$$L =< \psi|i\partial_t|\psi > - < \psi|H|\psi > -\lambda(< \psi|\psi > -1)$$

[b] Work out the Hartree-Fock equations taking into account explicit dependence of the Hamiltonian and the wave functions on the spin of the electrons assuming the special case of two electrons. Specifically, assume that the Hamiltonian has the form

$$H = H_1 + H_2 + V_{12}$$

where

$$H_k = (-1/2m)\nabla_k^2 - Ze^2/r_k + ge(sigma_k, B(t, r_k))/2m, k = 1, 2,$$

$$V_{12} = e^2/r_{12} + K(\sigma_1, \sigma_2),$$

Consider each of the following antisymmetric trial wave functions

$$|\psi >= (psi_1(r_1)\psi_2(r_2) - \psi_2(r_1)\psi_1(r_2))| + + >,$$

$$|\psi >= (psi_1(r_1)\psi_2(r_2) - \psi_2(r_1)\psi_1(r_2))| - - >,$$

$$|\psi >= (psi_1(r_1)\psi_2(r_2) - \psi_2(r_1)\psi_1(r_2))(| + - > +| - + >),$$

$$|\psi >= (\psi_1(r_1)\psi_2(r_2) + \psi_2(r_1)\psi_1(r_2))(| + - > -| - + >)$$

Note that the Hilbert space of each of the electrons is

$$L^2(\mathbb{R}^3) \otimes \mathbb{C}^2$$

Here $(1/2)\sigma_1 = (1/2)(\sigma_{1x}, \sigma_{1y}, \sigma_{1z})$ and $(1/2)\sigma_2 = (1/2)(\sigma_{2x}, \sigma_{2y}, \sigma_{2z})$ are identical copies of the spin vector observable each component acting on $\mathbb{C}^2$. Show explicitly using this calculation, the presence of spin-exchange energy between the two electrons coming as a term in each of the one particle differential equations from the component $K(\sigma_1, \sigma_2)$ in the interaction potential V_{12} apart from the usual smeared charge electrostatic interaction potential coming from the term e^2/r_{12} in V_{12}

8.7 Application of the extended Kalman filter to estimating the states of the particles in the Hartree-Fock approximation

From the above, we derive the time dependent Hartree-Fock equations after taking into account the constraint

$$1 =< \psi|\psi >= N! \sum_{\sigma} \Pi_k < \psi_k|\psi_{\sigma k} >= N! det((< \psi_k|\psi_m))$$

This constraint is incorporated by means of a time dependent Lagrange multiplier $\lambda(t)$ which contributes an extra term

$$-\lambda(t)(N! \sum_{\sigma} \Pi_k < \psi_k|\psi_{\sigma k} > -1)$$

Express the Hartree-Fock equations for the two electron case as

$$i\partial_t \psi_k(t, r_k) = F_k(\psi_1, \bar{\psi}_1, \psi_2, \bar{\psi}_2, t, r_k), k = 1, 2$$

Now add noise to the rhs with Ito correction terms to make the evolution non-linear unitary. Consider the problem in which the average value of an observable X is measured with measurement noise:

$$dz(t) =< \psi_t|X|\psi_t > dt + dv(t)$$

where

$$\psi_t = \psi_{1t} \otimes \psi_{2t} - \psi_{2t} \otimes \psi_{1t}$$

Express the Hartree-Fock equations as well as the measurement equations in terms of the real and imaginary parts of $\psi_k, k = 1, 2$ and construct the extended Kalman filter for estimating on a dynamical basis, the ψ_{1t}, ψ_{2t}. Work out the special case when the Hamiltonian is the Schrodinger Hamiltonian by spatially discretizing each of the one electron wave functions and arranging their spatial samples in the form of a finite dimensional state vector.

8.8 Newtonian cosmology, evolution of inhomogeneities in a background homogeneous and isotropic space

The velocity field is given by Hubble's law plus a small perturbation. The density field is given by a uniform in space density plus a small perturbation and likewise the gravitational potential is given by the formula for the potential of a uniform mass density in space plus a small perturbation. The Navier-Stokes, equation of continuity and the Poisson equation govern the dynamics of the space-time dependent velocity perturbation, density perturbation and the gravitational potential pertrubation. These equations are linearized and transformed into ordinary differential equations in time by assuming a plane wave form for these pertrubations with wave length proportional to the radius of the sphere, ie, of the expanding universe.

Newtonian cosmology, evolution of inhomogeneities in a background homogeneous and isotropic space: The universe is a sphere of radius $R(t)$ filled with a definite amount of matter M so that its density at time t is $\rho_0(t) = 3M/4\pi R(t)^3$. Small inhomogeneous and anisotropic perturbations cause the density to get shifted by $\rho_1(t,r)$. Likewise, the unperturbed velocity field described in terms of the comoving spatial coordinates r_1 $(r(t) = R(t)r_1)$ is the trajectory of a particle having constant comoving coordinate r_1 is given by $dR(t)r_1/dt = R'(t)r_1 = (R'(t)/R(t))R(t)r_1 = (R'(t)/R(t))r = v_0(t,r)$. Small inhomogneous and anisotropic perturbations shift this velocity field by $v_1(t,r)$. Likewise, the unperturbed gravitational field is that determined by a uniform density $\rho_0(t)$ within the sphere. It is given by $g_0(t,r) = -4\pi G\rho_0(t)r/3$. Due to inhomogneities and anisotropicity, it gets perturbed by $g_1(t,r)$. The dynamical equations of our universe are (a) The hydrodynamical equations, the equation of continuity and the Poisson equation relating gravity to hte distribution of matter, corresponding to a velocity field $R'(t)r/R(t) + v_1(t,r)$, the density field $\rho_0(t) + \rho_1(t)$ and the gravitational field $g_0(t,r) + g_1(t,r)$. These equations are with Hubble's constant $H(t) = R'(t)/R(t)$,

$$(\rho_0(t)+\rho_1)(H(t)r+v_1,\nabla)(H(t)r+v_1)+H'(t)r+\partial_t v_1)$$

$$= -f'(\rho_0(t))\nabla\rho_1+(\rho_0(t)+\rho_1)(g_0+g_1)---(1)$$

$$div((\rho_0+\rho_1)(H(t)r+v_1)) + \rho_0'(t) + \partial_t\rho_1 = 0 --- (2)$$

$$\nabla.(g_0+g_1) = -4\pi G(\rho_0+\rho_1), \nabla\times g_1 = 0 --- (3)$$

Note that $\nabla \times g_0 = 0$ since g_0 is proportional to the radial vector r. Equating terms of zeroth order of smallness in the first equation gives

$$\rho_0(t)(H(t)^2 r + H'(t)r) = \rho_0(t)g_0(t,r)$$

This gives

$$R''(t)/R(t) = -4\pi G/3$$

or

$$R''(t) = -4\pi GR(t)/3$$

which is precisely the equation of motion of a particle sitting on the surface of the sphere of radius $R(t)$ subject to the gravitational pull of all the matter within the sphere. Equating terms of first order of smallness in the first equation gives

$$\rho_0(t)(H(t)(r,\nabla)v_1(t,r)+H(t)(v_1(t,r),\nabla)r+\partial_t v_1(t,r))$$

$$= -f'(\rho_0(t))\nabla\rho_1(t,r)+\rho_0(t)g_1(t,r)+\rho_1(t)g_0(t,r)$$

Here, we make use of the equation of state of the matter field $p = f(\rho)$. Next, the zeroth order term in the second equation gives

$$3\rho_0(t)H(t) + \rho_0'(t) = 0$$

or equivalently,

$$3\rho_0(t)R'(t)/R(t) + \rho_0'(t) = 0$$

which integrates to

$$\rho_0(t)R(t)^3 = constt = 3M/4\pi$$

as expected, namely the total mass of matter within the sphere is a constant. Next, the zeroth order term in the third equation is

$$div g_0 = -4\pi G\rho_0(t)$$

which is clearly satisfied by $g_0 = -4\pi G\rho_0(t)r/3$. The first order terms in the second equation gives

$$H(t)div(\rho_1(t,r)r) + \rho_0(t)div(v_1(t,r)) + \partial_t v_1(t,r) = 0$$

or

$$3H(t)\rho_1(t,r) + H(t)(r,\nabla)\rho_1(t,r) + \rho_0(t)div(v_1(t,r)) + \partial_t v_1(t,r) = 0$$

Finally, the first order term in the third equation is

$$div g_1(t,r) = 4\pi G\rho_1(t,r)$$

We arrange the three first order equations in sequential order below:

$$\rho_0(t)(H(t)(r,\nabla)v_1(t,r)+H(t)(v_1(t,r),\nabla)r+\partial_t v_1(t,r))$$
$$= -f'(\rho_0(t))\nabla\rho_1(t,r)+\rho_0(t)g_1(t,r)+\rho_1(t)g_0(t,r)---(4),$$

$$H(t)\rho_1(t,r) + H(t)(r,\nabla)\rho_1(t,r) + \rho_0(t)div(v_1(t,r)) + \partial_t v_1(t,r) = 0 --- (5),$$

$$div g_1(t,r) = 4\pi G\rho_1(t,r) --- (6)$$

These are five equations for the five functions $v(t,r) = (v_i(t,r))_{i=1}^3, \rho_1(t,r), g_1(t,r)$. We can reduce these pde's in the space-time variables into ode's in just the time variable by making the substitution

$$v_1(t,r) = v_1(t).exp(iq.r/R(t)), \rho_1(t,r)$$
$$= \rho_1(t).exp(iq.r/R(t)), g_1(t,r) = g_1(t).exp(iq.r/R(t))$$

where q wave vector of the matter and gravity waves, the corresponding wavelength is $2\pi R(t)/|q|$ which expands proportionally to the radius of of the universe.

Exercise:Carry out the process of deriving ode's for $v_1(t), \rho_1(t), g_1(t)$ with q as a parameter vector.

8.9 Problems in commutative algebra

R is a ring and M a maximal ideal of R. Show that R/M is a field by showing that if $x \in R, x \notin M$, then there exists $y \in R$ such that $xy - 1 \in M$. In fact, since M is a maximal ideal, if $x \notin M$, then $< x, M >= R$ an hence there exist $y \in R$ such that $xy + m = 1$ for some $m \in M$. Conclude the proof from here.

Let Y be a variety corresponding the polynomial ring $K[x_1, ..., x_n]$ where K is a field. In other words, there exists a set $T \subset K[x_1, ..., x_n]$ such that $Y = Z(T)$=the set of common zeroes of T in K^n. Let $I(Y)$ denote the ideal in $K[x_1, ..., x_n]$ corresponding to Y, ie, $I(Y)$ is the set of all $p \in K[x_1, ..., x_n]$ that vanish on Y. Then, $T \subset I(Y) = I(Z(T))$. Show that $I(E)$ is an ideal in $K[x_1, ..., x_n]$ for any $E \subset K^n$. Prove Hilbert's Nullstellensatz: If I is an ideal in $K[x_1, ..., x_n]$, then $I(Z(I))$=radical of I=the set of all $f \in K[x_1, ..., x_n]$ such that $f^m \in I$ for some positive integer m. Give an example by taking I to be that generated by $(x_j - a_j)^{m_j}, j = 1, 2, ..., r$ for some fixed $r \leq n$ and some fixed $a_1, ..., a_r \in K$ where the $m_j's$ are positive integers and showing that $Z(I)$ is the set of all points in K^n whose first r coordinates are $a_1, ..., a_r$ and the remaining $n - r$ coordinates are arbitrary elements of K. Show further that any $f \in I(Z(I))$ iff f can be expressed as $(x_1 - a_1)f_1 + ... + (x_r - a_r)f_r$ for some $f_1, ..., f_r \in K[x_1, ..., x_n]$. Show finally using the multinomial theorem that for $m \geq m_1 + ... + m_r$ we have that f^m is expressible as $(x_1 - a_1)^{m_1} g_1 + ... + (x_r - a_r)^{m_r} g_r$ for some $g_1, ..., g_r \in K[x_1, ..., x_r]$.

[3] In the context of the previous problem, define $E \subset A^n$ to be closed (in the Zariski topology) if $E = Z(T)$ for some $T \subset K[x_1, ..., x_n]$. Show that if Y_1 and Y_2 are arbitrary subsets of K^n, then $I(Y_1 \cup Y_2) = I(Y_1) \cap I(Y_2)$. Also show that if $T_1, T_2 \subset K[x_1, ..., x_n]$, then $Z(T_1.T_2) = Z(T_1) \cup Z(T_2)$ where $T_1.T_2 = \{f_1 f_2 : f_1 \in T_1, f_2 \in T_2\}$. Also show that $Z(\bigcup_a T_a) = \bigcap_a Z(T_a)$. Deduce thus the Zariski topology is a well defined topology.

In the context of the previous two problems, let Y be a variety. Then, show that $Z(I(Y)) = \bar{Y}$ where the closure is w.r.t to the Zariski topology. First note that by the definition of a closed set, $Z(I(Y))$ is a closed subset of K^n. Now, let $x \in Y$. Then $I(Y)$ vanishes at x and hence $x \in Z(I(Y))$. Thus $Y \subset Z(I(Y))$. This shows that $\bar{Y} \subset Z(I(Y))$. Further, suppose $C = Z(T)$ is any closed set containing Y, where we can assume that T is an ideal. Then, T vanishes on $Z(T)$ and hence also on Y. Hence $T \subset I(Y)$. It follows that $Z(I(Y)) \subset Z(T) = C$. It follows that the intersection of all the closed sets containing Y also contains $Z(I(Y))$. Hence, we deduce that $Z(I(Y)) \subset \bar{Y}$.

Chapter 9

Quantum Mechanics, Stochastic Control, Supersymmetry, and Cohomology Applications

9.1 Topics in automata applied to fluid models

Exercise: Write down the Navier-Stokes and continuity equation for the velocity field and the density field of a compressible fluid given the equation of state $p = g(\rho)$ in nonlinear state vector form:

$$\partial_t \psi(t,r) = F(\psi(t,r), \nabla_r \otimes \psi(t,r), \nabla_r \otimes \nabla_r \otimes \psi(t,r), f(t,r))$$

where $f(t,r)$ is the external force density and $\psi(t,r)$

$$= [\rho(t,r), v(t,r)^T]^{T\, isthestatevector.Nowconsideranapproximationinw}_{hichthefluidvelocityvectorateachspatialpixelcanmovealongonlyoneofNdirectionsinspaceateachspatialpixelpoint}$$

Denote these possible directions in $\mathbb{R}^3$ by $e_k, k = 1, 2, ..., N$. Thus choose a random vector $v(t, r_k)$ for each k and each time t that its direction can assume only the unit vector values $e_1, ..., e_N$ with probabilities $p(t,k,j), j = 1, 2, ..., N$. Derive differential equations for $p(t,k,j)$ and $\rho(t, r_k)$ by using the Navier-Stokes equation and the equation of continuity. Specifically, assume that $v(t, r_k)$ can be expressed as $\phi(t,k)e_j$ with a probability of $p(t,k,j)$ so the average velocity at (t, r_k) is given by

$$< v(t, r_k) >= \phi(t,k). \sum_{j=1}^{N} p(t,k,j) e_j$$

Likewise, the average value of $(v, \nabla)v(t, r_k)$ can be approximated. Then by taking the expectation of the Navier-Stokes equation and the equation of continuity, we can derive differential equations for $p(t,k,j), \rho(t, r_k)$.

9.2 String theoretic synthesis of probability densities

First we write down the string action in terms of the metric and the gauge field interacting with the metric. Using these equations of motion, we calculate the pde satisfied by the string propagator in terms of the metric and gauge fields. From this string propagator parameterized by the metric and gauge fields, we compute using the vertex function of string theory, the transition probability amplitude of the string making a transition from an initial state to a final state in infinite time using a standard formula of string theory. More generally, we can compute using the Hamiltonian operator of a string and the vertex func-

DOI: 10.1201/9781003728627-9

tion, the probability amplitude of the string making a transition from a given initial state to a final state in finite time t. The magnitude square of this transition probability is then matched to a given probability density functional that varies with time using the stochastic gradient algorithm with adaptive control parameters being those present in the metric and gauge fields. This gives us a quantum neural network for tracking probability density functionals. Moreover, the adaptation constant can also be controlled in accordance with the prediction error energy involved in predicting a given signal based on the quantum average of an observable calculated using the string transition probability density functional. This scheme would guarantee density tracking while simultaneously ensuring that the mean of an observable w.r.t this density is highly correlated with a given density functional.

Consider a Bosonic string having tree amplitude

$$< \phi_2|\Delta V(k_2, z_2)\Delta V(k_1, z_1)\Delta|\phi_1 >= M(k_1, z_1, k_2, z_2, \phi_2, \phi_1)$$

where

$$\Delta = (L_0 - 1)^{-1} = \int_0^1 z^{L_0 - 2}dz$$

is the string propagator with

$$L_0 = \sum_{n\geq 1} \alpha(-n).\alpha(n)$$

and $V(k, z)$ is the vertex function defined by

$$V(k, z) = exp(ik.X(z)), X(z) = -i\sum_{n\neq 0} \alpha^\mu(n)z^n/n + p^\mu z + x^\mu$$

Can this amplitude be realized via Hamiltonian dynamics of the string ? Writing $z = exp(i(\tau - \sigma))$, we see that $X^\mu(z) = X^\mu(\tau, \sigma)$ indeed satisfies the differential equation

$$(\partial_\tau^2 - \partial_\sigma^2)X^\mu(\tau, \sigma) = 0$$

This differential equation is derived from the Lagrangian density

$$L = (1/2)\partial_\tau X^\mu.\partial_\tau X_\mu - (1/2)\partial_\sigma X^\mu.\partial_\sigma X_\mu$$

The corresponding Hamiltonian density is given by

$$H = (1/2)\partial_\tau X^\mu.\partial_\tau X_\mu + (1/2)\partial_\sigma X^\mu.\partial_\sigma X_\mu = \sum_{n\geq 1} \alpha(-n).\alpha(n) = L_0$$

Thus,

$$X^\mu(\tau, \sigma) = exp(-i\tau L_0)X^\mu(0, \sigma).exp(i\tau L_0)$$

The propagator is defined in the way of conventional way of quantum mechanics: In the time domain,

$$K(t - t') = U(t - t')\theta(t - t'), U(t) = exp(-itH)$$

Since

$$i\partial_t U(t) = HU(t),$$

it follows that

$$i\partial_t K(t) = iU'(t)\theta(t) + iU(t)\delta(t) = HK(t) + i\delta(t)I$$

or equivalently,

$$(i\partial_t - H)K(t) = i\delta(t)I$$

which is the conventional differential equation for the propagator. In the frequency domain, this is equivalent to taking

$$K(E) = \int K(t)exp(iEt)dt = i(E - H)^{-1}$$

When we take $E = 1, H = L_0$, then the above string propagator is recovered. For example, take the massive Klein-Gordon field with Lagrangian

$$L(\phi, \partial_\mu \phi) = (1/2)\partial_\mu \phi.\partial^\mu \phi - m^2\phi^2/2$$

The field Hamiltonian density is

$$H = (1/2)(\partial_t \phi)^2 + (1/2)(\nabla \phi)^2 + m^2\phi^2/2$$

The field equation is

$$\partial_\mu \partial^\mu \phi + m^2\phi = 0$$

or equivalently,

$$(\partial_t^2 - \nabla^2 + m^2)\phi = 0$$

The field propagator

$$\Delta(x, x') = < T(\phi(x).\phi(x')) >$$

in view of the equal time bosonic canonical commutation relations

$$[\phi(t, r), \partial_t \phi(t, r')] = i\delta^3(r - r')$$

satisfies the differential equation

$$(\partial_t^2 - \nabla^2 + m^2)\Delta(x, x') = i\delta^4(x - x')$$

or equivalently in the frequency domain,

$$\Delta(E) = (E^2 + \Delta^2 - m^2)^{-1}$$

Writing the first quantized particle Hamiltonian as

$$H_1 = (-\Delta^2 + m^2)^{1/2}$$

we see that according to first quantized quantum mechanics, the propagator would satisfy

$$(E - H_1)\Delta(E) = i.I$$

However, according to the second quantized field theory, it instead satisfies,

$$(E + H_1).(E - H_1)\Delta(E) = iI$$

ie,

$$\Delta(E) = i(E^2 - H_1^2)^{-1} = i(E^2 + \Delta^2 - m^2)^{-1}$$

Again consider the Dirac propagator defined according to second quantized field theory:

$$(\gamma^\mu p_\mu - m)\psi(x) = 0, p_\mu = i\partial_\mu,$$

$$\Delta(x,x') =< T(\psi(x).\psi(x')^*) >$$

in view of the equal time Fermionic anticommutation relation

$$\{\psi(t,r), \psi(t,r')^*\} = i\gamma^0\delta^3(r-r')$$

satisfied the differential equation

$$(\gamma^\mu i\partial_\mu - m)\Delta(x,x') = \delta^4(x-x')$$

or equivalently, in the frequency domain

$$\Delta(E) = (E - (\alpha, -i\nabla) - \beta m)^{-1} = (E - H_1)^{-1}$$

which indeed coincides with the result of first quantized quantum mechanics once we note that the first quantized Dirac Hamiltonian is given by

$$H_1 = (\alpha, -i\nabla) + \beta m$$

Note that the Dirac field Lagrangian density is

$$L = \psi^*\gamma^0(i\gamma^\mu\partial_\mu - m)\psi$$

and the corresponding Hamiltonian density is

$$H = \psi^*((\alpha, -i\nabla) + \beta m)\psi = \psi^* H_1 \psi$$

Now in order to justify an expression of the form

$$< \phi_2|V(k_1,z_1)\Delta.V(k_2 m z_2)\Delta...\Delta V(k_M, z_M)|\phi_1 >$$

as a string quantum amplitude, we look at the case of a quantum field theory with unperturbed Hamiltonian H_0 and perturbation $V(t)$ so that the total Hamiltonian is $H(t) = H_0 + V(t)$. The transition amplitude is given by

$$< \phi_2|U(\infty, -\infty)|\phi_1 >$$

where $U(t,s)$ is the evolution operator

$$U(t,s) = T\{exp(-i\int_s^t H(u)du)\}$$

This can be evaluated using the Dyson series in the interaction picture as

$$lim_{s\to-\infty,t\to\infty}\sum_{n\geq 0}(-i)^n\int_{s<u_n<...<u_1<t} < \phi_2|U_0(t)V_1(u_1)...V_1(u_n)U_0(-s)|\phi_1 > du_1...du_n$$

where

$$V_1(t) = U_0(t)^* V(t) U_0(t), U_0(t) = exp(-itH_0)$$

The nth term in this series is the same as

$$< \phi_2 | lim_{s,t} \int_{s<t_n<...<t_n<t} U_0(t-t_1)V(t_1)U_0(t_1-t_2)V(t_2)...U_0(t_{n-1}-t_n)V(t_n)U_0(t_n-s)dt_1...dt_n|\phi_1 >$$

Now the free particle propagator in the frequency domain, as we saw above is

$$\Delta(E) = (E - H_0)^{-1} = \int_0^\infty U_0(t) exp(iEt)dt$$

It follows that for $t \geq 0$,

$$U_0(t) = \int_{\mathbb{R}} \Delta(E).exp(-iEt)dE$$

and hence assuming $V(t) = V$ to be time independent, we have

$$I =< \phi_2 | \int_{s<t_n<...<t_n<t} U_0(t-t_1)VU_0(t_1-t_2)V...VU_0(t_n-s)dt_1...dt_n|\phi_1 >$$

$$= \int_{s<t_n<...<t_1<t} < \phi_2|\Delta(E_0)V\Delta(E_1)V...V\Delta(E_n)|\phi_1 > exp(i(E_0(t-t_1)+$$

$$E_1(t_1-t_2)+...+E_n(t_n-s)))dt_1...dt_n dE_0...dE_n$$

Defining

$$u_0 = t - t_1, u_1 = t_1 - t_2, ..., u_{n-1} = t_{n-1} - t_n,$$

the region of $\{t_j\}$-integration becomes the simplex

$$u_0 < t - s, u_0 + u_1 < t - s, ..., u_0 + ... + u_{n-1} < t - s, u_0, ..., u_{n-1} > 0$$

and hence

$$I = \int_{0<u_0<t-s, 0<u_0+u_1<t-s,...0<u_0+...+u_{n-1}<t-s} < \phi_2|\Delta(E_0)V\Delta(E_1)V...V\Delta(E_n)|\phi_1 >$$

$$exp(i(E_0u_0+E_1u_1+...+E_{n-1}u_{n-1}+E_n(t-s-u_0-...-u_{n-1}))du_0...du_{n-1}dE_0...dE_n$$

In the limit as $t \to \infty$ and $s \to -\infty$, this becomes

$$\int_{0<u_0,...,u_{n-1}<\infty} < \phi_2|\Delta(E_0)V\Delta(E_1)V...V\Delta(E_n)|\phi_1 > exp(i(E_0u_0+E_1u_1+...$$

$$+E_{n-1}u_{n-1}+E_n(t-s-u_0-...-u_{n-1}))du_0...du_{n-1}dE_0...dE_n$$

We can control the string dynamics by a gauge field by adding to its Lagrangian an extra gauge term, so that the total string action becomes

$$\int (1/2)\eta^{ab} X^\mu_{,a} X_{\mu,b} d\tau.d\sigma + \int B_{\mu\nu}(X)\epsilon^{ab} X^\mu_{,a} X^\nu_{,b} d\tau.d\sigma$$

In that case, the string propagator gets modified by the additional gauge field interaction. To determine this modification we first write down the string field equations by setting the variation of the above action w.r.t X^μ to zero to obtain

$$\eta^{ab} X_{\mu,ab} - 2(B_{\mu\nu}(X)\epsilon^{ab} X^\nu_{,b})_{,a} + B_{\rho\sigma,\mu}(X)\epsilon^{ab} X^\rho_{,a} X^\sigma_{,b} = 0$$

which simplifies to

$$\eta^{ab}X_{\mu,ab} + \epsilon^{ab}(B_{\rho\nu,\mu}(X) - 2B_{\mu\nu,\rho}(X))X^{\rho}_{,a}X^{\nu}_{,b}) = 0$$

or equivalently defining

$$B_{\mu\rho,\nu} + B_{\rho\nu,\mu} + B_{\nu\mu,\rho} = H_{\mu\rho\nu}$$

we can write the above equation as

$$\Box X_{\mu} + H_{\mu\rho\nu}(X)(dX^{\rho} \wedge dX^{\nu})/d\tau.d\sigma = 0$$

the two dimensional analog of the geodesic equation of a particle in general relativity, or equivalently the two dimensional analog of the equation of motion of a charged particle in an electromagnetic field. Here, we are assuming that the metric on the string world sheet is Minkowskian flat. However, it can be readily generalized to the non-flat case to give

9.3 Stochastic optimal control with fine tuning of the parameters based on the large deviation principle

The state is a Markov process with prescribed infinitesimal generator dependent upon an input process and a set of free parameters. We first design the optimal state feedback function as the input process based on the stochastic Hamilton-Jacobi algorithm that minimizes the expected value of the cost integral defined as a distance measure between the state process and a desired non-random state process. With this input prescribed, the state becomes a Markov process with known generator dependent only upon the free parameters. We fine tune these parameters so that the resulting cost integral is minimized, or we can tune these parameters so that the probability of the cost functional exceeding a given threshold is a minimum. The probability of this cost function exceeding the threshold can be computed in many cased by assuming that the state process satisfies a stochastic differential equation driven by Brownian motion, Poisson noise or more generally a combination of the two, ie, Levy noise and assuming that this driving noise is weak so that the large derivation rate function can be used to approximate this probability. An alternate way to design the parameters is to calculate using the large deviation principle, the rate function for the empirical distribution of the Markov process using standard formulae in large deviation theory and then to maximize the probability that this empirical distribution assumes a given distribution by minimizing the rate function of the empirical distribution evaluated at the given distribution.

$x(t)$ is the state vector of the system at time t and $u(t)$ is the control input vector. $x(t)$ is assumed to be a Markov process with infinitesimal generator

$K_t(x, y|u(t), \theta)$. Consider the problem of minimizing the average cost function from time s to time T:

$$V(s, T, x, u) = \mathbb{E}[\int_s^T L(x(t), u(t))dt|x(s) = x]$$

with respect to $u(t), s \leq t \leq T$ subject to the condition that $u(t)$ must be of the instantaneous state feedback form

$$u(t) = F(t, x(t))$$

Denote

$$min(V(s, T, x, u) : u(ds), t \leq s \leq T) = V^*(s, T, x)$$

Then, we can write

$$V^*(s, T, x) = min_{u(s)}(L(x, u(s))ds + \mathbb{E}(V^*(s + ds, T, x(s + ds))|x(s) = x))$$

$$= min_u(L(x, u)ds + V^*(s, T, x) + ds.\partial_s V^*(s, T, x) + ds \int K_s(x, y|u, \theta)V^*(s, T, y)dy)$$

or equivalently, we obtain the stochastic Bellman-Hamilton-Jacobi equation

$$\partial_s V^*(s, T, x) + min_u[L(x, u) + \int K_s(x, y|u)V^*(s, T, y)dy] = 0$$

Roughly speaking, the algorithm for calculating $u(s)$ is as follows. Assume that $u(s + ds), V(s + ds, T, x)$ have been determined. Then, determine $u(s)$ as a function of $x(s)$ as follows:

$$u(s) = F(s, x(s)), F(s, x) = argmin_u[L(x, u) + \int K_s(x, y|u, \theta)V^*(s + ds, T, y)dy]$$

and determine $V^*(s, T, x)$ as

$$V^*(s, T, x) = V^*(s + ds, T, x) + ds.[L(x, F(s, x)) + \int K_s(x, y|F(s, x), \theta)V^*(s + ds, T, y)dy]$$

Suppose that for all parameters θ in a given set Θ, we have calculated the feedback function $F(s, x) = F(s, x, \theta)$. Then we wish our solution $x(t)$ to be as close as possible to a desired non-random state $x_d(t)$ over the time duration $[0, T]$. For that we must minimize the cost function

$$\mathbb{E} \int_0^T C(x(t), x_d(t))dt$$

w.r.t θ. To do so, we must determine the pdf $p(t, x)$ of $x(t)$ at each time t. Equivalently writing

$$\mathbb{E}[C(x(t), z)|x(s) = y] = G(t, s, z, y), t \geq s$$

we see that G satisfies the backward Kolmogorov differential equation

$$G(s,s,z,y) = C(y,z)$$

and

$$G(t,s,z,y) = \mathbb{E}(G(t,s+ds,z,x(s+ds))|x(s)=y)$$

$$= G(t,s,z,y)+ds[\partial_s G(t,s,z,y)+\int K_s(y,v|F(s,y),\theta)G(t,s,z,v)dv]$$

or equivalently,

$$\partial_s G(t,s,z,y)+\int K_s(y,v|F(s,y),\theta)G(t,s,z,v)dv = 0, t \geq s, G(t,t,z,y) = C(y,z)$$

For each $\theta \in \Theta$, we thus get the average cost function

$$G(t,0,x_d(t),y|\theta) = \mathbb{E}(C(x(t),x_d(t))|x(0)=y)$$

and we minimize

$$\int_0^T \mathbb{E}(C(x(t),x_d(t))|x(0)=y)dt = \int_0^T G(t,0,x_d(t),y|\theta)dt$$

w.r.t θ. This gives us a fine tuning of the stochastically controlled system. Note that $u(t)$ can be regarded as a local control parameter while θ is a global control parameter.

9.4 Feynman's theory of automata as applied to the physical world around us

First to fluid dynamics modeled using discrete particles moving along the different sides of a regular hexagon, then the continuous time Navier-Stokes dynamics can be projected onto this discrete model. Likewise in quantum mechanics, there is a definite probability amplitude of a particle making a transition from each of the sides of a hexagon into a neighbouring hexagon and one can formulate a discrete state Schrodinger equation which in the limit when of zero hexagon size will converge to the standard Schrodinger equation for a particle moving in a vector potential and a scalar potential.

Feynman's theory of automata as applied to the physical world around us: Consider a fluid, it is described by a velocity field say $v_x(t,x,y), v_y(t,x,y)$ in two dimensional space. We partition 2-D space into hexagons. Denote these hexagons by $\Delta_k, k = 1,2,....$ Choose a given hexagon Δ_k with centre coordinates (x_k, y_k). Let d denote the length of any side of the hexagon. The fluid velocity at (x_k, y_k) is $v_k(t) = v(t, x_k, y_k)$. We consider its projection along the direction perpendicular to the jth side of this hexagon. Specifically, the normal directions to the six sides of the hexagon make angles $\theta(k) = 2\pi/12 + 2k\pi/6, k = 0,1,...,5$

with the x axis. So the discretized velocity field at (x_k, y_k) along these six directions is $v_{kj} = v_k.cos(\theta(j)), j = 0, 1, ..., 5$. So if we regard two dimensional space as a discrete spatial automata consisting of hexagons with the fluid molecules replaced by ball bearings within these hexagons, then we should be actually writing down ordinary differential equations for the time functions $v_{kj}(t), k = 1, 2, ..., j = 0, 1, ..., 5$. This idea of discrete automata can also be immediately applied to quantum mechanics. Let $\psi_k(t)$ denote the probability amplitude for the particle to be within Δ_k and let $\Delta_{kj}, j = 0, 1, ..., 5$ label the six neighbouring hexagons of Δ_k, so that Δ_{kj} and Δ_k have common side whose normal makes an angle θ_j with the x-axis. Let $\psi_{kj}(t)$ denote the probability amplitude for the particle to be within the hexagon Δ_{kj}. Then assuming only transitions between neighbouring hexagons are possible, it follows that the differential equations should be of the form

$$ih\psi_k'(t) = \sum_{j=0}^{5} a(k, j)\psi_{kj}(t), k = 1, 2, ...$$

Now when we partition two dimensional space into hexagons with the first hexagon having its centre at $(0, 0)$, then the centre points of the different hexagons are precisely at the lattice sites $\sum_{j=0}^{5} n_j \mathbf{a}_j$ where the $n_j's$ are arbitrary integers and the $\mathbf{a}_j's$ are the lattice vectors defined by

$$\mathbf{a}_j = (R.cos(\theta_j), R.sin(\theta_j)), j = 0, 1, ..., 5$$

with R being the distance between the centre's of neighbouring hexagons, given by

$$R = 2.d.sec(\pi/6) = 4d/\sqrt{3}$$

where d is the side-length of a hexagon. We write

$$\psi(t, n_0, ...n_5) = \psi(t, \mathbf{n}), \mathbf{n} = (n_0, ..., n_5)$$

for the amplitude of the particle to be located at the hexagon having centre coordinates $\sum_j n_j \mathbf{a}_j$. Then, Schrodinger's equation can be expressed as

$$ih\partial_t\psi(t, \mathbf{n}) = \sum_j a(\mathbf{n}, j)\psi(t, \mathbf{n} + \mathbf{e}_j) + b(\mathbf{n})\psi(t, \mathbf{n})$$

where

$$\mathbf{e}_j = (\delta_{j0}, ..., \delta_{j5}), j = 0, 1, ..., 5, b(\mathbf{n}) = \sum_j a(\mathbf{n}, j) = 0$$

$a(\mathbf{n}, j)/ih$ is the probability amplitude per unit time of the particle making a transition from the hexagon with central coordinates $\sum_{k=0}^{5} n_k \mathbf{a}_k + \mathbf{a}_j$ to the hexagon with central coordinates $\sum_{k=0}^{5} n_k \mathbf{a}_k$. $b(\mathbf{n})$ is the probability per unit time of the particle located at the hexagon indexed by $\mathbf{n}$ not to make any transition at all.

By making a second order Taylor approximation of the form

$$\mathbf{a}_j = Rf_j, f_j = (cos(\theta_j), sin(\theta_j))$$

and

$$\psi(t, \mathbf{n}) = \psi(t, R\sum_j n_j f_j), x = R\sum_j n_j f_j$$

$$\psi(t, \mathbf{n}+e_j) = \psi(t, x+Rf_j) \approx \psi(t,x) + R(f_j, \nabla)\psi(t,x) + (1/2)R^2(f_j,\nabla)^2\psi(t,x)$$

$$b(\mathbf{n})\psi(t, \mathbf{n} + \sum_j a(n,j)\psi(t, n+e_j) \approx$$

$$(\sum_j a(n,j)+b(\mathbf{n}))\psi(t,x)+R\sum_j a(n,j)(f_j,\nabla)\psi(t,x)+(R^2/2)\sum_j a(n,j)(f_j,\nabla)^2\psi(t$$

$$= R\sum_j a(n,j)(f_j,\nabla)\psi(t,x) + (R^2/2)\sum_j a(n,j)(f_j,\nabla)^2\psi(t,x)$$

Defining the 2×2 matrix

$$\mathbf{Q}(n) = R^2\sum_j a(n,j)f_j f_j^T$$

and the 2×1 vector

$$\mathbf{q}(n) = R\sum_j a(n,j)f_j$$

we can write the above approximation as

$$\psi(t+n+e_j) = (\mathbf{b}(\mathbf{n}), \nabla) + (1/2)Tr(\mathbf{Q}(\mathbf{n})\nabla.\nabla^T)\psi(t,x)$$

and hence our Schrodinger equation in the continuum limit approximates to

$$i\partial_t\psi(t,\mathbf{x}) = (\mathbf{b}(x), \nabla)\psi(t,x) + (1/2)Tr(\mathbf{Q}(\mathbf{x})\nabla.\nabla^T)\psi(t,x)$$

The associated Hamiltonian operator is therefore

$$H = (-1/2)(\mathbf{p}^T\mathbf{Q}(\mathbf{x})\mathbf{p} + i\mathbf{b}(\mathbf{x})^T\mathbf{p}, \mathbf{p} = -i\nabla_x$$

This Hamiltonian corresponds to that of a charged particle of charge $-e$ moving in two dimensions with mass matrix

$$\mathbf{M}(x) = -\mathbf{Q}(\mathbf{x})^{-1},$$

magnetic vector potential vector

$$\mathbf{A}(\mathbf{x}) = -ie^{-1}\mathbf{Q}(\mathbf{x})\mathbf{b}(\mathbf{x})$$

and electric potential

$$\Phi(x) = (e/2)\mathbf{A}(\mathbf{x})^T\mathbf{M}(\mathbf{x})^{-1}\mathbf{A}(\mathbf{x})$$

so that the Hamiltonian can be expressed as

$$H = (1/2)(\mathbf{p} + e\mathbf{A}(\mathbf{x}))^{\mathbf{T}}\mathbf{M}(\mathbf{x})^{-\mathbf{1}}(\mathbf{p} + \mathbf{eA}(\mathbf{x})) - \mathbf{e\Phi}(\mathbf{x})$$

This model shows that a charged particle moving in an electromangnetic field and having an anisotropic mass matrix can be obtained as a continuum limit of a discrete model corresponding to an automata theory.

9.5 The variational formulation of the mode computation in a waveguide having inhomogeneous and anisotropic permittivity and permeability

The variational formulation of the generalized Helmholtz equation for the fields can be combined with the finite element method to derive a numerical algorithm for computing the modes in such a guide. The idea behind the finite element method is to partition the guide cross section into triangles and express the field within each triangle as an affine linear function of the xy coordinates in terms of the vertex field values and then by substituting into the action functional after partitioning the set of triangle vertices into boundary vertices and interior vertices, derive the modes by solving a generalized eigenvalue problem with the matrices having a nonlinear dependence upon the mode e

Assume that $\epsilon(x,y), \mu(x,y)$ are 3×3 matrices that are functions of ω, x, y only. Assuming the z-dependence of the fields as $exp(-\gamma z)$, we can set up the Maxwell curl equations and solve for the transverse components in terms of the longitudinal components:

$$\nabla_\perp E_z \times z - \gamma.\hat{z} \times E_\perp = -j\omega(\mu_\perp.H_\perp + \mu_{1z}H_z)$$

$$\nabla_\perp H_z \times z - \gamma.\hat{z} \times H_\perp = j\omega(\epsilon_\perp.E_\perp + \epsilon_{1z}E_z)$$

where

$$\mu_{1z} = [\mu_{xz}, \mu_{yz}]^T, \epsilon_{1z} = [\epsilon_{xz}, \epsilon_{yz}]^T$$

$$\nabla_\perp \times E_\perp = -j\omega(\mu_{2z}^T H_\perp + \mu_{zz}H_z)\hat{z},$$

$$\nabla_\perp \times H_\perp = j\omega(\epsilon_{2z}^T E_\perp + \epsilon_{zz}E_z)\hat{z},$$

where

$$\mu_{2z} = [\mu_{zx}, \mu_{zy}]^T, \epsilon_{2z} = [\epsilon_{zx}, \epsilon_{zy}]^T$$

Solving the transverse equations gives us

$$E_\perp = M_1(\gamma,x,y)\nabla_\perp E_z + M_2(\gamma,x,y)\nabla_\perp H_z + m_3(\gamma,x,y)E_z + m_4(\gamma,x,y)H_z$$

$$H_\perp = N_1(\gamma,x,y)\nabla_\perp E_z + N_2(\gamma,x,y)\nabla_\perp H_z + n_3(\gamma,x,y)E_z + n_4(\gamma,x,y)H_z$$

and finally substituting into the longitudinal equations gives us generalized two dimensional Helmholtz equations in the form

$$K_{1ij}(\gamma,x,y))\partial_i\partial_j\psi(\gamma,x,y) + K_{2i}(,x,y)\partial_i\psi(,x,y) + K_3(\gamma,x,y)\psi(\gamma,x,y) = 0$$

where the indices i, j run over $\{1,2\}$, ie, $\partial_1 = \partial_x, \partial_2 = \partial_y$. K_{1ij}, K_{2i}, K_3 are 2×2 matrix valued functions and

$$\psi = [E_z, H_z]^T$$

is a 2×1 vector valued function. In the perturbation approximation of weak inhomogeneity and anisotropy, these equations assume the form

$$(\nabla_\perp^2 + h_0^2)\psi(x,y) + \delta.[F_{ij}(h_0^2,x,y)\partial_i\partial_j + G_i(h_0^2,x,y)\partial_i + K(h_0^2,x,y)]\psi(x,y) = 0$$

This field equation taking into account the boundary conditions, can be derived from a variational principle by introducing an auxiliary dual field $\phi(x,y)$ taking values in $\mathbb{C}^2$ and with the action

$$-\int_D(\phi(x,y)^*,(\nabla_\perp^2+h_0^2)\psi(x,y)+\delta.[F_{ij}(h_0^2,x,y)\partial_i\partial_j+G_i(h_0^2,x,y)\partial_i+K(h_0^2,x,y)]\psi(x,y))dxdy$$

$$=\int_D(\nabla_\perp\phi(x,y)^*,\nabla_\perp\psi(x,y))dxdy-\delta.\int_D(\phi(x,y)^*,[F_{ij}(h_0^2,x,y)\partial_i\partial_j\psi(x,y))dxdy$$

$$-\int_D(\phi(x,y)^*,(G_i(h_0^2,x,y)\partial_i+K(h_0^2,x,y))\psi(x,y))dxdy-h_0^2\int(\phi(x,y)^*,\psi(x,y))dxdy$$

If this functional is extremized w.r.t ϕ^*,ψ and h_0^2, then it results in the modal equations for h_0^2. More generally, we consider a linear partial differential operator $L(h_0^2,x,y)$ parametrized by the modal value h_0^2 and with coeffcients dependent on x,y. We wish to solve the equation

$$L(h_0^2,x,y)\psi(x,y) = 0$$

for the modal eigenvalue, taking into account the boundary conditions on $\psi(x,y)$. This is possible by extremizing the functional

$$\int(\phi(x,y)^*,L(h_0^2,x,y)\psi(x,y))dxdy = \int(\phi,L(h_0^2)\psi)dxdy$$

Implementing this variational principle using the finite element method amounts to partitioning the cross section of the guide into little triangles $\Delta_k, k = 1,2,...,N$ and within each triangle Δ_k, we express $\psi(x,y)$ as a linear combination

$$\psi(x,y) = \psi_{k1}\chi_{k1}(x,y) + \psi_{k2}\chi_{k2}(x,y) + \psi_{k3}\chi_{k3}(x,y)$$

where $_{kj}(x,y), j = 1,2,3$ are basis functions for the kth triangle so as to satisfy

$$\chi_{kj}(x_{km},y_{km}) = \delta_{jm}, j,m = 1,2,3$$

with $(x_{km},y_{km}), m = 1,2,3$ being the vertices of Δ_k. Here ψ_{kj} is the value $\psi(x_{kj},y_{kj})$ of ψ at the jth vertex of Δ_k. Then the discretized form of the above functional can be expressed as

$$\phi^*Q(h_0^2)\psi$$

where $Q(h_0^2)$ is a matrix of size $3N \times 3N$ and ψ,ϕ are respectively column vectors of sizes $3N \times 1$ whose components are respectively (ψ_{kj}) and (ϕ_{kj})

The boundary constraints whether Dirichlet or Neumann or mixed, can all be expressed as linear constraints

$$B\psi = 0$$

where B is of size $p \times 3N$ where p is the number of boundary pixels. Taking these constraints into account using a Lagrange multipier vector λ of size $p \times 1$ gives the function to be extremized as

$$Re(\phi^* Q(h_0^2)\psi) - 2Re(\lambda^* B\psi)$$

Extremizing w.r.t ϕ^* gives us

$$Q(h_0^2)\psi = 0 - - - (1)$$

Extremizing w.r.t ψ^* gives us

$$Q(h_0^2)^*\phi - B^*\lambda = 0$$

Extremizing w.r.t λ^* gives the constraints

$$B\psi = 0$$

This method will not generally work since the field equations (1) cannot be applied to the boundary pixels since these are constrained by mixed boundary conditions. The correct way is therefore to partition ψ into two parts, the interior pixel part ψ_I and the boundary pixel part ψ_B. Thus,

$$\psi = [\psi_I^T, \psi_B^T]^T$$

The field equations are applicable to the interior pixel part keeping ψ_B fixed, say zero if we use the Dirichlet conditions. For the general case, the boundary conditions $B\psi = 0$ can be expressed as

$$B_1\psi_I + B_2\psi_B = 0$$

where B_1, B_2 each have the same number of rows equal to the number of boundary pixels. The field equations have the form

$$L_{11}(h_0^2)\psi_I + L_{12}(h_0^2)\psi_B = 0$$

where

$$L = \begin{pmatrix} L_{11} & L_{12} \\ L_{21} & L_{22} \end{pmatrix}$$

Taking into account the boundary conditions, the field equations for ψ_I are expressible as

$$(L_{11}(h_0^2) + L_{12}(h_0^2)B_2^{-1}B_1)\psi_I = 0$$

which gives the modes h_0^2 as solutions to

$$det(L_{11}(h_0^2) + L_{12}(h_0^2)B_2^{-1}B_1) = 0$$

We can also consider a discretized action of the form $\psi^* Q(h_0^2)\psi$ with $Q(h_0^2)$ Hermitian with the boundary conditions $B\psi = 0$. This problem can be solved in the same way by partitioning the set of pixels into an interior part and a boundary part. It can however also be solved wihout such a paritioning by simply using Lagrange multipliers to account for the constraints. The resulting function to be extremized is

$$\psi^* Q(h_0^2)\psi - 2Re(\lambda^* B\psi)$$

The variational method gives

$$Q(h_0^2)\psi - B^*\lambda = 0, B\psi = 0$$

so

$$\psi = Q(h_0^2)^{-1}B^*\lambda, BQ(h_0^2)^{-1}B^*\lambda = 0$$

so the modal equations are

$$det(BQ(h_0^2)B^*) = 0$$

9.6 A problem on convolutional channels with twisting:

We define a group of linear transformations of the mixed state of a quantum system by means of its quantum characteristic function. This group of transformations involves first scaling the quantum characteristic function, then applying a transformation that takes Gaussian states to Gaussian states, followed finally by multiplication by the classical characteristic function of an infinitely divisible probability distribution. The corresponding transformation on the state is obtained by inversion of the quantum characteristic function using properties of the Weyl operator and the Galuber-Sudarshan non-orthogonal resolution of the identity in Boson Fock space.

Let ρ be the state of a quantum system in a finite dimensional Hilbert space. Let $a = (a_1, ..., a_n)$ be annihilation operators and $a^* = (a_1^*, ..., a_n^*)$ be the corresponding creation operators so that

$$[a_i, a_j] = 0, [a_i, a_j^*] = \delta(ij)$$

Let

$$z = (z_1, ..., z_n) \in \mathbb{C}^n, z^*.a = \sum_i \bar{z}_i a_i, z.a^* = \sum_i z_i a_i^*$$

Consider the Weyl operator

$$W(z) = exp(z^*.a - z.a^*)$$

Note that

$$W(z)^* = W(-z) = W(z)^{-1}$$

We wish to compute

$$W(z)W(u)$$

for $z, u \in \mathbb{C}^n$ in terms of $W(z+u)$. let

$$W(t(z+u)) = F(t)W(tu)$$

Then, clearly,

$$F'(t)W(tu)+F(t)W(tu)(u^*a-ua^*) = W(t(z+u))((z+u)^*a-(z+u)a^*)$$

$$= F(t)W(tu)((z+u)^*a-(z+u)a^*)$$

so that

$$F'(t) = F(t)W(tu)(z^*a - za^*)W(-tu) = F(t)exp(tad(u^*a - ua^*))(z^*a - za^*)$$

$$= F(t)((z^*a - za^*) + t(-u^*z + z^*u)) = F(t)((z^*a - za^*) + 2itIm(< z|u >))$$

where $< z|u >= z^*u$. Integrating this equation gives us

$$F(t) = exp(t(z^*a - za^*)).exp(itIm < z|u >= W(tz).exp(itIm < z|u >)$$

so that we obtain the following result which is one of the forms of the Weyl commutation relations:

$$W(z+u) = exp(i.Im < z|u >)W(z)W(u)$$

This shows that $z \to W(z)$ is a projective unitary representation of the Abelian group $\mathbb{C}^n$ in the Hilbert space $\Gamma_s(\mathbb{C}^n) = L^2(\mathbb{R}^n)$. Now consider the following quantum Fourier transform of the state ρ:

$$\hat{\rho}(z) = Tr(\rho W(z))$$

We seek to invert this equation, ie, express ρ as a linear superposition of $\hat{\rho}(z), z \in \mathbb{C}^n$. Note that in view of the Galuber-Sudarshan non-orthogonal resolution of the identity,

$$\pi^{-n} \int |\phi(z) >< \phi(z)|dzd\bar{z} = I, |\phi(z) >= |e(z) > exp(-|z|^2/2)$$

we can write

$$\rho = C \int |\phi(u) >< \phi(u)|\rho dud\bar{u}$$

so

$$\hat{\rho}(z) = C \int < \phi(u)|\rho.W(z)|\phi(u) > dud\bar{u}$$

Now, let

$$W(tz) = F(t)exp(tz^*a)$$

so

$$F'(t)exp(tz^*a) + F(t)z^*aexp(tz^*a) = F(t)exp(tz^*a)(z^*a - za^*)$$

so

$$F'(t) = -F(t)exp(-tad(z^*a))(za^*) = -F(t)(za^* - t|z|^2)$$

so

$$F(t) = exp(-tza^*).exp(-t^2|z|^2/2)$$

ie

$$W(z) = exp(-|z|'2)exp(-za^*).exp(z^*a)$$

Thus,

$$W(z)|e(u) >= exp(-|z|^2/2)exp(z^*u)exp(-za^*)|e(u) >$$

Now observe that

$$|e(u) >= \sum_n u^n(a^{*n}/n!)|0 >= exp(u.a^*)|0 >$$

and hence

$$W(z)|e(u) >= exp(-|z|'2+ < z|u >)|e(u-z) >$$

Thus,

$$) < \phi(u)|\rho.W(z)|\phi(u) >=< e(u)|\rho|e(u-z) > exp(-|z|^2/2+ < z|u >$$

$$+ < u|u-z > -|u|^2) =< e(u)|\rho|e(u-z) > exp(-|z|^2/2+2iIm < z|u >)$$

So

$$exp(|z|^2/2)\hat{\rho}(z) = C\int < e(u)|\rho|e(u-z) > exp(2iIm < z|u >)dud\bar{u}$$

Now,

$$\hat{\rho}(z-u) = Tr(\rho W(z-u)) = Tr(\rho W(z)W(-u))exp(-iIm < z|u >)$$

$$= exp(-iIm < z|u >)Tr(\rho W(z)W(u)^*)$$

It follows that the Kernel

$$K(z,u) = \hat{\rho}(z-u)exp(iIm < z|u >)$$

is positive definite on $\mathbb{C}^n \times \mathbb{C}^n$. This is a necessary condition for $\hat{\rho}(z)$ to be a quantum characteristic function. Now writing $\xi = R(z) = [x^T, y^T]$ where $z = x + iy$ and $u = v + iw, R(u) = [v^T, w^T]^T = \eta$, we write $\hat{\rho}(\xi)$ and $\hat{\rho}(\eta)$ for $\hat{\rho}(z)$ and $\hat{\rho}(u)$. Now consider the function

$$(T\hat{\rho})(\xi) = \hat{\rho}(Axi).exp(-\xi^T M\xi/2)\phi(\xi)$$

where $\hat{\rho}$ is a quantum characteristic function and A, M are matrices with M real symmetric and ϕ some function on $\mathbb{R}^{2n}$. We ask the question: Under what conditions on A, M, ϕ, is $T\hat{\rho}$ also a quantum characteristic function ? Noting that $Im <z|u>= y - y^T v + x^T w = \xi^T J\eta$ where $J = \begin{pmatrix} 0 & I \\ -I & 0 \end{pmatrix}$, for $T\hat{\rho}$ to be a quantum characteristic function, we require that $K(\xi,\eta) = (T\hat{\rho})(\xi - \eta).exp(i\xi^T J\eta)$ be a positive definite kernel on $\mathbb{R}^{2n} \times \mathbb{R}^{2n}$. Now

$$K(\xi,\eta) = (T\hat{\rho})(\xi - \eta).exp(i\xi^T J\eta)$$

$$= \hat{\rho}(A\xi - A\eta)exp(i\xi^T J\eta).exp(\xi^T M\eta).exp(-\xi^T M\xi/2).exp(-\eta^T M\eta/2).\phi(\xi - \eta)$$

$$= \hat{\rho}(A\xi - A\eta)exp(i\xi^T A^T JA\eta)exp(i\xi^T J\eta).exp(\xi^T (M - iA^T JA)\eta).exp(-xi^T M\xi/2).$$

$$.exp(-\eta^T M\eta/2).\phi(\xi - \eta) --- (1)$$

Since $\hat{\rho}(\xi - \eta)exp(i\xi^T J\eta)$ is positive definite, it follows that $\hat{\rho}(A\xi - A\eta)exp(i\xi^T A^T JA\eta)$ is also positive definite. So by using Schur's theorem that the element wise product of positive definite matrices is positive definite, it follows from (1) that $K(\xi,\eta)$ will be positive definite if $\phi(\xi - \eta)$ is positive definite and $exp(\xi^T (M - iA^T JA)\eta)$ is also positive definite. A necessary and sufficient condition for the latter is that $M - iA^T JA$ be a positive definite matrix. So we obtain a sufficient condition for $T\hat{\rho}$ to be a quantum characteristic function. Now assume that T_1, T_2 are two such operations corresponding to (M_1, A_1, ϕ_1) and (M_2, A_2, ϕ_2) respectively. We write $T_k = T(M_k, A_k, \phi_k), k = 1, 2$. Thus, if $\hat{\rho}$ is quantum characteristic function, then so is $T(M_2, A_2, \phi_2)oT(M_1, A_1, \phi_1)\hat{\rho}$. We ask the question, can we write $T(M_2, A_2, \phi_2)oT(M_1, A_1, \phi_1) = T(M, A, \phi)$ for some M, A, ϕ. If so, then we could talk about a group of twisted quantum convolutional channels. Now,

$$T(M_1, A_1, \phi)\hat{\rho}(\xi) = \hat{\rho}(A_1\xi)\phi_1(\xi)exp(-\xi^T M_1\xi/2),$$

$$T(M_2, A_2, \phi_2)oT(M_1, A_1, \phi_1)\hat{\rho}(\xi) = (T(M_1, A_1, \phi_1)\hat{\rho})(A_2\xi).\phi_2(\xi).exp(-\xi^T M_2\xi/2)$$

$$= \hat{\rho}(A_1 A_2\xi)\phi_1(A_2\xi)\phi_2(\xi)exp(-\xi^T (M_2 + A_2^T M_1 A_2)\xi/2) = T(M, A, \phi)\hat{\rho}(\xi)$$

where

$$\phi = (\phi_1 oA_2).\phi_2, M = M_2 + A_2^T M_1 A_2, A = A_1 A_2$$

In this way, we are able to talk about a group of twisted quantum convolutional channels whose elements consist of triplets (M, A, ϕ) satisfying the above mentioned constraints with the group composition law being given by

$$(M_1, A_1, \phi_1)o(M_2, A_2, \phi_2) = (M_2 + A_2^T M_1 A_2, A_1 A_2, (\phi_1 oA_2).\phi_2)$$

Now consider a one parameter group $(M_t, A_t, \phi_t), t \in \mathbb{R}$ of twisted convolutional channels. We require that

$$(M_t, A_t, \phi_t)o(M_s, A_s, \phi_s) = (M_{t+s}, A_{t+s}, \phi_{t+s})$$

for all $t, s \in \mathbb{R}$. It follows using the above composition law for the channels that

$$M_{t+s} = M_s + A_s^T M_t A_s, A_{t+s} = A_t A_s, \phi_{t+s} = (\phi_t oA_s)\phi_s$$

The second equation gives

$$A_t = exp(tK), K = (dA_t/dt)|_{t=0}$$

The first then gives on differentiating w.r.t s at $s = 0$,

$$dM_t/dt = L + K^T M_t + M_t K, L = (dM_s/ds)|_{s=0}$$

which solves to give

$$M_t = \int_0^t exp((t-s)K^T)L.exp((t-s)K)ds = \int_0^t exp(sK^T)L.exp(sK)ds$$

To obtain ϕ_t, we first recall that $\phi_t(\xi - \eta)$ is positive definite and so apart from a positive real multiple, must be the characteristic function of a probability distribution. We now prove that it must be the characteristic function of an infinitely divisible distribution. Since $\hat{\rho}$ must be a quantum characteristic function before and after transformation, its value at $\xi = 0$ must be unity and hence the value of ϕ_t at $\xi = 0$ must also be unity. Thus ϕ_t must be the characteristic function of a probability distribution for every $t \in \mathbb{R}$. Now, let $t = nh$ where n is a positive integer. Then, the equation $\phi_{t+s} = (\phi_t o A_s)\phi_s$ implies

$$\phi_{nh} = \phi_h(\phi_{(n-1)h} o A_h) = \phi_h.(\phi_h o A_h).(\phi_{(n-2)h} o A_h^2) = ... = [\Pi_{k=0}^{n-1}(\phi_h o A_h^k)]\phi_0 o A_h^n$$

Since ϕ_h, ϕ_0 are characteristic functions of probability distributions, so are $\phi_h o A_h^k$ and $\phi_0 o A_h^n$ and since n is an arbitrary positive integer, it follows from the above formula that ϕ_t is the characteristic function of an infinitely divisible probability distribution function. We observe that $\phi_0 = 1$ and ϕ_t satisfies the pde

$$d\phi_t/dt = (\nabla\phi_t(\xi))^T K\xi + \phi_t(\xi)\psi(\xi), \psi(\xi) = d\phi_s(\xi)/ds|_{s=0}$$

It also satisfies

$$d\phi_s(\xi)/ds = \psi(A_s\xi)$$

These equations are respectively obtained by differentiating $\phi_{t+s} = (\phi_t o A_s)\phi_s$ w.r.t s and t at $s = 0$ and at $t = 0$.

9.7 Decomposition of a tensor product of representations of two groups G_1, G_2, or equivalently, decomposition of a tensor product of representations of two mutually commuting subgroups of a group into irreducibles with application to the decomposition of the standard representation of $GL(n,V) \times S_n$ in $V^{\otimes n}$

Given two groups G_1, G_2, consider unitary representations π_1, π_2 of these two groups and consider the unitary representation $(g_1, g_2) \to \pi_1(g_1) \otimes \pi_2(g_2)$ of the group $G_1 \times G_2$. Decomposing this representation π of $G = G_1 \times G_2$ into irreducibles, we get

$$\pi(g_1, g_2) = \bigoplus_{\alpha \in \hat{G}_1} U_\alpha(g_1) \otimes V_{Q(\alpha)}(g_2)$$

where $U_\alpha, \alpha \in \hat{G}_1$ are the irreducible unitary representations of G_1 and for each $\alpha \in \hat{G}_1$, $\beta = Q(\alpha) \in \hat{G}_2$ determines an irreducible unitary representation V_β of G_2. Let $\mathcal{A}(G_k) = A_k$ denote the group algebra of $G_k, k = 1, 2$. Then if $U_\alpha(g_1)$ is of size $d_1(\alpha) \times d_1(\alpha)$ and $V_\beta(g_2)$ is of size $d_2(\beta) \times d_2(\beta)$, it follows that the minimal central projection $P_1(\alpha)$ corresponding to $\alpha \in \hat{G}_1$ has the form $diag[0, 0, ..., I_{d_1(\alpha)}, 0, 0, ...]$ and likewise for $P_2(\beta)$. It is clear that since $\pi(A_1) = \pi(A_1 \times \{e_2\})$ is in the commutant $\pi(\{e_1\} \times A_2\})'$ of $\pi(\{e_1\} \times A_2\}$ and hence $U_\alpha \times V_{Q(\alpha)}(A_1 \times \{e_2\}\})$ is in the commutant $U_\alpha \times V_{Q(\alpha)}(\{e_1\} \times A_2)'$. From the fact that the minimal central projections of a representation commute with the representation of the group algebra and the fact that the commutant of an irreducible representation of the group algebra consists of scalar multiples of the corresponding minimal central projection, it follows that the minimal central projections of the representation π of G are $I_{d(\alpha)} \otimes I_{d(Q(\alpha)}$ or more precisely, $diag[0, 0, ..., I_{d(\alpha)} \otimes I_{d(Q(\alpha)}, 0, 0, ...]$. Note that $U_\alpha \times V_{Q(\alpha)}(A_1 \times \{e_2\}\})$ equals precisely the commutant of $U_\alpha \times V_{Q(\alpha)}(\{e_1\} \times A_2)$ and likewise $U_\alpha \times V_{Q(\alpha)}(\{e_1\} \times A_2)$ equals precisely the commutant of $U_\alpha \times V_{Q(\alpha)}(A_1 \times \{e_2\})$. This fact can be used to compute the decomposition of the representation $\pi_1 \times \pi_2$ of the group $SU(m) \times S_n$ in $V^{\otimes n}$ where $V = \mathbb{C}$ and

$$\pi_1(g_1)(v_1 \otimes ... \otimes v_m) = g_1 v_1 \otimes .. \otimes g_1 v_n, \pi_2(g_2)(v_1 \otimes ... \otimes v_n) = v_{g_2^{-1}1} \otimes ... \otimes v_{g_2^{-1}n}$$

m for $g_1 \in SU(m), g_2 \in S_n$.

Remark 1: $(U_\alpha \times V_{Q(\alpha)})(A_1 \times \{e_2\}) = U_\alpha(A_1) \otimes I_{d_2(Q(\alpha))}$ equals precisely all matrices of the form $X_1 \times I_{d_2(Q(\alpha)}$ where X_1 ranges over all $d_1(\alpha) \times d_1(\alpha)$ matrices and likewise $(U_\alpha \times V_{Q(\alpha)})(\{e_1\} \times A_2) = I_{d_1(\alpha)} \otimes V_{d_2(Q(\alpha)}(A_2)$ equals precisely all matrices of the form $I_{d_1(\alpha)} \otimes X_2$ as X_2 ranges over all $d_2(Q(\alpha)) \times$

$d_2(Q(\alpha))$ matrices. From this fact, it follows at once that each one here is the commutant of the other.

Remark 2: The decomposition

$$\pi(g_1, g_2) = (\pi_1 \times \pi_2)(g_1, g_2) = \bigoplus_{\alpha \in \hat{G}_1} U_\alpha(g_1) \otimes V_{Q(\alpha))}(g_2)$$

of the representation $\pi_1 \times \pi_2$ of $G_1 \times G_2$ has the characteristic feature that if $\mathcal{H}_{1\alpha}$ is the representation space of U_α and $\mathcal{H}_{2\beta}$ is the representation space of $V_{\beta)}$, so that $\mathcal{H}_\alpha = \mathcal{H}_{1\alpha} \otimes H_{2Q(\alpha)}$ is the representation space of the irreducible representation $U_\alpha \times V_{Q(\alpha)}$ of $G = G_1 \times G_2$, then if $e_\alpha \in H_\alpha$, we have that

$$< \bigoplus_\alpha e_\alpha | \pi(g_1, g_2) | \bigoplus_\alpha e_\alpha >= \sum_\alpha < e_\alpha | U_\alpha(g_1) \otimes V_{Q(\alpha)}(g_2) | e_\alpha >$$

Note also that if for some v in the representation space of π such that v falls in the range of $I_1 \otimes V_{2\beta}(g_2)$ for some $g_2 \in G_2$ and some $\beta \in \hat{G}_2$ and such that $u = (P_{1\alpha} \otimes I_2)v \neq 0$ where $P_{1\alpha}$ is the minimal central projection corresponding to U_α in the decomposition of π_1, then it must necessarily follow that $\beta = Q(\alpha)$ and hence that $u = (P_{1\alpha} \otimes P_{2Q(\alpha)})v$, where $P_{2\beta)}$ is the minimal central projection corresponding to V_β in the decomposition of π_2. This fact helps us to determine which irreducible representation $Q(\alpha)$ of π_2 is paired up with the irreducible representation α of π_1 in the decomposition of $\pi = \pi_1 \times \pi_2$ (Reference: Barry Simon, Representations of finite and compact groups, American Mathematical Society). It follows from this observation that if all groups are semisimple and that if a vector v in the representation space $\mathcal{H} = \bigoplus_{\alpha \in \hat{G}_1} (U_\alpha \times V_{Q(\alpha)})$ of π is a non-zero weight vector of the representation $g_2 I_1 \otimes V_\beta(g_2)$ of G_2 for some $\beta \in \hat{G}_2$ and if $(P_{1\alpha} \otimes I_2)v \neq 0$, then necessarily $\beta = Q(\alpha)$.

Example: Consider the actions of $SU(m)$ and S_n on $T_n(\mathbb{C}^m) = (\mathbb{C}^m)^{\otimes n}$ defined as above. These two actions commute and hence we get a unitary representation of $SU(m) \times U(n)$ in $T_n(\mathbb{C}^m)$. Now the irreducible representations of S_n are obtained by taking a Young Tableaux F having in all n boxes. Let r_k denote the number of boxes in the kth row of this Tableaux for $k = 1, 2, ..., m_1$. We are assuming here that the jth column of the Tableaux has m_j boxes, $j = 1, 2, ..., r_1$. Let $e(F)$ denote the central projection of the group algebra corresponding to the representation of S_n defined by F. Let V denote the representation of S_n in $T^n(\mathbb{C}^m)$. Then, $V(e(F))$ equals the central projection of the representation V corresponding to the representation F. Now denote by U the representation of $SU(m)$ in $T^n(\mathbb{C}^m)$.

9.8 Microstrip line analysis

Microstrip line analysis: The electromagnetic field is generated by a single non-vanishing component $A_z(\omega, r)$ of the magnetic vector potential. It satisfies the Helmholtz equation

$$(\nabla^2 + k^2)A_z = 0, k^2 = \omega^2 \epsilon \mu$$

The electric field components are given by

$$E_z = -j\omega A_z, E_x = E_y = 0$$

Note that we are working in the Coulomb gauge in which $V = 0$ and $divA = 0$. The latter gives

$$\partial_z A_z = 0$$

and hence $A_z = A_z(\omega, x, y)$ is independent of z. The boundary conditions imply that E_z vanishes at $x = 0, a$ and at $y = 0, b$. The magnetic field components are

$$B_x = \partial_y A_z, B_y = -\partial_x A_z, B_z = 0$$

The boundary conditions on the magnetic field are that B_x should vanish when $x = 0, a$ and B_y should vanish when $y = 0, b$. In view of these remarks, it becomes clear that

$$(\partial_x^2+\partial_y^2+k^2)A_z(\omega, x, y) = 0, \partial_y A_z = 0, x = 0, a, \partial_x A_z = 0, y = 0, b,$$
$$A_z = 0, x = 0, aory = 0, b$$

It follows then this boundary value problem is equivalent to

$$(\partial_x^2 + \partial_y^2 + k^2)A_z(\omega, x, y) = 0, A_z = 0, x = 0, aory = 0, b,$$

This has the solution

$$A_z(\omega, x, y) = C.sin(m\pi x/a)sin(n\pi y/b), k^2 = ((m\pi/a)^2 + (n\pi/b)^2)$$

Another model: The microstrip line consists of only two perfectly conducting surfaces: $x = 0, a$. The magnetic vector potential consists of only one non-zero component $A_x(\omega, x, y, z)$. It satisfies the Helmholtz equation

$$(\nabla^2 + k^2)A_x = 0, k^2 = \omega^2\epsilon\mu$$

The magnetic field ande electric fields are given by

$$B = curlA, curlcurlA = curlB = j\omega\epsilon E$$

or equivalently,

$$j\omega\epsilon E = \nabla(\partial_x A_x) + k^2 A$$

or equivalently, in component form,

$$j\omega\epsilon E_x = (\partial_x^2 + k^2)A_x, j\omega\epsilon E_y = \partial_x\partial_y A_x, j\omega\epsilon E_z = \partial_x\partial_z A_x,$$

$$B_x = 0, B_y = \partial_z A_x, B_z = -\partial_y A_x$$

The boundary conditions that E_y, E_z vanish at $x = 0, a$ (Note that B_x being identically zero, trivially satisfies the other boundary condition that it should vanish when $x = 0, a$) give

$$\partial_x A_x = 0, x = 0, a$$

Note that if in addition, the surfaces $y = 0, b$ are also perfect conductors, then, in addition, we must impose the boundary conditions

$$B_y = 0, y = 0, b, E_x = 0, y = 0, b, E_z = 0, y = 0, b$$

which amount to requiring that

$$A_x = 0, y = 0, b$$

Thus in this case, we obtain the modes as

$$h^2 = \gamma^2 + \omega^2\epsilon\mu = (m\pi/a)^2 + (n\pi/b)^2,$$

$$A_x = C.cos(m\pi x/a)sin(n\pi y/b)exp(-\gamma z)$$

9.9 Problems in supersymmetry superstring theory related to classical and quantum stochastics

Construction of the canonical quantum noise processes in the Hudson-Parthasarathy quantum stochastic calculus theory using Fock space methods combined with the algebra of an infinite sequence of standard independent quantum harmonic oscillators. Construction of the coherent states/exponential vectors using the eigenstates of an infinite sequence of independent quantum harmonic oscillators.

Derivation of super-commutation rules in quantum stochastic calculus based on the "grading operator" $(-1)^{\Lambda_t(H)} = exp(i\pi\Lambda_t(H))$.

Long exact sequence of cohomology groups derived from an exact sequence of modules and complexes associated with modules. It should be noted that in quantum field theory, cohomology groups derived from the covariant differential map on the exterior algebra play an important role in determining properties of the zero modes of the Dirac-Yang-Mills operator on a vector bundle of a differentiable manifold. Zero modes corresponding to positive/left and negative/right chirality zero mass Fermions lead to the chiral anomaly as the difference between the dimensions of the zero modes corresponding to the two chiralities being expressed as a topological invariant via a formula known as the Atiyah-Singer index theorem. It is thus natural to attempt to prove a purely algebraic version of this theorem for an abstract sequence of differential maps on a complex. This would be in tune with Richard Feynman's philosophy that the physical is discrete at the minute level and not continuous so rather than modeling the physical world in terms of differentiable manifolds, we should use algebraic structures such a modules, algebras and the differential of a map from an algebra into a module.

The relationship between nilpotent radical of a ring and its prime ideals.

An example of prime ideals and maximal ideals.

The Zariski closed irreducible subset of a ring corresponding to a prime ideal and corresponding to a prime ideal generated by a single irreducible polynomial in a polynomial ring. The scheme of a ring as its spectrum, ie, the set of all the prime ideals of the ring, the closure of a single irreducible element of the ring.

Consider the quantum noise processes $\Lambda^a_b(t), a, b \geq 0$ in the Hudson-Parthasarathy quantum stochastic calculus theory where $\Lambda^0_0(t) = t$ and if ϵ^a_b is one if $a = b \geq 1$ and zero otherwise, then these processes obey the quantum Ito formula

$$d\Lambda^a_b(t).d\Lambda^c_d(t) = \epsilon^a_d d\Lambda^c_b(t)$$

so that $\Lambda^a_0(t) = A_a(t), \Lambda^0_a(t) = A_a(t)^*, a \geq 1$ are respectively annihilation and creation processes and $\Lambda^a_b, a, b \geq 1$ are conservation processes satisfying

$$dA_a(t).d\Lambda^c_a(t) = dA_c(t), a \geq 1, c \geq 1, d\Lambda^a_c(t).dA_a(t)^* = dA_c(t)^*$$

Note that

$$(\Lambda^a_b(t))^* = \Lambda^b_a(t), a, b \geq 0$$

Give an explicit construction of the quantum stochastic processes $\Lambda^a_b(t)$ using Boson Fock space methods.

Hint: One way is to first construct $A_a(t), t \geq 1$ using the Weyl translation operators on the Boson Fock space $\Gamma_s(\mathbb{R}_+ \otimes \mathbb{C}^d)$ by the rule

$$W(v)|e(u)>= exp(-|v|^2/2- < v, u >)|e(u + v) >$$

and letting $p(v)$ be the generator of the one parameter unitary subgroup $t \rightarrow W(tv)$ and then using an appropriate linear combination of $p(v)$ and $p(iv)$ to construct $a(v)$ and finally defining $A_k(t) = a(\chi_{[0,t]}e_k)$ where $e_k, k = 1, 2, ..., d$ is an orthonormal basis for $\mathbb{C}^d$ and $A_0(t) = t = A_0(t)^*$. and Writing $d\Lambda^a_b(t) = dA_b(t)^* dA_a(t)/dt$. Derive the quantum Ito formula

$$dA_a(t)dA_b(t)^* = \delta_{ab}dt, a, b \geq 1$$

and hence deduce that

$$d\Lambda^a_b(t).d\Lambda^c_d(t) = dA_b(t)^* dA_a(t).dA_d(t)^* dA_c(t)/dt^2$$
$$= \delta_{ad}.dA_b(t)^* dA_c(t)/dt = \delta_{ad} d\Lambda^c_b(t), a, b, c, d \geq 1$$

and also that

$$dA_a(t).d\Lambda^b_c(t) = dA_a(t)dA_c(t)^* dA_b(t)/dt = \delta_{ac}dA_b(t), a, b, c \geq 1$$

etc. Another way to construct the $\Lambda^a_b, a, b \geq 1$ is to choose an onb $f_n(t), n \geq 1$ for $L^2(\mathbb{R}_+)$ and to define the operators

$$a(n, k) = a(f_n \otimes e_k), n \geq 1, k = 1, 2, ..., d$$

Note that $\{f_n(t)e_k\}_{n,k}$ is an onb for $L^2(\mathbb{R}_+ \otimes \mathbb{C}^d)$. Then prove that the $a(n,k)$ mutually commute and

$$[a(n,k), a(m,l)^*] = \delta(n,m)\delta(k,l) - - - (1)$$

Now define

$$A_k(t) = \sum_n a(n,k) < f_n|\chi_{[0,t]} >= \sum_n a(n,k) \int_0^t f_n(s)ds$$

and for any $d \times d$ matrix H, define

$$\Lambda(H,t) = \sum_{n,m,k,l} a(n,k)^* a(m,l) < f_n \otimes e_k|\chi_{[0,t]}H|f_m \otimes e_l >$$

and compute

$$[A_k(t), A_m(s)^*], [A_k(t), \Lambda(H,s)]$$

using the commutation relations (1). Finally use the following identity for orthonormal bases:

$$\sum_{n,k} |f_n \otimes e_k >< f_n \otimes e_k| = I$$

or equivalently,

$$\sum_{n,k} f_n(t)\bar{f}_n(s)|e_k >< e_k| = \delta(t-s)I_d$$

Remark:

$$\Lambda(H,t) = \sum_{n,m} a(n,k)^* a(m,l) H_{kl}| < f_n|\chi_{[0,t]}|f_m >, H_{kl} =< e_k|H|e_l >$$

Consider the operator

$$F = \sum_{k,l} H_{kl} a(k)^* a(l), [a(k), a(l)^*] = \delta(k,l), [a(k), a(l)] = 0.$$

Assuming that H is a normal matrix, it has the orthogonal spectral representation

$$H_{kl} = \sum_r \lambda(r) w_{kr} \bar{w}_{lr}$$

where $W = ((w_{kr}))$ is a $d \times d$ unitary matrix. Equivalently,

$$H = WDW^*, D = diag[\lambda_1, ..., \lambda_d], \lambda_j \in \mathbb{C}$$

It follows that

$$F = \sum \lambda_r b(r)^* b(r), b(r) = \sum_l \bar{w}_{lr} a(l)$$

Thus, since W is unitary,

$$[b(r), b(s)^*] = \delta(r,s), [b(r), b(s)] = 0$$

It is clear then that

$$b(r)|e(u)>=\sum_l \bar{w}_{lr}a(l)|e(u)>=\sum_l \bar{w}_{lr}u(l)|e(u)>=(W^*u)_r|e(u)>$$

From the definition of the coherent vector,

$$|e(u)>=\sum_n (u^n/\sqrt{n!})|n>$$

we have

$$a(r)|e(u)>=u(r)|e(u)>$$

and if $|f(u)>$ is the coherent vector for $\{b(r)\}$, we have

$$b(r)|f(u)>=u(r)|f(u)>$$

Therefore, we obtain

$$|e(u)>=|f(W^*u)>$$

Further, from the fact that

$$a(r)^*a(r)|m>=m(r)|m>$$

so

$$exp(i\sum_r \mu(r)a(r)^*a(r))|m>=exp(i\mu.m)|m>, \mu.m=\sum_r \mu(r)m(r)$$

and hence

$$exp(i\sum_r \mu(r)a(r)^*a(r))|e(u)>=|e(exp(iD(\mu))u)>, D(\mu)$$
$$=diag(\mu(1),...,\mu(d)), u=((u(r)))_{r=1}^d$$

and likewise for b with respect to its number state and its coherent vector. We conclude that

$$exp(isF)|e(u)>=|f(exp(iD(\lambda))W^*u)>=|f(W^*.exp(iH)u)>=|e(exp(iH)u)>$$

Now coming back to the original problem involving time, we write

$$F=\Lambda(H,t)=\sum_{n,m,k,l} a(n,k)^*a(m,l)<f_n\otimes e_k|\chi_{[0,t]}\otimes H|f_m\otimes e_l>$$

and applying the above argument to $\chi_[[0,t]\otimes H$ in place of H and to $\{a(n,k)\}$ in place of $\{a(k)\}$ and to $\{f_n\otimes e_k\}$ in place of $\{e_k\}$ and to $L^2(\mathbb{R}_+)\otimes\mathbb{C}^d$ in place of $\mathbb{C}^d$ (ie, extending the argument to infinite dimensions), we get

$$exp(is.\Lambda(H,t))|e(u)>=|e(exp(is.\chi_{[0,t]}Hu)>$$

which is the way in which we originally defined the conservation process $\Lambda(H,t)$ via "Weyl rotations".

In the context of the previous problem, consider a $d \times d$ matrix H and define

$$\Lambda_t(H) = \sum_{a,b=1}^{d} H_{ab}\Lambda_b^a(t)$$

Show that

$$exp(i\Lambda_t(H))|e(u) >= |e(exp(iH)\chi_{[0,t]}u + \chi_{(t,\infty)}u) >$$

Deduce that for $s \leq t$,

$$dA_a(s)exp(i\Lambda_t(H))|e(u) >= (exp(iH)u(s))_a ds.exp(i\Lambda_t(H))|e(u) >$$

and

$$exp(i\Lambda_t(H)).dA_a(s)|e(u) >= u_a(s)ds.exp(i\Lambda_t(H))|e(u) >$$

Deduce that

$$exp(i\Lambda_t(H)).\sum_b (exp(iH))_{ab}dA_b(s) = dA_a(s).exp(i\Lambda_t(H)), s \leq t, a = 1, 2, ..., d$$

Let $0 \to A \to A' \to A'' \to 0$ be an exact sequence of modules. Let for $i = 1, 2, ...$ and any module B,

$$B \to T^1(B) \to T^2(B) \to ...$$

be a complex with the differential map $d_i : T^i(B) \to T^{i+1}(B), i = 0, 1, ..., T^0(B) = B$, i.e, $d_{i+1}od_i = 0, i = 0, 1, ...$ where the $T^{i's}$ are left exact functors. Define the cohomology groups

$$H^i(B) = Ker(d_{i+1})/Im(d_i)$$

Show that we have a long exact sequence

$$0 \to A \to A' \to A'' \to T^1(A) \to T^1(A') \to T^1(A'') \to T^2(A) \to ...$$

ie, the general term is

$$T^i(A) \to T^i(A') \to T^i(A'') \to T^{i+1}(A) \to T^{i+1}(A') \to T^{i+1}(A'') \to ...$$

Show that the intersection of all the prime ideals of a ring equals the nil radical of the ring, ie, the set of all nilpotent elements of the ring. The ring is assumed to be commutative.

Let a be an ideal of a ring R and define $V(a)$ to be the set of all prime ideals of R that contain a. The topology on $SpecR$ (the set of all prime ideals in R) is defined by taking the $V(a)'s$ as the closed sets where a is an ideal. Thus, a subset E of $SpecR$ is open iff it is of the form $V(a)^c$ for some ideal

a in R. Show that this is a well defined topology, ie, the union of open sets and finite intersection of open sets are indeed open. Taking R to be the ring of polynomials in n variables $R = K[x_1, ..., x_n]$ where K is a field, show that if $a_1, ..., a_r \in K, r \leq n$, then

$$p = (x_1 - a_1)R + ... + (x_r - a_r)R$$

is a prime ideal. Show that this prime ideal is a maximal ideal iff $r = n$.

[6] Let $f \in K[x_1, ..., x_n] = R$ be an irreducible polynomial. Show that $fR = (f)$ is a prime ideal. Define η to be the intersection of all the prime ideals in R that contain f. Show that the closure of $\{\eta\}$ in $SpecR$ equals $V(\eta) = V((f))$. Note that the closure of a set $E \subset SpecR$ equals the intersection of all the sets of the form $V(a)$ where a is an ideal such that $E \subset V(a)$. Note also that $E \subset V(a)$ means that E consists of precisely those prime ideals p that contain a. This problem shows that the closure of $\{\eta\}$ is the set of all prime ideals that contain f and one such ideal is precisely $\eta = (f)$ which is the intersection of all the prime ideals containing f and which therefore the smallest prime ideal containing f.

9.10 Problems in superstring theory

Vertex function in string theory arising from perturbations in the gravitational metric.

Conserved supercurrent associated with local supersymmetry transformations of the superstring action, energy-momentum tensor of a superstring and its conservation associated with the invariance of its action and translations of the world-sheet coordinates or equivalently, associated with the change in the superstring Lagrangian by a total world-sheet space-time differential under an infinitesimal translation.

How to create local supersymmetry in a superstring from global supersymmetry by adding extra gravitino terms.

Gauge mediation of supersymmetry breaking via messenger superfields

Explain the meaning of the vertex function in string theory and its application to the computation of string transition amplitudes using the string propagator. Also explain how the graviton vertex function arises by considering the string action

$$\int g_{\mu\nu}(X)\eta_{ab}X^{\mu}_{,a}X^{\nu}_{,b}d^2\sigma$$

and assuming

$$g_{\mu\nu}(X) = \eta_{\mu\nu} + f_{\mu\nu}(X),$$

where $f_{\mu\nu}(X)$ is a plane gravitational wave

$$f_{\mu\nu}(X) = \xi_{\mu\nu}exp(ik.X), X = (X^{\mu}(\tau, \sigma))$$

The notation $d^2\sigma$ stands for $d\tau.d\sigma$. Regarding the interaction term

$$\int exp(ik.X)\xi_{\mu\nu}\eta_{ab}X^{\mu}_{,a}X^{\nu}_{,b}d^2\sigma$$

as a vertex, we can calculate the matrix elements for the transition amplitudes.

Write down explicitly the superstring action and define appropriate supersymmetry transformations under which the superstring action is invariant. Evaluate the components of the energy-momentum tensor of the superstring. Show that if we replace global supersymmetry transformations by local supersymmetry transformations, then when the equations of motion hold, a certain supercurrent is conserved. To do this, you must use the fact that the variation of the superstring action under an infinitesimal change of the Boson and Fermion string amplitudes is zero when the equations of motion hold and in particular, under this condition, it is also zero under infinitesimal local superstring transformations. Derive the explicit formulas for the conserved supercurrent.

In order that the superstring action be invariant under local supersymmetry transformations, not merely under global ones, we must add additional terms to the action, introducing in the process a gravitino field. Derive the resulting superstring action and specify the explicit local supersymmetry transformations under which it is invariant.

Explain the notion of gauge mediation of supersymmetry breaking via messenger superfields. Basically, we consider a set $\{\Phi_n, \bar{\Phi}_n\}$ of left left Chiral messenger superfields and another set S_n of left-Chiral superfields in the supersymmetry breaking sector. The corresponding supersymmetry breaking superpotential is given by

$$f(S_n, \Phi, \Phi^*) = \sum_n \lambda(n) S_n \Phi_n \bar{\Phi}_n$$

Note that supersymmetry will be broken when the $\mathcal{F}$-part of S_n has a non-zero vacuum expectation value. The F-component of f contributes to the supersymmetry breaking Lagrangian. The above Lagrangian contributes a term that is quadratic in the scalar components ϕ_n and ϕ_n of Φ_n and hence generates scalar masses provided that the F-component F_n of S_n has a non-vanishing expected value in vacuum. The scalar mass term arising from this superpotential is $\sum_n \lambda(n) < F_n > \phi_n\bar{\phi}_n$. Likewise, the term in the superpotential that contributes to the Fermion masses in the superfield Φ_n arises when the scalar component s_n of S_n has a non-zero expectation value. This Fermion mass term is quadratic in ψ_{Ln} and $\bar{\psi}_{Ln}$ where ψ_{Ln} is the term in Φ_n that is linear in θ_L. This Fermion mass term is $\sum_n \lambda(n) < s_n > \psi_{Ln}\bar{\psi}_{Ln}$. Now we wish to investigate how the supersymmetry breaking superpotential f communicates via the Left Chiral messenger superfields $\Phi_n, \bar{\Phi}_n$ a corresponding supersymmetry breaking in the gauge superfields. The idea is that these messenger superfields interact with the gauge superfield $V = t_A V^A(x,\theta)$ via an interaction Lagrangian $a[\sum_n \Phi_n^* exp(-V)\Phi_n]_D + b\sum_n[\bar{\Phi}_n^* exp(-V)\bar{\Phi}_n]$. One of the terms in the first part is for example $\phi_n^* t_A t_B \phi_m \eta_{\mu\nu} V_\mu^A(x) V_\nu^B(x)$ (This comes from the D-term because V_μ^A appears as a quadratic term in θ in the Lagrangian and the product of two terms, each being quadratic in θ is a term that is of fourth degree in θ. When supersymmetry is broken by the messenger fields, ϕ_n as we saw acquires masses, ie, there is a mass term $\sum_n \lambda(n) < s_n > \phi_n \bar{\phi}_n$ coming from the supersymmetric breaking sector and when this term is added to the term $\phi_n^* t_A t_B \phi_m \eta_{\mu\nu} V_\mu^A(x) V_\nu^B(x)$ coming from the observable sector, we see using the path integral formalism that an integration gives a propagator correction term for the gauge field V_μ^A. Specifically, such a term arises from

$$\int exp(iS_0(V)+\sum_n \lambda(n) < s_n > \phi_n\bar{\phi}_n+\phi_n^* t_A t_B \phi_m \eta_{\mu\nu} V_\mu^A(x)V_\nu^B(x))V_\alpha^C(x)V_\beta^D(x).DvD\phi.D\bar{\phi}$$

In general, this mechanism of propagator correction can be seen from the standpoint of Gaussian integrals as follows: Without messenger field terms, the propagator of the gauge field V is given by

$$\int exp(iS_0(V))VV^T dV$$

where $S_0(V)$ is a fourth degree polynonmial in V which reduces to a second degree polynomial when the gauge field is Abelian. After the messenger field term interacts with this gauge field, this propagator gets corrected to

$$\int exp(iS_0(V) + i\phi^T(Q_0 + Q_1(V))\phi)VV^T.d\phi.dV$$

where $Q_1(V)$ is quadratic in V and ϕ is the scalar component of the messenger superfield. Integrating over ϕ first gives the corrected gauge field propagator as

$$\int exp(iS_0(V))det(Q_0 + Q_1(V))^{-1}VV^T dV$$

Now we must explore the situation of how this perturbation in the gauge superfield propagator induces a corresponding perturbation in the matter super field propagators in the observable sector. Let χ denote the matter superfields in the observable sector. These matter superfields interact with the with the gauge superfields V of the observable sector via an interaction of the form

$$[\chi^*.exp(-V).\chi]_D$$

It is clear from this formula that corrections to the V-propagator will induce corrections to the χ propagator. Specifically, the χ-propagator can be computed as

9.11 Problems on waveguides for quantum communication systems

Consider a waveguide of length d, with a quantum electromagnetic field applied at both its ends. The transverse field components are expressible in terms of the longitudinal components. The longitudinal field components are expressible as superposition of modal basis functions of x and y multiplied by exponential functions of z. The coefficients in these field expansions are photon creation and annihilation operators as can be seen by computing the energy of the field within the guide and expressing it as a quadratic form in the linear combination coefficients. The fact that these coefficients have sinusoidal variations in time is a consequence of the fact that the fields satisfy the wave equations. The frequencies of variation are the discrete oscillation frequencies. When the field is in a coherent state, calculate the average $s(21)$ parameters, ie, the ratio of the quantum average of a given component of a transverse component of the reflected electric field at the output to that of a given transverse component of the incident electric field at the input subject to the condition that the same component of the incident transverse electric field at the output is zero.

Consider a charged Bosonic string interacting with the em field within a waveguide. Write down the Hamiltonian of the string with this interaction. Set up the Euler-Lagrange equations for this string with the em interaction in the form

$$L = (1/2)\int_0^L [\partial_t X^\mu.\partial_t X_\mu - (1/2)\partial_\sigma X^\mu.\partial_\sigma X_\mu]d\sigma - a\int_0^L \partial_t X^\mu.A_\mu(X)d\sigma$$

Here, $A_\mu(X)$ is the covariant electromagnetic four potential within the guide. The sum of this string Lagrangian and the field Lagrangian within the guide is the total Lagrangian. Set up the field equations. Explain how communiation through the guide using a string can be carried out by perturbing the string at the transmitter end and measuring the resulting perturbation in the em field at the receiving end. Likewise, we can perturb the electromagnetic field at the transmitter end and measure its effect on a string at the receiver end. In addition to this, we can have the string interacting with a string gauge field with skew-symmetric potential $B_{\mu\nu}(X)$ and field tensor $B_{\mu\nu,\rho}+B_{\nu\rho,\mu}+B_{\rho\mu,\nu} = H_{\mu\nu\rho}(X)$. This gauge field has action $\int H_{\mu\nu\rho}(X)H^{\mu\nu\rho}(X)d^dX$ and when this string gauge field interacts with a string $X_0^\mu(t,\sigma)$, the interaction Lagrangian is $\int_0^L \epsilon_{ab}B_{\mu\nu}(X_0)X_{0,a}^\mu X_{0,b}^\nu d\sigma$

Generalize [2] to the case of superstrings. Specifically, let $X^\mu(t,\sigma)$ be the Bosonic string and $\psi^\mu(t,\sigma)$ the Fermionic string. The Lagrangian in the absence of any interactions is

$$L_0 = \int_0^L (1/2)\eta_{ab}(\partial_a X^\mu \partial_b X_\mu)d\sigma + \int_0^L \psi^{\mu T}\rho^0\rho^a\partial_a\psi_\mu d\sigma$$

The interaction Lagrangian between the gauge field $H_{\mu\nu\rho}(X)$ and the Bosonic string is as before. Dirac's equation for the Fermionic string in the absence of interactions is obtained from this Lagrangian as

$$\rho^a\partial_a\psi^\mu = 0$$

In the presence of a Fermionic gauge potential $A_a(t, \sigma, X)$, this equation is

$$\rho^a(\partial_a + iA_a(t, \sigma, X))\psi^\mu = 0$$

The corresponding modification to the Fermionic Lagrangian is

$$\psi^{\mu T}\rho^0\rho^a(\partial_a + iA_a(t, \sigma, X))\psi$$

The presence of this other gauge field will lead to a modification in the dynamics of the Bosonic string as well as in the dynamics of the field $H_{\mu\nu\rho}$. We then also require a Lagrangian of the gauge field A_a. We shall for simplicity assume that $A_a(t, \sigma)$ does not depend upon the Bosonic string coordinates explicitly. Then, its presence in the Fermionic string dynamics will not affect the Boson string dynamics. As in electromagnetism, we assume that its Lagrangian is given by

$$L_A = K\int F_{ab}F^{ab}d\sigma, F_{ab} = A_{b,a} - A_{a,b}$$

Problem: Write down the equations of the superstrings and fields involved and explain how quantum communication can be carried out with this setup. Specifically, express the solutions to the gauge potentials $A_a(t, \sigma), B_{\mu\nu}(X)$ as superpositions of plane waves with coefficients being Bosonic creation and annihilation operators plus perturbation terms coming from the interactions of these fields with the superstring and likewise express the solution to the superstring equations as superpositions of plane waves with coefficients being Bosonic and Fermionic creation and annihilation operators plus perturbation terms coming from the interactions of this superstring withe the gauge potentials. By introducing classical perturbations in the gauge potentials, show how one can control the statistics of the superstrings in a tensor product of Bosonic and Fermionic coherent states.

Consider a quantum transmission line with differential equations given by

$$\partial_z v(t, z) + Ri(t, z) + L\partial_t i(t, z), \partial_z i(t, z) + Gv(t, z) + C\partial_t v(t, z) = 0$$

In the spatial Fourier series domain with

$$v(t, z) = \sum_n v_n(t)exp(j2\pi nz/d), i(t, z) = \sum_n i_n(t)exp(j2\pi nz/d)$$

these equations are

$$(j2\pi n/d)v_n(t) + Ri_n(t) + Li_n'(t) = 0, (j2\pi n/d)i_n(t) + Gv_n(t) + Cv_n'(t) = 0, n \in \mathbb{Z}$$

For each n, these equations represent a damped harmonic oscillator. Derive these equations in the quantum context from a Hamiltonian with Lindblad operators. Consider the effect of random line loading with quantum noise $A_n(t), A_n(t)^*$, so that they become

$$Ldi_n(t) + (Ri_n(t) + (j2\pi n/d)v_n(t))dt = \sigma_1 ndA_n(t) + \sigma_{2n}dA_n(t)^* - - - -(1)$$

$$Cdv_n(t) + (Gv_n(t) + (j2\pi n/d)i_n(t))dt = \rho_{1n}dA_n(t) + \rho_{2n}dA_n(t)^* - - - -(2)$$

where $A_n(t), A_n(t)^*$ satisfy the quantum Ito formula of

Hudson and Parthasarathy:

$$dA_n(t).dA_m(t)^* = \delta_{nm}dt$$

Derive these quantum stochastic differential equations
from the Hudson-Parthasarathy
noisy Heisenberg dynamics by introducing a unitary evolution $U_n(t)$ for each n with

$$dU_n(t) = (-(iH_n + L_nL_n^*/2)dt + L_ndA_n(t) - L_ndA_n(t)^*)U_n(t)$$

and choosing an observable X_n with its Heisenberg evolution being

$$j_t(X_n) = U_n(t)^*X_nU_n(t) = X_n(t)$$

so that

$$dj_t(X_n) = j_t(i[H_n, X_n] + \theta_{0n}(X_n))dt + j_t(\theta_{1n}(X_n))dA_n(t) + j_t(\theta_{2n}(X_n))dA_n(t)^*$$

Show that by appropriate choice of H_n, X_n and with the structure maps $\theta_{kn}, k = 0, 1, 2$ defined in terms of the Lindblad operators, we can realize the qsdes (1),(2).

Consider a gauge superfield $V(x, \theta)$. Let $\Omega(x, \theta)$ be a left Chiral superfield. Explain how by means of a gauge transformation, ie appropriate choice of $\Omega(x, \theta)$, $V(x, \theta)$ can be brought into the Wess-Zumino gauge, ie with vanishing C, ω, M, N components. Note that the gauge transformation is

$$V(x, \theta) \rightarrow exp(i\Omega(x, \theta))V(x, \theta)exp(-i\Omega(x, \theta)^*)$$

or equivalently, if the gauge transformation is infinitesimal,

$$V \rightarrow V + i(\Omega - \Omega^*)$$

Consider a non-Abelian gauge superfield $exp(t_AV^A(x, \theta))$. Explain how using a gauge transformation

$$exp(t_AV^A) \rightarrow exp(it_A\Omega^A).exp(t_AV^A).exp(-it_A\Omega^{A*}) - - - (1)$$

this gauge superfield can be brought into the Wess-Zumino gauge. Here Ω^A are left-Chiral superfields. After this gauge superfield has been brought into the Wess-Zumino gauge, show that the effect of restricted gauge transformations,. ie, with $\Omega^A = \Lambda^A(x_+)$, is to transform the components of V^A, namely V_μ^A, Λ^A and D^A in accordance to a usual gauge transformation of gauge fields in accordance with the infinitesimal gauge parameter function $\Lambda^A(x)$. Consider now a gauge superfield V^A in the Wess-Zumino gauge. Define the left Chiral superfield

$$W_L = t_AW_{LA} = D_R^T\epsilon D_R.exp(t_AV^A).D_Lexp(-t^AV_A)$$

Show that under a gauge transformation as in (1) with $\Omega^A(x,\theta)$ being an arbitrary left Chiral superfield, W_L transforms covariantly, ie,

$$W_L \to exp(it_A\Omega^A).W_L.exp(-it^A\Omega^A)$$

Hence show that $[W_{LA}^T\epsilon W_{LA}]_F$ is a supersymmetric Lagrangian density that is also gauge invariant. Note that we are assuming the t_A to be Hermitian with $Tr(t_At_B) = \delta_{AB}$. Show that this Lagrangian generalizes that of a non-Abelian gauge field by including terms involving the gaugino field λ^A as well as the auxiliary field D^A.

Supersymmetric Maxwell equations: We can write $W_{LA} = D_R^2 S_A$ where S_A is some superfield because $D_R W_{LA} = 0$ since W_{LA} is left chiral and $D_R^3 = 0$. Then,

$$S = \int [W_{LA}^T\epsilon W_{LA}]_F d^4x = \int\int [(D_R^2(S_A)^T\epsilon D_R^2(S_A)]_F d^4x =$$

$$= \int [D_R^2(D_R^2 S_A^T.\epsilon.S_A)]_F d^4x = \int [D_R^2 S_A^T\epsilon.S_A]_D d^4x = \int D_R^2 S_A^T\epsilon S_A d^4x d^4\theta$$

on using integration by parts in superspace. By setting the variational derivative of S w.r.t the superfield S_A to zero, derive the super-non-Abelian Maxwell equations for W_{LA}.

General superfield equations for Chiral fields in the absence of gauge interacgtions. Let Φ be a left Chiral superfield. Then a supersymmetric action functional for the components of this superfield is given by

$$S = \int [K(\Phi^*,\Phi)]_D d^4x + \int [f(\Phi)]_F d^4x$$

f is called the superpotential. Suppose we express Φ in terms of its components as

$$\Phi(x,\theta) = \Phi(x_+,\theta_L) = \phi(x_+) + \theta_L^T\gamma^5\epsilon\psi(x_+) + F(x_+)\theta_L^T\epsilon\theta_L$$

where

$$x_+^\mu = x^\mu + \theta_L^T\epsilon\gamma^\mu\theta_R$$

Note that

$$F(x_+)\theta_L^T\epsilon\theta_L = F(x)\theta_L^T\epsilon\theta_L$$

because the product of three $\theta_L' s$ is zero. Note that any left Chiral superfield can be expressed as a function of x_+, θ_L and conversely, every function of these two must be a left Chiral superfield because of the characterizing property of a left Chiral superfield that it is annihilated by D_R. Note that

$$D_R = (1-\gamma^5)D/2 = \epsilon\partial_{\theta_R} + \gamma^\mu\theta_L\partial_\mu$$

and hence x_+^μ and θ_L are both annihilated by D_R. The above action functional in the case when the Kahler potential K is bilinear in Φ^*, Φ can be shown to give a generalized version of the sum of the Klein-Gordon action for the complex scalar field $\int(\partial_\mu\phi^*.\partial^\mu\phi)$ plus the action of the Dirac field with a mass dependent upon the scalar field and the superpotential f and an auxiliary Lagrangian for the auxiliary field F. The total Lagrangian density is in this case,

$$\mathcal{L} = \partial_\mu\phi^*\partial^\mu\phi + \psi^T\gamma^5\epsilon(i\gamma^\mu\partial_\mu + f''(\phi))\psi + f'(\phi)F + F^2/2$$

It is clear that this Lagrangian is linear-quadratic in the auxiliary field F. Thus while evaluating the path integral for computing propagators of the scalar and Dirac fields, an integration of the auxiliary field is equivalent to setting the auxiliary field equal to its value at the stationary point of the action w.r.t the auxiliary field. This stationary point is

$$F = -f'(\phi)$$

Note that since the fields ϕ, F will be assumed to be real and ψ to be a Majorana spinor field, Φ^* is obtained by replacing θ_L by θ_R in Φ and ψ by $\gamma^5\epsilon\gamma^0\psi$ and hence the modified Lagrangian of the scalar and Dirac fields is given by

$$\mathcal{L} = (1/2)(\partial_\mu\phi.\partial^\mu\phi - f'(\phi)^2) + \psi^T\gamma^5\epsilon(i\gamma^\mu\partial_\mu + f''(\phi))\psi$$

In the special case, when f is a quadratic function, we recover the sum of Lagrangians of the Klein-Gordon field and the Dirac field with constant mass. In this general case, however, we see that any change in the dynamics of the scalar field will affect the dynamics of the Dirac field and vice-versa. This correlated property can be used to control the probability density of the Dirac wave function by making changes in the dynamics of the Klein-Gordon scalar field via control potentials. Of course, the introduction of such control potentials will break the supersymmetry of the theory.

Chapter 10

Algebraic Geometry, Superfield Theory, Electroweak Symmetry, and Stochastic Calculus

10.1 Problems and remarks in algebraic geometry

A prime divisor of a scheme X is a subscheme Y of codimension 1 in X. Give an example from polynomial varieties illustrating how this definition agrees with our intuitive definition of a curve.

Solution: Let $X = specK[x_1, ..., x_n]$ the polynomial ring in n variables over a field K. Then, an example of a prime divisor of X is $V(f)$ where f is an irreducible/prime polynomial. Note that if a is an ideal of $R = K[x_1, ..., x_n]$, then, $V(a)$ is the set of all prime ideals containing a. Since (f) is a prime ideal, $V(f)$ is therefore, the set of all prime ideals which contain f as an element. By definition of the topology on a scheme, the closed sets of X are precisely sets of the form $V(a)$ where a ranges over all the ideals of R. Since f is prime, $V(f)$ is a closed irreducible set. We claim that $V(f)$ has codimension 1 in X. Indeed, $V(f) \subset V(0) = X$ shows that $CodimV(f) \geq 1$. However, if $V(f) \subset Z$ where Z is a closed irreducible subset of X, then $Z = V(p)$ is a prime ideal and $p \subset (f)$. Since p is a prime ideal and f is a prime polynomial, this is possible only if $p = (f)$ unless $p = (0)$. This proves that $Z = (f)$ or (0) and confirms our claim that $codimV(f) = 1$ and hence that $Y = V(f)$ is a prime divisor of X. Note that if f and g are two distinct prime polynomials, then $V((f, g))$ is a closed subset of X but is not a prime divisor of X because it does not have codimension one in X. Indeed, $V((f, g)) \subset V(f) \subset V(0) = X$ and $V((f, g)) \subset V(g) \subset V(0)$ show that codim $V(f, g) \geq 2$. It is easy to see that in fact codim $V((f, g)) = 2$

Now let $f_1, ..., f_k$ be distinct prime polynomials in R. Consider the variety $V = Z(f_1, ..., f_k)$. $t(V)$ is the scheme whose elements are all the closed irreducible subsets of V. Suppose, we extend $f_1, ..., f_k$ to a set $f_1, ..., f_n$ of distinct prime polynomials in R. Then,

$$((f_1, ..., f_k)) \subset ((f_1, ..., f_{k+1})) \subset ...((f_1, ..., f_n))$$

and it is clear that $((f_1, ..., f_r)), r = k, k + 1, ..., n$ are all prime ideals in R. Then, it follows immediately that

$$Z(f_1, ..., f_n) \subset Z(f_1, ..., f_{n-1}) \subset ... \subset Z(f_1, ..., .f_k)$$

DOI: 10.1201/9781003728627-10

are all irreducible subsets of V. It is then clear that

$$dimV = dimZ(f_1, ..., f_k) = n - k$$

Hence, V is a curve iff $dimV = 1$ iff $k = n - 1$. On the other hand, we have $t(V) =$ the set of all closed irreducible subsets of $V = Z(f_1, ..., f_k)$. We clearly have that a set C is a closed, irreducible subset of V iff $C = Z(p)$ where p is a prime ideal containing $(f_1, ..., f_k)$. In particular, $Z(f_1, ..., f_r), r = k, k+1, ..., n$ are all closed irreducible subsets of V. It becomes evident that

$$t(Z(f_1, ..., f_n)) \subset t(Z(f_1, ..., f_{n-1})) \subset ... \subset t(Z(f_1, ..., f_k)) = t(V)$$

If C is a closed irreducible subset of V, then $t(C)$ is a closed irreducible subset of $t(V)$. Indeed, C being a closed and irreducible subset of V implies that $C = Z(p)$ where p is a prime ideal and $Z(p) \subset Z(f_1, ..., f_k)$, ie, p is a prime ideal containing $f_1, ..., f_k$.

What do you understand by a scheme in the context of the spectrum of a ring R, ie $X = specR$, the topological space of all prime ideals in R with the closed sets being precisely all the sets of the form $V(a)$, the set of all prime ideals in R that contain an ideal a. Show that this defines a valid topology on X, namely this set of closed sets is indeed closed under finite unions and arbitrary intersections. Specifically, by using the definition of a prime ideal, show that $V(a.b) = V(a) \cup V(b)$ for any two ideals a and b and that $V(\sum_\alpha a_\alpha) = \bigcap_\alpha V(a_\alpha)$ for any family of ideals $\{a_\alpha\}_\alpha$ in R. Now to complete the definition of a scheme, we must a sheaf of regular functions on X. So, given any open subset U of X, define $O_X(U)$ to be the set of all functions $f : U \to \bigcup_{p \in U} R_p$ where for any prime ideal p, R_p is the localization of R at p, ie, R_p consists of formal elements of the form a/b where $a, b \in R$ and $b \notin p$. Two elements $a/b, c/d$ of this form are said to be equal in R_p if for some $q \notin p$, we have $q(ad - bc) = 0$. Note that the set of all elements of R that are not contained in p is a multiplicative system, ie, closed under multiplication by the definition of a prime ideal. The definition of a/b as an equivlalence class satisfies all the nice properties that fractions should satisfy. Specifically, if $a/b = c/d$ and $c/d = e/f$ then $a/b = e/f$ and $(b/c).(a/b) = a/c$. The function f should satisfy: $f(p) \in R_p$ for all $p \in U$ such that for each $p \in U$, there should exist an open neighbourhood V of p in U for which if $q \in V$, then $f(q) = a/b$ for some $a/b \in R_q$, ie $a \in R, b \notin q$. In other words, f should be locally constant. $O_X(U)$ as U ranges over all open subsets of $X = specR$, forms a sheaf of rings. Its local ring at some $p \in X$ is given by $O_{X,p}$ which is isomorphic to R_p. An element of $O_{X,p}$ can be identified with the equivalence class of all pairs (f, U) wbere U is an open set in X containing p, $f \in O_X(U)$ with the equivalence relation that $(f, U) = (g, V)$ iff f and g agree on some open subset W of $U \cap V$ containing p. Note that in Sheaf theory, $O_{X,p}$ is called the stalk of the sheaf O_X at p.

10.2 More problems in supersymmetry

Spontaneous breaking of supersymmetry by gravitational effects leading to the massless scalar, Dirac, gauge and gaugino fields acquiring masses. The cosmological constant gives negative energy to the vacuum. This must be compensated by a positive energy of gravitino's in the vacuum in order that the vacuum has zero energy as required by conditions for supersymmetry.

Conservation of supersymmetry current as required by the invariance of the action under global supersymmetry and the equations of motion, ie, zero variation of the action under local supersymmetry transformations as implied by the equations of motion.

Conservation of supercurrent in superfield space, supersymmetric matter and gauge Lagrangians leading to interactions between the Bosonic and Fermionic components of the superfield, computing the corrected propagators of the component matter and gauge fields due to these interactions using the differential equation approach and using the creation-annihilation operator theoretic approach based on perturbative expansions of the solutions to the field equations.

More on the conservation of supercurrent in superfield space.

Estimate the masses of the scalar field, the Dirac field, the gauge and gaugino fields produced when supersymmetry is spontaneously broken by gravitational effects, ie, when the components of the metric superfield acquire nonzero vacuum expectation values.

If L is the Lagrangian of a supersymmetric field theory, ie, L changes by a perfect differential under global supersymmetry transformations, then under global supersymmetry transformations with parameter α, L will change by $\alpha^T\partial_\mu J^\mu(x)$ and under therefore under local supersymmetry transformations with parameter $\alpha(x)$, L will therefore change by

$$\delta L(x) = (\partial_\mu\alpha(x))^T N^\mu(x) + \alpha(x)^T\partial_\mu J^\mu(x)$$

When the equations of motion hold, the variation of the action under any infinitesimal variation of the component superfields must be zero, ie,

$$0 = \delta\int L(x)d^4x = \int \delta L(x)d^4x = \int(\partial_\mu\alpha(x))^T N^\mu(x) + \alpha(x)^T\partial_\mu J^\mu(x))d^4x$$

and therefore, from the arbitrariness of the local parameter $\alpha(x)$, it follows that we must have the conservation law

$$\partial_\mu(-N^\mu(x) + J^\mu(x)) = 0$$

ie the supersymmetry current

$$S^\mu(x) = J^\mu(x) - N^\mu(x)$$

is conserved.

Derive an expression for the supercurrent as a superfield whose conservation contains both energy-momentuum conservation and conservation of the supersymmetry current and besides these, also other conservation laws.

hint: Consider the left Chiral superfield Φ and the following supersymmetric action for its component superfields:

$$I = \int [K(\Phi^*, \Phi)]_D d^4x + 2Re \int [f(\Phi)]_F d^4x$$

Since Φ is a Chiral superfield, it can be expressed as

$$\Phi = D_R^2 S$$

for some superfield S where $D_R^2 = D_R^T \epsilon D_R$. We can write

$$I_1 = \int [K(\Phi^*, \Phi)]_D d^4x = \int K(\Phi^*, \Phi) d^4x d^4\theta = \int K(D_L^2 S^*, D_R^2 S) d^4x d^4\theta$$

Its variation w.r.t S^* is given by

$$\delta I_1/\delta S^* = \int \delta[K(D_L^2 S^*, D_R^2 S)/\delta\Phi^*] D_L^2 \delta S^* d^4x d^4\theta$$

$$= \int D_L^2 \delta[K(D_L^2 S^*, D_R^2 S)/\delta\Phi^*] \delta S^* d^4x d^4\theta$$

Likewise, we have

$$I_2 = 2Re \int [f(\Phi)]_F d^4x = \int [f(D_R^2 S)]_F d^4x = 2Re \int [D_R^2 g(S)]_F d^4x$$

$$= 2Re \int g(S) d^4x d^4\theta$$

where if

$$f(\Phi) = \sum_{n \geq 1} c(n) \Phi^n$$

then

$$g(S) = \sum_n c(n) S (D_R^2 S)^{n-1}$$

= The variation of I_2 w.r.t S^* is given by noting that $\Phi^* = D_L^2 S^*$,

$$\delta I_2/\delta S^* = \frac{\delta}{\delta S^*} \int [\bar{f}(D_L^2 S^*)]_F d^4x = \int [(\delta \bar{f}(\Phi^*)/\delta\Phi^*) D_L^2 \delta S^*]_F d^4x$$

$=\int [D_L^2[(\delta \bar{f}(Phi^*)/\delta\Phi^*)\delta S^*]]_F d^4x because$ D$_L$ acting on any function of Φ^* is zero since Φ^* is right Chiral. Now we can write

$$\int [D_L^2[(\delta \bar{f}(Phi^*)/\delta\Phi^*)\delta S^*]]_F d^4x = \int [\delta \bar{f}(\Phi^*)/\delta\Phi^*] \delta S^* d^4x d^4\theta$$

Thus, the variational principle

$$\delta(I_1 + I_2)/\delta S^* = 0$$

yields the superfield equations

$$D_L^2 \frac{\delta K(\Phi^*, \Phi)}{\delta \Phi^*} + \delta \bar{f}(\Phi^*)/\delta \Phi^* = 0$$

Equivalently, taking the adjoint of this equation or equivalently applying the variational principle $\delta(I_1 + I_2)/\delta S = 0$ yields

$$D_R^2 \frac{\delta K(\Phi^*, \Phi)}{\delta \Phi} + \frac{\delta f(\Phi)}{\delta \Phi} = 0$$

These are to be solved subject to the constraint that Φ is a left Chiral field. Suppose now we introduce a gauge superfield V and derive from it the left Chiral spinor gauge superfield

$$W_L = D_R^2 exp(t.V) D_L exp(-t.V)$$

Then the left Chiral field $W_L^\epsilon W_L$ is easily seen to be gauge invariant and hence we can introduce in addition to the matter Lagrangian interacting with the gauge field, $[\Phi^*.exp(-t.V)\Phi]_D + 2.Re[f(\Phi)]_F$, the gauge field gauge invariant Lagrangian

$$a_1.Re(W_L^T \epsilon W_L]_F + a_2.Im(W_L^T \epsilon W_L]_F$$

is easily seen that this action for the matter and gauge fields is invariant under super-gauge transformations

$$\Phi \to exp(it.\Omega)\Phi, exp(t.V) \to exp(it.\Omega).exp(t.V).exp(-it.\Omega^*)$$

where Ω is an arbitrary left Chiral superfield, apart from being a supersymmetric action. Further, an expansion of this action in terms of its component fields resembles that of the scalar KG field action plus the Dirac field action plus the non-Abelian gauge field (in the U(1) cade the electromagnetic field) action with the standard mutual interactions apart from containing gaugino and finally auxiliary field terms, ie, corresponding to the F component of Φ and the D component of V. After this expansion, while evaluating the effective action using the Feynman path integral, we set the auxiliary fields F, D equal to their values in terms of the other components at which the action is stationary. This is because the auxiliary fields F, D appear in in linear-quadratic degees in the total action. On doing so, we find that the mass of the Dirac field is a function of the scalar field via the superpotential function f. We can therefore hope to control the probability density of the Dirac field by controlling the scalar field via parameters introduced into the superpotential. It should be noted that the Dirac field mass depends only on the second derivative $f''(\phi)$ whereas the scalar KG mass depends on the first derivative $f'(\phi)$. Hence, if we take $f(\phi) = c_1\phi + c_2\phi^2$, then $f''(\phi) = 2c_2$ while $f'(\phi) = c_1 + 2c_2\phi$. If we control therefore only the parameter c_1 in the superpotential, we would not be able to control the Dirac field. Apart from interactions between the non-abelian gauge field with the matter and Dirac fields in the form of gauge convariant derivatives, we also have a similar interaction between the non-Abelian gauge field with the gaugino via the gauge covariant derivative. Further, there is a triple interaction term between the Dirac field, the scalar KG field and the gaugino field coming from the term $[\Phi^* exp(t.V)\Phi]_D$. Note that the gaugino field is a component of the gauge superfield V while the scalar and Dirac fields are components of the left Chiral matter superfield Φ. We can compute the effect of these interactions on the propagators of the scalar field, the Dirac field, the non-Abelian gauge field and the gaugino fields either by writing down their differential equations and using perturbation theory to solve them or else by expanding the free fields in the absence of interactions in terms of creation and annihilation operators in the momentum domain and then substituting these expansions into the interaction terms and evaluating the corrections to the operator expansions of the fields via perturbation theory applied to the field equations in the presence of the interactions. The resulting corrections to the fields will be higher degree polynomials in the creation and annihilation operators and we can compute the corrected propagators via taking vacuum expectation values of products of creation and anihilation operator fields in the momentum domain.

Taking $K(\Phi^*, \Phi) = \Phi^*\Phi$, the above superfield equations are

$$D_L^2\Phi + \delta\bar{f}(\Phi^*)/\delta\Phi^* = 0, D_R^2\Phi^* + \delta f(\Phi)/\delta\Phi = 0$$

Note that these equations are consistent because the product of three $D_L's$ ($D_R's$) is zero and hence D_L (D_R) annihilates $D_L^2\Phi$ ($D_R^2\Phi^*$) and hence $D_L^2\Phi$ ($D_R^2\Phi^*$) is a right (left) Chiral superfield as is $\delta\bar{f}(\Phi^*)/\delta\Phi^*$ ($\delta f(\Phi)/\delta\Phi$).

Note that

$$D = \gamma^\mu\theta.\partial_\mu + \gamma^5\epsilon.\partial_\theta$$

so since

$$\gamma^5\gamma^\mu = -\gamma^\mu\gamma^5$$

it follows that

$$D_L = (1+\gamma^5)D/2 = -\gamma^\mu\theta_R.\partial_\mu + \gamma^5\epsilon.\partial_{\theta_L},$$

$$D_R = (1-\gamma^5)D/2 = \gamma^\mu\theta_L.\partial_\mu - \gamma^5\epsilon.\partial_{\theta_R},$$

and hence

$$\{D_L, D_R^T\} = [\gamma^\mu\epsilon(1-\gamma^5)/2 + (1+\gamma^5)\epsilon\gamma^{\mu T}/2]\partial_\mu = (1+\gamma^5)\gamma^\mu\epsilon\partial_\mu$$

where we have used the fact that

$$(\epsilon\gamma^\mu)^T = -\epsilon\gamma^\mu, \epsilon^T = -\epsilon, \gamma^\mu\gamma^5 = -\gamma^5\gamma^\mu$$

Likewise,

$$\{D_R, D_L^T\} = (1-\gamma^5)\gamma^\mu\epsilon\partial_\mu$$

(Note that if A, B are two vector operators with component operators A_a, B_a, then by $\{A, B\}$, we mean the matrix operator whose $(a,b)^{th}$ element is $\{A_a, B_b\} = A_aB_b + B_bA_a$). These two identities agree with the fact that

$$\{D, \bar{D}\} = \{D, D^T\}\gamma^5\epsilon = (\{D_L, D_R^T\} + \{D_R, D_L^T\})\gamma^5\epsilon = 2\gamma^\mu\partial_\mu$$

Note that

$$\{D_L, D_L^T\} = \{D_R, D_R^T\} = 0.$$

since θ_L anticommutes with θ_R, ∂_{θ_L} anticommutes with ∂_{θ_R} and ∂_{θ_L} anticommutes with ∂_{θ_R}. Note that

$$\{\partial_\theta, \theta^T\} = I$$

Note further that since D_L anticommutes with itself and likewise for D_R, it follows that

$$D_L D_L^T = D_L^T \epsilon D_L \epsilon = \bar{D}_L D_L \epsilon = D_L^2 \epsilon$$

where D_L^2 is an abbreviation for $D_L^T \epsilon D_L$. and likewise,

$$D_R D_R^T = D_R^2 \epsilon$$

Define the supercurrent

$$\Theta^\mu = i(\Phi^* \partial^\mu \Phi - \Phi \partial^\mu \Phi^*) + a.\bar{D}\Phi^* \gamma^\mu D\Phi$$

where Φ is a left-Chiral superfield. It is clear that Θ^μ is a non-Chiral real superfield since

$$(\Phi^* \partial^\mu \Phi)^* = \Phi \partial^\mu \Phi, (\Phi \partial^\mu \Phi^*)^* = \Phi^* \partial^\mu \Phi,$$

$$(\bar{D}\Phi^* \gamma^\mu D\Phi)^* = (D_R^T \Phi^* \epsilon \gamma^\mu D_L \Phi)^* =$$

$$D_L^T \Phi (\epsilon \gamma^\mu)^* D_R \Phi^* = D_L^T \Phi (\epsilon \gamma^\mu D_R \Phi^* = D_R^T \Phi^* \epsilon \gamma^\mu D_L \Phi$$

We can equivalently write

$$\Theta^\mu = \Phi^* \partial^\mu \Phi - \Phi \partial^\mu \Phi^* + a.\bar{D}_R \Phi^* \gamma^\mu D_L \Phi$$

where

$$\bar{D} = D^T \gamma^5 \epsilon$$

Then, $D = D_L + D_R, D_R = D_L^*$ and

$$\gamma_\mu D_L \Theta^\mu = \gamma_\mu \Phi^* \partial^\mu D_L \Phi - \gamma_\mu D_L \Phi \partial^\mu \Phi^* + a.Z$$

where since Φ and Φ^* mutually commute so that $D_L \Phi$ and $D_R \Phi^*$ anticommute and hence $D_L D_L^T \Phi^*$ and $D_R Phi$ commute, and $\epsilon \gamma^\mu$ is an antisymmetric matrix,

$$Z = \gamma_\mu D_L (D_R^T Phi^* \epsilon \gamma^\mu D_L \Phi) = \gamma_\mu (D_L D_R^T \Phi^* \epsilon \gamma^\mu D_L \Phi + D_L D_L^T \Phi \epsilon \gamma^\mu D_R \Phi^*)$$

$$Nowsince$$

D_{La} anticommutes with D_{Lb}, we can write

$$D_{La} D_{Lb} = C.\epsilon_{ab}$$

for some operator C. Thus,

$$D_L^2 = D_L^T \epsilon D_L = \epsilon_{ab} D_{La} D_{Lb} = CTr(\epsilon \epsilon^T) = 2C$$

Also

$$D_L D_R^T \Phi^* = \{D_L, D_R^T\} \Phi^* = (1 + \gamma^5) \gamma^\mu \epsilon \partial_\mu \Phi^*$$

Thus,

$$Z = \gamma_\mu ((1 + \gamma^5) \gamma.\partial \epsilon \Phi^* \epsilon \gamma^\mu D_L \Phi + D_L^2 \Phi \gamma^\mu D_R \Phi^*)$$

$$= (\gamma.\partial) \Phi^* D_L \Phi + D_L^2 \Phi.D_R \Phi^*$$

where in simplifying the first term, we have used the fact that γ^5 anticommutes with γ^μ and hence with $\gamma.\partial$ and that $(1+\gamma^5)D_L = 2D_L$. Also,

$$\gamma_\mu\Phi^*\partial^\mu D_L\Phi - \gamma_\mu D_L\Phi\partial^\mu\Phi^*$$

$$= \Phi^*(\gamma.\partial)D_L\Phi - \gamma_\mu\partial^\mu\Phi^* D_L\Phi$$

$$= \Phi^*(\gamma.\partial)D_L\Phi - \gamma.\partial\Phi^* D_L\Phi$$

$$(\gamma.\partial)D_L = (\gamma.\partial)(1+\gamma^5)D_L = (1-\gamma^5)\gamma.\partial.D_L$$

$$= \{D_R, D_L^T\}\epsilon.D_L$$

and hence,

$$\Phi^*(\gamma.\partial)D_L\Phi = \Phi^*\{D_R, D_L^T\}\epsilon D_L\Phi$$

$$= \Phi^* D_R D_L^T \epsilon D_L\Phi + \Phi^* D_{Lb} D_{Ra}\epsilon(b,c)D_{Lc}\Phi$$

$$= \Phi^* D_R D_L^2\Phi + \Phi^* D_{Lb}\epsilon(b,c)\{D_{Ra}, D_{Lc}\}\Phi$$

$$= \Phi^* D_R D_L^2\Phi + \Phi^* D_{Lb}\epsilon(b,c)((1-\gamma^5)\epsilon)_{ac}\Phi$$

$$= \Phi^* D_R D_L^2\Phi + \Phi^*(1-\gamma^5)\epsilon.D_L\Phi = \Phi^* D_R D_L^2\Phi$$

since

$$(1-\gamma^5)D_L = 0$$

Note:

$$\gamma.\partial\Phi^* D_L\Phi = \{D, \bar{D}\}\Phi^*.D_L\Phi$$

Now, since $D_L\Phi^* = 0$, it follows that

$$\{D, \bar{D}\}_{ab}\Phi^* = \{D_L + D_R, \bar{D}_L + \bar{D}_R\}_{ab}\Phi^* = (D_{La}D_{Rb} + \bar{D}_{Lb}D_{Ra})\Phi^*$$

This result is however, not needed. We obtain from the above equations that the term $\gamma.\partial\Phi^* D_L\Phi$ cancels out and we get

$$\gamma^\mu D_L\Theta^\mu =$$

$$D_L^2\Phi.D_R\Phi^* + \Phi^* D_R D_L^2\Phi$$

Now making use of the superfield equations derived earlier, ie

$$D_L^2\Phi = -\delta\bar{f}(\Phi^*)/\delta\Phi^*$$

we get

$$\gamma^\mu D_L\Theta^\mu = c.(D_R\Phi^*).\delta\bar{f}(\Phi^*)/\delta\Phi^* + \Phi^*.D_R.\delta f(\Phi^*)/\delta\Phi^*$$

$$= D_R((\delta\bar{f}(\Phi^*)/\delta\Phi^*)\Phi^* + (c-1)\bar{f}(\Phi^*))$$

This result is however, not needed. We obtain from the above equations that the term $\gamma.\partial\Phi^* D_L\Phi$ cancels out and we get

$$\gamma^\mu D_L \Theta^\mu =$$

$$D_L^2\Phi.D_R\Phi^* + \Phi^* D_R D_L^2\Phi$$

Now making use of the superfield equations derived earlier, ie

$$D_L^2\Phi = -\delta\bar{f}(\Phi^*)/\delta\Phi^*$$

we get

$$\begin{aligned}\gamma^\mu D_L\Theta^\mu &= c.(D_R\Phi^*).\delta\bar{f}(\Phi^*)/\delta\Phi^* + \Phi^*.D_R.\delta f(\Phi^*)/\delta\Phi^* \\ &= D_R((\delta\bar{f}(\Phi^*)/\delta\Phi^*)\Phi^* + (c-1)\bar{f}(\Phi^*))\end{aligned}$$

10.3 Quantum neural networks based on the interaction of SU(2) gauge fields with a matter field in Dirac's equation: Controlling the vacuum expectation values of the Higgs field so as to generate appropriate probability distributions corresponding to the Fermionic matter field wave functions

Quantum neural networks based on the interaction of SU(2) gauge fields with a matter field in Dirac's equation. We shall design a quantum neural network based on the electroweak theory of Wienberg, Salam and Glashow to track a probability density in space-time. Specifically, we consider the total action for the electron-Lepton matter field interacting with an $SU(2)$ gauge field and also with a Higgs doublet scalar field. The $SU(2)$-gauge Symmetry is broken when the Higgs field acquires non-zero vacuum expectation values causing thereby the gauge field to acquire masses. Further, we also add to the total action an SU(2)-invariant Yukawa interaction between the electron-Lepton matter field with the Higgs field and when the Higgs field acquires non-zero vacuum expectation values so that the symmetry is broken, the electron-Leptonic field also acquires mass. By controlling the vacuum expectation values of the Higgs doublet field, we get control over the mass of the electron-Lepton field and thereby control over the probability distribution associated with the electron-Lepton field. Simultaneously since the acquisition of vacuum expectation values of the Higgs field leads to the gauge field acquiring masses, we get control over the gauge field via control of the vacuum expectation values of the Higgs field and since the gauge field interacts with the electron-Lepton matter field, this also generates control over the probability density associated with the electron-Lepton matter field wave function.

Let $A_\mu(x) = A^a_\mu(x)\tau_a$ be the gauge field that interacts with the matter field described by Dirac's wave function $\psi(x) \in \mathbb{C}$. $\tau_a, a = 1, 2, 3$ are taken to be the standard Hermitian generators of $\int u(2)$, namely the Pauli spin matrices:

$$\tau_1 = \begin{pmatrix} 0 & 1 \\ 1 & 0 \end{pmatrix}, \tau_2 = \begin{pmatrix} 0 & -i \\ i & 0 \end{pmatrix},$$

$$\tau_3 = \begin{pmatrix} 1 & 0 \\ 0 & -1 \end{pmatrix}$$

Thus, any element of the group $SU(2)$ can be expressed as

$$g = exp(i(\sum_{a=1}^{3} t_a\tau_a)), t_a \in \mathbb{R}$$

The Yang-Mills matter wave equation with interaction with the gauge field taken into account is given by

$$\gamma^\mu(i\partial_\mu + eA^a_\mu\tau_a)\psi(x) = 0$$

We can also consider a mass term in this equation arising from a spontaneously broken symmetry in a coupling of the matter field with the Higgs scalar field. This interaction has the form

$$(\psi(x)^* \otimes \psi(x))^T A\phi(x)$$

where $\phi(x) = [\phi_1(x), \phi_2(x)]^T$ is a Higgs doublet whose wave equation in the absence of interactions is derived from the Lagrangian

$$(1/2)[(\partial^\mu - ieA^\mu(x))\phi(x)]^*[(\partial_\mu - ieA_\mu(x))\phi(x)]$$

It should be noted that when we take into account also the gauge field Lagrangian

$$(-1/4)F^a_{\mu\nu}F^{a\mu\nu}$$

where

$$F^a_{\mu\nu} = \partial_\mu A^a_\nu - \partial_\nu A^a_\mu - eC(abc)A^b_\mu A^c_\nu$$

with $C(abc)$ the structure constants of the above Lie algebra generators:

$$[\tau_a, \tau_b] = iC(abc)\tau_c$$

Specifically, using the standard commutation relations for the Pauli matrices

$$[\tau_1, \tau_2] = 2i\tau_3, [\tau_2, \tau_3] = 2i\tau_1, [\tau_3, \tau_1] = 2i\tau_2$$

so

$$C(123) = C(231) = C(312) = 2 = -C(213) = -C(321) = -C(132)$$

then these gauge bosons also acquire masses by virtue of the Higgs bosons acquiring vacuum expectation values. Specifically, if we write $< \phi_a(x) >= \alpha(a), a = 1, 2, < \phi_a(x)^*\phi_b(x) >= \beta(ab), a, b = 1, 2$. Then the mass term in in the gauge Lagrangian becomes

$$(e^2/2)A^a_\mu A^{\mu b}Tr(\tau_a\tau_b.\beta)$$

In the absence of Higgs correlation in the vacuum state, we can write $\beta(ab) = \alpha(a)^*\alpha(b)$ and we can write

$$Tr(\tau_a\tau_b\beta) = 2|\alpha|^2\delta_{ab} = 2(|\alpha(1)|^2 + |\alpha(2)|^2)\delta_{ab}$$

and the gauge field mass term Lagrangian becomes

$$e^2|\alpha|^2 A_\mu^a A^{\mu a}$$

Likewise the electron field mass term after the Higgs boson acquires vacuum expectation values is given by

$$(\psi^* \otimes \psi)^T A < \phi(x) >= (\psi^* \otimes \psi)^T A\alpha$$

Note that $(\psi^* \otimes \psi)_T A\phi$ to be invariant under $SU(2)$, we require that

$$(g^* \otimes g)_T Ag = A\forall g \in SU(2)$$

ie A should be preserved by $SU(2)$.

10.4 Problem on Z_n graded quantum stochastic calculus for $n \geq 2$. Supercommutation relations with $\mathbb{Z}_n$ grading can be used to describe the dynamical evolution of quantum fields corresponding to n kinds of particles as a generalization of standard supersymmetry corresponding to $n = 2$, ie, between Bosons and Fermions

Choose integers $r_0 = -1 < 0 < r_1 < r_2 < ... < r_{n-1} < r_n = N$. Here, N is a positive integer. Let $0 \leq \alpha, \beta \leq N$. If $r_{l-1} < \alpha \leq r_l$ and $r_{m-1} < \beta \leq r_m$ for some $l, m \in \{0, 1, ..., n\}$, then define $\sigma(\alpha, \beta) = l - m(modn)$. We let $E(\alpha, \beta)$ denote the $N + 1 \times N + 1$ matrix whose $(\alpha, \beta)^{th}$ entry is a one and the other entries are all zero. We set

$$\sigma(E(\alpha, \beta)) = \sigma(\alpha, \beta)$$

Note that

$$\sigma(\alpha, \beta) \in \{0, 1, ..., n - 1\}$$

We note in particular that if $r_{l-1} < \alpha \leq r_l$, then

$$\sigma(\alpha, 0) = l - 1$$

while if $r_{m-1} < \beta \leq r_m$, then

$$\sigma(0, \beta) = -(m - 1)$$

Thus, for any α, β, we have

$$\sigma(\alpha, \beta) = \sigma(\alpha, 0) + \sigma(0, \beta)$$

Example (Z_2-graded quantum stochastic calculus): Let $n = 2$ and $r_1 = r, r_2 = N$. Thus, $1 \leq r < N$. Then, if $0 \leq \alpha, \beta \leq N$, we get as a special case of the above,

$$\sigma(\alpha, \beta) = 1 - 1 = 0, 0 \leq \alpha, \beta \leq r,$$

$$\sigma(\alpha, \beta) = 0 - 1 = 1(mod2), 0 \leq \alpha \leq r, r < \beta \leq N,$$

$$\sigma(\alpha, \beta) = 1 - 0 = 1r < \alpha \leq N, 0 \leq \beta \leq r,$$

$$\sigma(\alpha, \beta) = 1 - 1 = 0, r \leq \alpha, \beta \leq N$$

Now let $\epsilon = ((\epsilon(\alpha, \beta)))_{0 \leq \alpha, \beta \leq N}$ denote the $N + 1 \times N + 1$ matrix for which $\epsilon(\alpha, \beta)$ is one if $1 \leq \alpha = \beta \leq N$ and zero otherwise. For two $N + 1 \times N + 1$ matrices A, B, we consider the product

$$C = A\epsilon B = AoB$$

For $\gamma = 0, 1, ..., n-1$, denote by $M(\gamma)$ the set of all $N+1 \times N+1$ matrices spanned by $E(\alpha, \beta)$ with $\sigma(\alpha, \beta) = \gamma$. We wish to prove that

$$M(\gamma).M(\gamma') \subset M(\gamma + \gamma')$$

To show this, we consider the product

$$E(\alpha, \beta)\epsilon.E(\mu, \nu)$$

Its $(u, w)^{th}$ entry $(u, w = 0, 1, ..., N)$ is given by

$$\sum_v E(\alpha, \beta)(u, v)\epsilon(v, v)E(\mu, \nu)(v, w)$$

$$= \epsilon(\beta, \mu)E(\alpha, \nu)(u, w)$$

Thus,

$$E(\alpha, \beta)oE(\mu, \nu) = E(\alpha, \beta)\epsilon.E(\mu, \nu) = \epsilon(\beta, \mu)E(\alpha, \nu)$$

Thus if $\beta \neq \mu$, there is nothing to prove while if $\beta = \mu$, then obviously,

$$\sigma(\alpha, \nu) = \sigma(\alpha, \beta) + \sigma(\beta, \nu) = \sigma(\alpha, \beta) + \sigma(\mu, \nu)$$

completing the proof. Now let

$$\omega = exp(2\pi i/N)$$

so that ω is a primitive N^{th} root of unity. Define the $N+1 \times N+1$ "grading matrix"

$$K = \sum_{\alpha=0}^{N} \omega^{\sigma(\alpha,0)} E(\alpha, \alpha)$$

and also define the $N \times N$ matrix

$$H = diag[\omega^{\sigma(j,0)} : j = 1, 2, ..., N] = \sum_{j=1}^{N} \omega^{\sigma(j,0)} F_j^j$$

where F_j^j is the $N \times N$ matrix with a one in the $(j, j)^{th}$ position and zeros at all the other positions. Also define the quantum stochastic process $G(t)$ by its action on exponential vectors as

$$G(t)|e(u) >= |e(Hu\chi_{[0,t]} + u\chi_{(t,\infty)} >$$

It is immediate that

$$\omega^{\sum_{j=1}^{N} \sigma(j,0)\Lambda_j^j(t)}|e(u) >= \Gamma(exp((2\pi i/N)\sum_{j=1}^{N} \sigma(j, 0)F_j^j))|e(u) >$$

$$= |e(Hu\chi_{[0,t]} + u\chi_{(t,\infty)} >= G(t)|e(u) >$$

Hence,

$$G(t) = \omega^{\sum_{j=1}^{N} \sigma(j,0)\Lambda_j^j(t)}$$

In particular, $G(t)$ is unitary. Specifically,

$$G(t)^{-1} = G(t)^* = \omega^{-\sum_{j=1}^{N} \sigma(j,0)\Lambda_j^j(t)} = \omega^{\sum_{j=1}^{N} \sigma(0,j)\Lambda_j^j(t)}$$

Remark: for $a, b \geq 1$,

$$\Lambda_b^a(t) = \lambda(F_b^a \chi_{[0,t]})$$

and hence

$$exp(x.\Lambda_b^a(t)) = \Gamma(exp(xF_b^a \chi_{[0,t]})), x \in \mathbb{C}$$

and more generally,

$$exp(\sum_{a,b=1}^{N} x(a,b)\Lambda_b^a(t)) = \Gamma(exp(X.\chi_{[0,t]})), X = \sum_{a,b=1}^{N} x(a,b)F_b^a = ((x(a,b)))$$

We observe the identity

$$KE(\mu,\nu)K^* = \sum_{\alpha,\beta=0}^{N} \omega^{\sigma(\alpha,0)+\sigma(0,\beta)} E(\alpha,\alpha)E(\mu,\nu)E(\beta,\beta)$$

$$= \omega^{\sigma(\mu,\nu)} E(\mu,\nu)$$

since

$$\bar{\omega} = \omega^{-1}, -\sigma(\alpha,0) = \sigma(0,\alpha), E(\alpha,\beta)E(\mu,\nu) = \delta(\beta,\mu)E(\alpha,\nu)$$

We observe that for $s \leq t$,

$$d\Lambda_b^a(s)G(t)|e(u)> = (Hu(s))_a dA_b(s)^* G(t)|e(u)>$$

$$= \omega^{\sigma(a,0)} u_a(s) dA_b(s)^* G(t)|e(u)>$$

one the one hand, and on the other,

$$G(t)d\Lambda_b^a(s)|e(u)> = u_a(s)G(t)dA_b(s)^*|e(u)>$$

so that

$$< e(v)|d\Lambda_b^a(s)G(t)|e(u)> = \omega^{\sigma(a,0)} u_a(s)\bar{v}_b(s)ds < e(v)|G(t)|e(u)>$$

and

$$< e(v)|G(t)d\Lambda_b^a(s)|e(u)> = u_a(s) < dA_b(s)G(t)^* e(v)|e(u)>$$

$$= u_a(s)(Hv(s))_b ds < e(v)|G(t)|e(u)> = u_a(s)\bar{v}_b(s)ds\omega^{\sigma(b,0)} < e(v)|G(t)|e(u)>$$

and hence we conclude that

$$\omega^{-\sigma(a,0)} d\Lambda_b^a(s)G(t) = \omega^{-\sigma(b,0)} G(t)d\Lambda_b^a(s), s \leq t$$

b or equivalently,

$$d\Lambda_b^a(s).G(t) = \omega^{\sigma(a,0)-\sigma(b,0)}G(t).d\Lambda_b^a(s), s \le t$$

or

$$d\Lambda_b^a(s).G(t) = \omega^{\sigma(a,b)}G(t).d\Lambda_b^a(s), s \le t$$

Note that these identities are also true if $a = 0$ or $b = 0$. Thus, for $s \le t$,

$$G(s)^{\sigma(a,b)}d\Lambda_b^a(s).G(t)^{\sigma(c,d)}d\Lambda_d^c(t) = \omega^{\sigma(a,b)\sigma(c,d)}G(s)^{\sigma(a,b)}G(t)^{\sigma(c,d)}d\Lambda_b^a(s)d\Lambda_d^c(t)$$

Now $G(s), s \ge 0$ forms a commuting family of operators and therefore, we obtain the following fundamental super-commutation relations: let

$$\xi_b^a(t) = \int_0^t G(s)^{\sigma(a,b)}d\Lambda_b^a(s)$$

Then, for $s < t$,

$$d\xi_b^a(s).d\xi_d^c(t) = \omega^{\sigma(a,b)\sigma(c,d)}d\xi_d^c(t)d\xi_b^a(s)$$

Moreover, for $s = t$, we obtain using Quantum Ito's formula,

$$d\xi_b^a(t)d\xi_d^c(t) = G(t)^{\sigma(a,b)+\sigma(c,d)}\epsilon_d^a d\Lambda_b^c(t)$$

Noting the obvious identity,

$$\sigma(a,b) + \sigma(c,d) = \sigma(a,d) + \sigma(c,b)$$

this becomes,

$$d\xi_b^a(t)d\xi_d^c(t) = G(t)^{\sigma(a,d)}\epsilon_d^a d\xi_b^c(t)$$

10.5 More remarks on algebraic geometry

Describing prime ideals of a quotient ring, example of a prime divisor in a scheme by analogy with prime polynomials in the polynomial ring in several complex variables.

Let A be a ring and let $\mathfrak{a}$ be an ideal in A. Consider the ring $A/\mathfrak{a}$ and its spectrum $Y = spec(A/\mathfrak{a})$. It is clear that Y can be regarded the set of all $\mathfrak{p} \in spec(A)$ that contain $\mathfrak{a}$. Note that $specB$, for any ring B, is by definition, the set of all prime ideals of B. Let E be any subset of A. Consider the set $E + \dashv = \{x + \mathfrak{a} : x \in E$. If this set is to be an ideal in $A/\mathfrak{a}$, then we must have $xy \in E + \mathfrak{a}$ for all $x \in A, y \in E$. Since $xy \in \mathfrak{a}$ for any $x \in A, y \in \mathfrak{a}$, in particular, this implies that as a subset of A, $E + \mathfrak{a}$ must be an ideal in A. Note that $E + \mathfrak{a} = E' + \mathfrak{a}$ in $A//\mathfrak{a}$ where $E' = E + \mathfrak{a}$ as a subset of A. This means that all the ideals in $A/\mathfrak{a}$ are of the form $F + \mathfrak{a}$ where F is an ideal in

A that contains $\dashv$. Now suppose $E + \mathfrak{a}$ is a prime ideal in $A/\mathfrak{a}$ where we can assume in view of the above that E is an ideal in A containing $\mathfrak{a}$. Then $u, v \in A$ $uv + \mathfrak{a} \in E + \dashv$ in $A/\mathfrak{a}$ implies $uv \in E$ in A since E contains $\mathfrak{a}$. By definition of a prime ideal, this should then imply that either $u + \mathfrak{a} \in E + \mathfrak{a}$ or $v + \mathfrak{a} \in E + \mathfrak{a}$ in $A/\mathfrak{a}$, or equivalently, since A contains $\mathfrak{a}$, we should have either $u \in E$ or $v \in E$. It follows from this argument that the prime ideals in $A/\mathfrak{a}$ are precisely ideals of the form $E + \mathfrak{a}$ where E is any prime ideal in A containing $\mathfrak{a}$.

If $f(z) = \frac{\Pi_{k=1}^{p}(z-a_k)^{m_k}}{\Pi_{k=1}^{q}(z-)^{n_k}}$ is a rational function of one complex variable z, then we define the divisor (f) corresponding to f as a formal sum of a_k with a weight of m_k and b_j with a weight of n_j, the sum being over $k = 1, 2, ..., p$ and $j = 1, 2, ..., q$. The underlying scheme X should be taken as $V(0)$, namely the space of all prime ideals in $\mathbb{C}[z]$. Each prime ideal is of the form $(z - a)$ for some $a \in \mathbb{C}$. We write this as

$$(f) = \sum_{k=1}^{p} m_k\{a_k\}\} - \sum_{j=1}^{q} n_j\{b_j\}$$

A specific term, for example $\{a_k\}$ should be read as corresponding to the subscheme $V((z-a_k))$ which consists of only one prime ideal, namely that generated by $z - a_k$. It is easy to see that $V(z-a_k) \subset V(0)$ and hence $codimV(z-a_k) = 1$. Likewise for $V(z - b_j)$. We have thus expressed the divisor (f) formally as an integer linear combination of subschemes, each of codimension one:

$$(f) = \sum_{k=1}^{p} m_k V(z - a_k) - \sum_{j=1}^{q} n_j V(z - a_j)$$

Now consider the scheme $X = spec\mathbb{C}[z_1, z_2]$ which comprises of all polynonmials over $\mathbb{C}$ in two complex variables. A scheme Y of codimension one in $X = V(0)$ can be constructed by taking a prime polynomial $f \in \mathbb{C}[z_1, z_2]$ and setting $Y = V((f))$. This can be seen from $V((f)) \subset V(0)$ and noting that $V(f)$ and $V(0)$ are irreducible subsets of X since $f, 0$ are prime polynomials. If we then define a prime divisor in X as a closed subscheme of codimension one, then $V(f)$ is a prime divisor provided that f is a prime polynomial in $\mathbb{C}[z_1, z_2]$. Now take a curve in $\mathbb{C}[z_1, z_2]$ of the form $f(z_1, z_2) = 0$ where f is a polynomial. Associate with this curve the subscheme $V(f)$ of X. If we are able to write f as a product of prime factors

$$f = \Pi_{k=1}^{d} p_k^{n_k}$$

where the $p_k's$ are prime polynomials, then the relation

$$V(z_1 - a, z_2 - b) \subset V(p_k) \subset V(f)$$

for any $k = 1, 2, ..., d$ where (a, b) is a zero of p_k, shows that $dimV(f) = 1$ since $(z_1 - a, z_2 - b)$ and (p_k) are both prime ideals. We see further that since $V(p_k) \subset V(0) = X$, that $codimV(p_k) = 1$ and hence by definition of a divisor on a scheme, $V(p_k)$ is a prime divisor. However, we cannot assert that $V(f)$ in general is a prime divisor since although $V(f) \subset V(0) = X$, $V(f)$ is not irreducible since f is not in general a prime polynomial.

10.6 Training a quantum neural network to generate an evolving family of probability densities based on information about the mean value of an observable in this evolving state

Suppose that the state evolution is described by the Lindblad equation (which is the Schrodinger-Liouville-Von-Neumann equation with noise)

$$\rho'(t) = -i[H(\phi(t)), \rho(t)] + \theta(\rho(t)|\phi(t))$$

where $\phi(t)$ are evolving parameters and the second term is the Lindblad noise term with the Lindblad operators being dependent upon the time varying parameter $\phi(t)$:

$$\theta(\rho|\phi) = (-1/2)(LL^*\rho + \rho LL^* - 2L^*\rho L), L = L(\phi), L^* = L(\phi)^*$$

The procedure: At time t, let the evolved state be $\rho(t)$ and the parameter be $\phi(t)$. Take an observable X and compute its expectation $Tr(\rho(t)X)$. Let $e(t)$ denote the optimum prediction error involved in predicting $Tr(\rho(t)X)$ based on $s(t-u), u = 0, \delta, 2\delta, ..., p\delta$, ie, by minimizing

$$\sum_{r=t-p}^{t} (Tr(\rho(r)X) - \sum_{k=0}^{p} a(k)s(r-k))^2$$

w.r.t $a(k), k = 0, 1, ..., p$. It is clear that the optimum $e(t) = Tr(\rho(t)X) - \sum_{k=0}^{p} a(k)s(t-k)$ can be expressed as a nonlinear function of $Tr(\rho(s)X), s = t, t-1, ..., t-p$ and $s(t-k), k = 0, 1, ..., 2p$. We now choose the position representation so that $\rho(t)$ is represented by the kernel $\rho(t, q, q')$. We adapt $\phi(t)$ to the next iteration step by the algorithm

$$\phi(t+1) = \phi(t) - \mu.\frac{\partial}{\partial\phi(t)}e(t+1)^2$$

where

$$e(t+1) = F(Tr(\rho(t+1-k)X), k = 0, 1, ..., p)$$

with

$$\rho(t+1) = \rho(t) - i\delta.[H(\phi(t)), \rho(t)] + \delta.\theta(\rho(t)|\phi(t))$$

This process of parameter adaption thus does not require any extra information about the pdf of the second process, it naturally generates the state and hence the pdf $\rho(t+1, q, q)$ at time $t+1$ by using information about the reference signal to which the mean of the second process is correlated. The hypothesis appears to be correct when the randomness in the second process is naturally generated due to quantum phenomena in the system.